Lecture Notes in Artificial Intelligence   9716

Subseries of Lecture Notes in Computer Science

More information about this series at http://www.springer.com/series/1244

Lyuba Alboul · Dana Damian
Jonathan M. Aitken (Eds.)

Towards Autonomous Robotic Systems

17th Annual Conference, TAROS 2016
Sheffield, UK, June 26 – July 1, 2016
Proceedings

 Springer

Editors
Lyuba Alboul
Sheffield Hallam University
Sheffield
UK

Jonathan M. Aitken
University of Sheffield
Sheffield
UK

Dana Damian
University of Sheffield
Sheffield
UK

ISSN 0302-9743 ISSN 1611-3349 (electronic)
Lecture Notes in Artificial Intelligence
ISBN 978-3-319-40378-6 ISBN 978-3-319-40379-3 (eBook)
DOI 10.1007/978-3-319-40379-3

Library of Congress Control Number: 2016940896

LNCS Sublibrary: SL7 – Artificial Intelligence

Printed on acid-free paper

This Springer imprint is published by Springer Nature
The registered company is Springer International Publishing AG Switzerland

Preface

These proceedings contain the papers presented at TAROS 2016, the 17th edition of the conference Towards Autonomous Robotic Systems, held in Sheffield, UK from the 28th to the 30th June 2016. The conference was held earlier in the year than previously, to coincide with the inaugural UK Robotics Week (25th June–1st July). TAROS 2016 included an academic conference, industry talks, robot demonstrations, and other satellite events organized by the University of Sheffield and Sheffield Hallam University, under the umbrella of Sheffield Robotics.

The TAROS series was initiated by Ulrich Nehmzow in Manchester in 1997 under the name TIMR (Towards Intelligent Mobile Robots). In 1999, Chris Melhuish and Ulrich formed the conference Steering Committee, which was joined by Mark Witkowski in 2003 when the conference adopted its current name. The Steering Committee has provided a continuity of vision and purpose to the conference over the years as it has been hosted by robotics centres throughout the UK. Under their stewardship, TAROS has become the UK's premier annual conference on autonomous robotics, while also attracting an increasing international audience. Sadly, Ulrich died in 2010, but his contribution is commemorated in the form of the "Ulrich Nehmzow Best Student Paper Award" sponsored by his family.

TAROS was originally intended as a forum enabling PhD students to have their first conference experience. While the conference has become larger and now attracts a wide range of papers from both senior and junior scientists, TAROS 2016 has kept to the ethos of providing an event that encourages new researchers. The IET sponsored a public lecture, presented by Prof Angelo Cangelosi entitled "Developmental Robotics for Embodied Language Learning".

For the 2016 edition, TAROS received 56 submissions. Two categories of papers were offered, full-length paper, of which 42 submissions were received, and short papers, of which we received 14 submissions. All the papers were reviewed by at least three members of our international Program Committee. Of the full-length paper submissions, 23 were accepted as full-length papers, corresponding to an acceptance rate of 41 %. Also included in this volume are 15 short papers, selected from the remaining 32 submissions, to be presented as posters and demonstrations. The publication of the collected papers from 2016 by Springer, in the *Lecture Notes in Artificial Intelligence* (LNAI) series, testifies to the high standard of the research reported here.

In conjunction with TAROS, Sheffield Robotics and the EPSRC UK-RAS network hosted an agricultural field robotics hackathon challenge at the Sheffield Robotics outdoor robotics proving ground in the Peak District National Park. The competition reflected an agricultural scenario, played out on real-world, challenging terrain. Teams developed software to control, as well as collect and process data from, air and ground robots. Tasks included high- and low-level surveying by fixed- and rotary-wing unmanned aerial systems, and ground-based sampling and intervention by wheeled rovers.

At TAROS 2011, the University of Sheffield and Sheffield Hallam University launched their city-wide initiative for robotics research—The Sheffield Centre for Robotics—which has evolved into the current Sheffield Robotics. This partnership has grown over the last 5 years to have a significant presence in the UK robotics community.

We take this opportunity to thank the sponsors for this year's conference: the IET Robotics and Mechatronics TPN for providing the Prize for the Most Promising Robot Application; Springer for providing the Best Paper Prize; the EPSRC UK-RAS Network for sponsoring one of the invited speakers, and the University of Sheffield for supporting the conference. In addition, we would like to thank the many people that were involved in making TAROS 2016 possible. On the organisational side this included Ana MacIntosh, Louise Caffrey, Ekaterina Netchitailova, James Law, and Iveta Eimontaite. We would also like to thank the authors who contributed their work and the members of the international Programme Committee, as well as the additional referees, for their detailed and considered reviews.

June 2016   Jonathan M. Aitken
                                                         Lyuba Alboul
                                                         Dana Damian
                                                         Tony Prescott
                                                       Jacques Penders

Organization

TAROS 2016 was organized jointly by the University of Sheffield and Sheffield Hallam University.

Conference Chairs

Tony Prescott	University of Sheffield, UK
Jacques Penders	Sheffield Hallam University, UK

Program Chairs

Lyuba Alboul	Sheffield Hallam University, UK
Dana Damian	University of Sheffield, UK
Jonathan M. Aitken	University of Sheffield, UK

Organizing Committee

Louise Caffrey	University of Sheffield, UK
Iveta Eimontaite	University of Sheffield, UK
James Law	University of Sheffield, UK
Ana MacIntosh	University of Sheffield, UK
Ekaterina Netchitailova	Sheffield Hallam University, UK

TAROS Steering Committee

Chris Melhuish	University of Bristol, UK
Mark Witkowski	Imperial College London, UK

Program Committee

Jonathan M. Aitken	University of Sheffield, UK
Lyuba Alboul	Sheffield Hallam University, UK
Rob Alexander	University of York, UK
Ronald Arkin	Georgia Institute of Technology, USA
Robert Babuska	Delft University of Technology, The Netherlands
Martin Beer	Sheffield Hallam University, UK
Nicola Bellotto	University of Lincoln, UK
Giuseppe Carbone	Sheffield Hallam University, UK
Dana Damian	University of Sheffield, UK
Alessandro Dinuovo	Sheffield Hallam University, UK
Clare Dixon	University of Liverpool, UK
Tony Dodd	University of Sheffield, UK
Stephane Doncieux	ISIR, France

Marco Dorigo	Université Libre de Bruxelles, Belgium
Matthew Dunnigan	Heriot-Watt University, UK
Mustapha Suphi Erden	Heriot-Watt University, UK
Richard French	University of Sheffield, UK
Antonios Gasteratos	Democritus University of Thrace, Greece
Dan Gladwin	University of Sheffield, UK
Roderich Gross	University of Sheffield, UK
Dongbing Gu	University of Essex, UK
Heiko Hamann	University of Paderborn, Germany
William Harwin	University of Reading, UK
Alan Holloway	Sheffield Hallam University, UK
Roberto Iglesias-Rodriguez	University of Santiago de Compostela, Spain
Joshua Jackson	University of Sheffield, UK
Yaochu Jin	University of Surrey, UK
Mike Jump	University of Liverpool, UK
Maarja Kruusmaa	Tallinn University of Technology, Estonia
Haruhisa Kurokawa	National Institute of Advanced Industrial Science and Technology, Japan
Fred Labrosse	Aberystwyth University, UK
Stanislao Lauria	Brunel University, UK
James Law	University of Sheffield, UK
Mark Lee	Aberystwyth University, UK
Marco Leo	Institute of Optics, Italy
Raul Marin	Jaume I University of Castellon, Spain
Owen McAree	University of Sheffield, UK
Chris Melhuish	University of Bristol, UK
Antonio Neves	DETI-IEETA-University of Aveiro, Portugal
Jacques Penders	Sheffield Hallam University, UK
Tony Prescott	University of Sheffield, UK
Rob Richardson	University of Leeds, UK
Marcos Rodrigues	Sheffield Hallam University, UK
Barbara Webb	University of Edinburgh, UK
Myra Wilson	Aberystwyth University, UK
Alan Winfield	University of the West of England, UK
Mark Witkowski	Imperial College London, UK
Ulf Witkowski	South Westphalia University of Applied Sciences, UK
Masaki Yamakita	Tokyo Tech, Japan

Additional Reviewers

Kaszubowski Lopes, Yuri	Pinciroli, Carlo
Massink, Mieke	Sayed, Muhammad
Ozdemir, Anil	Trifan, Alina

Contents

Active Compliance Control of the RED Hand: A PID Control Approach

Muhamad Faizal Abdul Jamil, Jamaludin Jalani[(⊠)], Afandi Ahmad, and Eddy Irwan Shah Saadon

Universiti Tun Hussen Onn Malaysia, Parit Raja, Batu Pahat, Johore, Malaysia
faizal-jamil@engineer.com,
{jamalj,afandia,eddy}@uthm.edu.my

Abstract. Robot with non-back-drivable actuators will appear stiff when in contact with the environment and human. This scenario is unsafe for the Human-Robot Interaction (HRI). In order to guarantee safety in HRI, the robot will be made "soft" such that a compliant control can be introduced. Apart from utilizing the proper mechanism design, the back drivability actuators can be achieved by a suitable choice of control. In particular, in this paper, a PID control is employed to achieve an active compliance control. The reference impedance model characteristics are exploited for which the system allows us to introduce a virtual mass-spring-damper system to adjust the compliant control level. The performance of the PID control will be tested on the RED Hand in the simulation. The results are recorded and analyzed for the thumb finger. The results show that the PID controller is capable of controlling the motion and position of the RED Hand. In addition, the compliance behavior for the RED Hand can be suitably adjusted based on the required compliant level.

Keywords: PID control · Robotic hand · Active compliant control · Model reference impedance control · Multi-fingered hand · Anthropomorphic hand · Force control

1 Introduction

Robot manipulators have been utilized in the industrial area to perform tasks such as pick-and-place, polishing, deburring and machining where in most cases their workspace are separated from human. However, a new generation of robots is designed to be incorporated with human. This has certainly exposed the human to the risk due to the intricate relationship between them. Clearly, this scenario involves physical contact with human and safety for the Human-Robot Interaction (HRI) cannot be compromised and must be guaranteed [1].

Robots with non-backdriveable actuators such as geared electric motors or hydraulic actuators appear to be stiff in many industrial robot applications [2]. Although, the stiffer actuator is preferred to improve the precision, stability and bandwidth of position control [3], the compliant actuator (i.e. none-stiff actuator) also has the advantages [4]. A stiff actuator can be described as its ability to move from one position to a specific position or location based on a predefined trajectory. On the other hand, a compliant actuator is

© Springer International Publishing Switzerland 2016
L. Alboul et al. (Eds.): TAROS 2016, LNAI 9716, pp. 1–7, 2016.
DOI: 10.1007/978-3-319-40379-3_1

regarded as deviations from its own equilibrium position, depending on the applied external force. Further description of compliant control is explained in [5].

In order to ensure safety in Human-Robot Interaction, the robot has to be "soft", and this can be achieved by introducing compliance feature to the robot. There are three types of robotic compliance, firstly, the passive compliance, which the linear springs is normally used [6]. Secondly, active compliance, where sensors and proper control algorithms are employed [7]. Thirdly, is the combination between active and passive compliance which is known as *hybrid compliance* [8].

In active compliance control, the behavior of the physical mass-spring-system can be realized by employing a suitable controller. The stiffness of the actuator is adjustable and spring-like behavior can be shown with the introduction of the external force sensor. This paper attempts to introduce a compliance control approach by using PID control scheme for the RED Hand. It is to note that, the PID control approach is considered a new control technique, particularly for a model reference active compliant control which will be applied for the RED Hand. The author found none of the available research, particularly for model reference active compliance control that employing the PID approach. However, the group in [9, 10] had proposed a PD for control technique for a stiff force control which can be used for our reference.

2 Mathematical Modeling

The mathematical dynamic modeling of the RED Hand is generally written as follows:

$$M(\dot{q})\ddot{q} + C(q, \dot{q})\dot{q} + G(q) = \tau \tag{1}$$

where M, C and G represent mass, velocity and gravity terms respectively. The torque vector τ represents the external actuating torques affecting each joint. Considering the effect of friction and stiction, Eq. (**1**) can be further written as

$$M(\dot{q})\ddot{q} + C(q, \dot{q})\dot{q} + G(q) + F = \tau \tag{2}$$

where F represents friction and stiction. Nevertheless, the effects of friction and stiction are not included in this simulation study [11] (i.e. only the complete RED Hand will be considered in the future which currently in progress).

3 PID Controller Design

The proposed control strategy comprises of an impedance reference model, a simulated RED Hand, PID control and sensor. The performance of the PID controller is tested in Simulink/MATLAB. Generally, a PID controller is written as follows:

$$u(t) = K_p e(t) + K_i \int_0^t e(t)dt + K_d \frac{de}{dt} \tag{3}$$

where $u(t)$ is the control output, K_p is the proportional gain, K_i is the integral gain, K_d is the derivative gain and e is the tracking error.

4 Compliance Control

The behavior of the reference model can be represented by the equation below:

$$M_s\ddot{q}_d + K_s\dot{q}_d + K_iq_d = -G_fH + M_s\ddot{q}_r + K_s\dot{q}_r + K_iq_r \tag{4}$$

The reference impedance model characteristics are defined by the mass matrix M_s, the damping matrix coefficient K_s and the stiffness coefficient matrix K_i. H is an externally sensed force q_r, which is the reference trajectory and q_d is the new demand to compensate the external force introduced via the input distribution gain, G_f. Furthermore, G_f is positive definite and H is an external force measurement, obtained via specially introduced sensors. Equation (4) can be further written as follows:

$$M_ss^2q_d(s) + K_ssq_d(s) + K_iq_d(s) = -G_fH + M_ss^2q_r(s) + K_ssq_r(s) + K_iq_r(s) \tag{5}$$

Then, Eq. (5) can be simplified as

$$q_r(s) - q_d(s) = \frac{-G_fH}{M_ss^2 + K_ss + K_i} \tag{6}$$

Since Eq. (6) is now a mass-spring-damper of a second order system, the following equation can be shown.

$$q_r(s) - q_d(s) = \frac{-G_fH}{M_ss^2 + 2\varsigma\omega_ns + \omega_n^2} \tag{7}$$

From (7) it can be found that M_s, K_s and K_i are used to adjust or fine-tune the level of compliance. For instance, if K_s is decreased, the robot becomes more compliant where $K_i = \omega_n^2$. The ω_n is a natural frequency. On the other hand, the robot hand can be very stiff if K_i is significantly increased. Similarly, the values of K_s can also affect the compliance level of the robotic hand where $K_s = 2\varsigma\omega_n$. The scalar ς is a damping ratio.

5 Performance Evaluation and Discussion

In this Section, two sets of simulation are carried out via Simulation 1 and Simulation 2. Simulation 1 allows us to investigate the effect of the compliant level by varying the value of K_i, while K_s is fixed. In contrast, Simulation 2 investigates the effect of compliant level when varying the value K_s, while K_i is unchanged. Here, only thumb finger will be tested due to its flexibility (i.e. adduction and abduction). The thumb has four (4) degrees of freedom while the rest of the fingers possess only three (3) degrees of freedom. In addition, the preliminary results show that both thumb and index fingers have produced a similar performance in terms of compliance level.

5.1   Simulation 1: Investigate the Effect of Varying the Value of K_i, While K_s is Fixed

In Simulation 1, the performance of tracking control and compliant level is investigated by increasing K_i. The values of K_i has been increased from 5×10^{-5}, 5, 50, and 500 while $K_s = 5 \times 10^{-3}$ is fixed. The result is shown in Figs. 1, 2, 3 and 4. The values of ς and ω_n are computed and shown. Moreover, the compliant level performances at position -22 rad are investigated. It can be observed that the robotic hand becomes less compliant as the value of K_i increased, with a fixed value of K_s. Overall performance shows that different compliant level has been successfully achieved starting from 2 s to 6.5 s, as shown in Figs. 1, 2, 3 and 4 when values of K_i increased from 5×10^{-5} to 10. It is also to note that the amplitude of the compliance level is higher when $K_i = 5 \times 10^{-5}$ and $K_s = 5 \times 10^{-3}$ which is approximately 110 rad (see Fig. 1). On the contrary, the amplitude becomes smaller (approximately 12 rad) as the gain of K_i increased as depicted in Fig. 4.

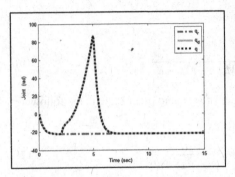

Fig. 1. $K_i = 5 \times 10^{-5}$, $K_s = 5 \times 10^{-3}$, $\omega_n = 7.07 \times 10^{-2}$ and $\zeta = 3.53 \times 10^{-2}$

Fig. 2. $K_i = 5$, $K_s = 5 \times 10^{-3}$, $\omega_n = 2.24$ and $\zeta = 1.12 \times 10^{-3}$

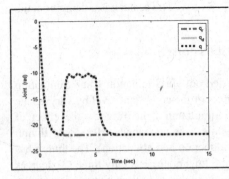
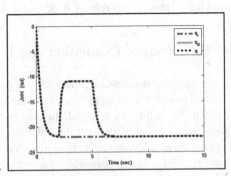

Fig. 3. $K_i = 50$, $K_s = 5 \times 10^{-3}$, $\omega_n = 7.07$ and $\zeta = 3.54 \times 10^{-4}$

Fig. 4. $K_i = 500$, $K_s = 5 \times 10^{-3}$, $\omega_n = 22.36$ and $\zeta = 1.12 \times 10^{-4}$

5.2 Simulation 2: Investigate the Effect of Varying the Value of K_s, While K_i is Fixed

Simulation 2 investigates the performance of the tracking control and compliant level by increasing K_s. The values of K_s have been increased from 5×10^{-3}, 5, 50 and 500 and $K_i = 5 \times 10^{-3}$ remains unchanged. The values of ς and ω_n are shown in Figs. 5, 6, 7 and 8.

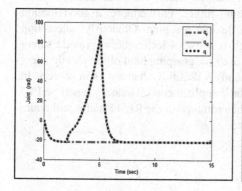

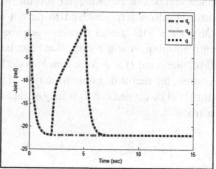

Fig. 5. $K_i = 5 \times 10^{-3}$, $K_s = 50 \times 10^{-3}$, $\omega_n = 7.07 \times 10^{-2}$ and $\zeta = 3.54 \times 10^{-4}$

Fig. 6. $K_i = 5 \times 10^{-3}$, $K_s = 5$, $\omega_n = 7.07 \times 10^{-2}$ and $\zeta = 35.36$

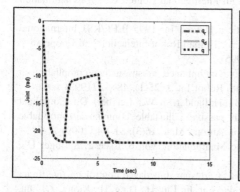

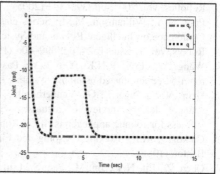

Fig. 7. $K_i = 5 \times 10^{-3}$, $K_s = 50$, $\omega_n = 7.07 \times 10^{-2}$ and $\zeta = 353.55$

Fig. 8. $K_i = 5 \times 10^{-3}$, $K_s = 500$, $\omega_n = 7.07 \times 10^{-2}$ and $\zeta = 3535.53$

It is observed that when gains K_s increased and K_i maintained, the values of ς increased while values of ω_n stays unchanged. It shows that by increasing the damping factor and maintaining the natural frequency, the hand becomes stiffer between the period of 2 s to 6.5 s (see Figs. 5, 6, 7 and 8). Nevertheless, tracking performance becomes normal where q follows q_d and q_r satisfactorily after 6.5 s. The amplitude of compliant level is gradually dropped from approximately 110 rad (see Fig. 5) to 11 rad (see Fig. 8) when K_s is increased.

6 Conclusions

In this paper, a model reference compliance control is accomplished by using a simple PID control scheme for the RED Hand. Simulation results show that different compliance levels can be achieved for the thumb finger. Furthermore, a good motion control is adequately achieved to permit a flexible grasping. Obviously, interaction control via a PID model reference approach is a useful technique to provide safety during grasping. A stiff robotic hand can be used for grasping hard objects while a less stiff robotic hand (i.e. a compliant robotic hand) is useful for handling soft objects. In the future, the real-time implementation of the compliant control strategy based on PID control will be carried out. It is to note that the prototype of the RED Hand is still under construction.

References

1. Lenarcic, J.: Should robots copy humans. In: Proceedings of IEEE International Conference on Intelligent Engineering Systems, pp. 101–106 (1997)
2. Sensinger, J.W., Weir, R.F.: Design and analysis of a non-backdrivable series elastic actuator. In: Proceedings of the 2005 IEEE 9th International Conference on Rehabilitation Robotics, vol. 2005, pp. 390–393 (2005)
3. Pratt, G.A., Williamson, M.M.: Series elastic actuators. In: 1995 IEEE/RSJ International Conference on Intelligent Robots and Systems', Human Robot Interaction and Cooperative Robots, vol. 1, issue 1524, pp. 399–406 (1995)
4. Wang, W., Loh, R.N.K., Gu, E.Y.: Passive compliance versus active compliance in robot-based automated assembly systems. Ind. Robot Int. J. **25**(1), 48–57 (1998)
5. Ham, V.R., Sugar, T.G., Vanderborght, B., Hollander, K.W., Lefeber, D.: Compliant actuator designs: review of actuators with passive adjustable compliance/controllable stiffness for robotic applications. IEEE Robot. Autom. Mag. **16**(3), 81–94 (2009)
6. Cutkosky, M.R.: Robotic Grasping and Fine Manipulation, vol. 6, issue 2. Springer US, Boston (1985)
7. Chen, Z., Lii, N.Y., Jin, M., Fan, S., Liu, H.: Cartesian impedance control on five-finger dexterous robot hand DLR-HIT II with flexible joint. In: Liu, H., Ding, H., Xiong, Z., Zhu, X. (eds.) ICIRA 2010, Part I. LNCS, vol. 6424, pp. 1–12. Springer, Heidelberg (2010)
8. Okada, M., Nakamura, Y., Hoshino, S.: Design of active/passive hybrid compliance in the frequencydomain-shaping dynamic compliance of humanoid shoulder mechanism. In: Proceedings of 2000 ICRA. Millennium Conference. IEEE International Conference on Robotics and Automation. Symposia Proceedings (Cat. No. 00CH37065), vol. 3 (2000)
9. Liu, H., Hirzinger, G.: Cartesian impedance control for the DLR Hand. In: 1999 IEEE/RSJ International Conference on Intelligent Robots and Systems, IROS '99, Kyongju, vol. 1, pp. 106–112 (1999)

10. Albu-Schaffer, A., Ott, C., Hirzinger, G.: A unified passivity-based control framework for position, torque and impedance control of flexible joint robots. Int. J. Robot. Res. **26**(1), 23–39 (2007)
11. Jalani, J., Herrmann, G., Melhuish, C.: Underactuated fingers controlled by robust and adaptive trajectory following methods. Int. J. Syst. Sci. **45**(2), 120–132 (2014)

A Wearable Automated System to Quantify Parkinsonian Symptoms Enabling Closed Loop Deep Brain Stimulation

Paolo Angeles[1]([✉]), Michael Mace[2], Marcel Admiraal[1], Etienne Burdet[2], Nicola Pavese[3], and Ravi Vaidyanathan[1]

[1] Department of Mechanical Engineering, Imperial College London, London SW7 2AZ, UK
paolo.angeles09@imperial.ac.uk
[2] Department of Bioengineering, Imperial College London, London SW7 2AZ, UK
[3] Department of Medicine, Imperial College London, London W12 0NN, UK

Abstract. This study presents (1) the design and validation of a wearable sensor suite for the unobtrusive capture of heterogeneous signals indicative of the primary symptoms of Parkinson's disease; tremor, bradykinesia and muscle rigidity in upper extremity movement and (2) a model to characterise these signals as they relate to the symptom severity as addressed by the Movement Disorder Society Unified Parkinson's Disease Rating Scale (MDS-UPDRS).

The sensor suite and detection algorithms managed to distinguish between the non-mimicked and mimicked MDS-UPDRS tests on healthy subjects ($p \leq 0.15$), for all the primary symptoms of Parkinson's disease. Future trials will be conducted on Parkinsonian subjects receiving deep brain stimulation (DBS) therapy. Quantifying symptom severity and correlating severity ratings with DBS treatment will be an important step to fully automate DBS therapy.

Keywords: Parkinson's disease therapy device · Quantification of Parkinson's disease symptoms · Rigidity model

1 Introduction

Parkinson's disease (PD) is a progressive disease that presents the gradual loss of both motor and non-motor functions. The primary motor symptoms of Parkinson's disease have been the main topics of research for a considerable amount of time and consist of tremor, bradykinesia and muscle rigidity. There is currently no cure but treatment can be administered in the form of oral medication or deep brain stimulation (DBS) during therapy sessions. The Movement Disorder Society Unified Parkinson's Disease Rating Scale (MDS-UPDRS) is an internationally recognised scale used by clinicians to evaluate and monitor PD-related disabilities and impairment through interview and observation. Part III of the MDS-UPDRS evaluates and scores the severity of the primary motor symptoms of the patient.

© Springer International Publishing Switzerland 2016
L. Alboul et al. (Eds.): TAROS 2016, LNAI 9716, pp. 8–19, 2016.
DOI: 10.1007/978-3-319-40379-3_2

To the author's knowledge, there has been no attempt to correlate quantifiable data of primary symptoms to these DBS parameters, i.e. automating DBS therapy. The sensor suite described in this study will look to eventually correlate the symptom severity to DBS parameters. As an initial step, this study focuses on the validation of the sensor suite on healthy subjects.

The system also has the potential to be used in conjunction with Parkinson's disease rehabilitation robots. By providing an objective and quantifiable rating for each symptom, the robot can then conduct an improved and custom procedure for more efficient rehabilitation.

The proposed experiment in this study will be conducted on healthy subjects who will undergo symptom severity tests without mimicked symptoms and with mimicked symptoms of Parkinson's disease. The objectives of this experiment are to record baseline results, from the non-mimicked tests, with the sensor suite (MDS-UPDRS = 0) for comparison purposes in future trials with Parkinson's subjects, and to determine specific models for quantifying primary symptoms of Parkinson's disease.

Rigidity is defined as increased muscle tone which is felt as resistance to passive movement. Mechanical impedance and viscoelastic properties (VEPs) of the affected limbs have been used in literature in an attempt to measure rigidity [1–4]. Mechanical impedance is defined as the vectorial sum of the viscous and elastic stiffness. VEPs are defined as the viscous stiffness and elastic stiffness. Impedance and VEPs metrics will be explored and compared in this study.

Bradykinesia restricts the speed of movement of limbs such as the rotation of the wrist. A gyroscopic sensor attached to a subject's forearm can be used to measure the rotational speed of the wrist [5,6]. Bradykinesia in theses studies was quantified using three parameters; the average angular velocity, the percentage of time the hand was active in a particular time window and the average range of rotation of the hand. Average angular velocity performed best in distinguishing the presence of bradykinesia and will therefore be utilised in identifying bradykinesia for this study.

The involuntary shaking experienced by a patient can be defined as tremor. Tremor has attempted to be quantified in PD patients using IMUs [7,8] to measure the accelerations of the affected limb. A predicted MDS-UPDRS tremor score was produced from linear regression of the peak powers of the accelerometers and gyroscopes. Peak power analysis of acceleration measurements will also be explored in the current study to identify tremor.

2 Methods

2.1 Recording Devices

A 6-axis force/torque transducer from ATI Industrial Automation was utilised for this experiment. The force transducer was attached to the wrist because it is the area of the forearm with least muscle mass and therefore least compliance. A joystick-like handle was attached to the force transducer to allow better control for the clinician of the subject's forearm during rigidity assessments. The force

Fig. 1. The device setup on a healthy subject

transducer was only used during the rigidity assessments. All forces and torques during the experiments were sampled at 1 kHz.

Inertial measurements units (IMUs) from x-io were attached to the forearm and upper arm to areas with the least muscle mass to minimise compliance. The IMUs included an accelerometer, a gyroscope and a magnetometer. IMU data were sampled at 64 Hz.

The data from all sensors were streamed to a laptop running a GUI program that displayed and recorded the data. A set-up on a subject is shown in Fig. 1. All data outputted from the sensors were validated before use in clinic.

2.2 Subjects

Four (three male, one female) healthy volunteers participated in the study. The mean age of the volunteers is 20 years old. The control subjects were neurologically and physically healthy. All subjects gave informed consent under the Ethics agreement of the Imperial College Research Ethics Committee.

2.3 Testing Protocol

Each subject was asked to sit on a chair with both arms rested. All sensors were then attached to the subject's arm and connected to the laptop for data collection. The testing process was verbally explained and physically demonstrated to each subject before tests began. The total time required from each subject, with 1 min rests between assessments, did not exceed 30 min. Each of the assessments described below are taken from the MDS-UPDRS. For this study, the author was the examiner.

Both of the subject's arms were tested, one after the other. Each subject also repeated each symptom assessment for a non-mimicked and mimicked test.

Subjects were shown by the author how best to mimic the symptoms. The author has attended Parkinson's disease clinics at Charing Cross Hospital, London.

Rigidity Assessment: Rigidity is only judged through slow passive motion by the examiner on the limb of the subject. Only the arms were examined during this study, specifically rigidity about the elbow. The subject started with their arm straight and relaxed, the examiner then moved the forearm into flexion and extension using the force sensor handle. The force sensor recorded any resistive forces to this motion. The IMUs were required to record both the angular velocity and displacement between the forearm and upper arm during the assessment. The force, angular velocity and angular displacement were used to model the amount of rigidity. This motion was repeated five times. This set of repetitions was then repeated a further two times with a 5 s break in-between each set.

Bradykinesia Assessment: Once the rigidity assessment was completed, the force sensor and handle were removed from the subject's arm leaving only the pair of IMUs. People with Parkinson's disease have much more difficulty and take a prolonged amount of time to pronate and supinate their wrists, an observation of bradykinesia. After a demonstration, the subject was asked to pronate and supinate their wrist for five repetitions. A 5 s break was taken before repeating the set of repetitions a further two times. The IMUs recorded the angular velocity during the assessment which was used to model the amount of bradykinesia.

Tremor Assessment: Three separate sub-assessments were conducted to quantify the three different types of tremor; postural, kinetic and rest tremor.

Postural tremor can occur when a subject is asked to hold a certain posture. Subjects were asked to hold out their arm for five seconds during the postural tremor assessment. The subject's arm had to be oriented such that their palms were facing downwards, wrists were straight and fingers were comfortably spread. Kinetic tremor can occur when a subject is asked to concentrate on moving a limb to a target location. Kinetic tremor assessment began by asking the subject to stretch out their arm to touch the examiner's finger with their index finger. The examiner then instructed the subject to move their finger to their own nose and back to the examiner's finger. The tremor in PD patients can be present throughout the motion or as the tremor reaches the intended target (examiner's finger or their own nose). The task was performed slowly to ensure no tremor was masked. Tremors can also being observed during a resting state, i.e. resting tremor. The subject was asked to sit comfortably in the chair, with their arms on the armrest and their fit supported by the floor. A break of 5 s was then given before repeating each set of repetitions a further two times. The IMUs were able to record accelerations to model the amount of tremor.

2.4 Analysed Parameters

To model and quantify rigidity, the torque and elbow angle data can be found with these sensor measurements to allow us to find impedance and VEPs of the arm. Angles between the forearm and upper arm were calculated by finding the difference in orientation of the two IMUs. The angular velocity was given from the gyroscopic readings.

Using the force from produced from the force transducer, a torque could be calculated from the product of the force and the moment arm (distance from the elbow pivot to force sensor placement). A torque model was presumed to have both elastic and viscous components and any components concerning acceleration were ignored due to the low frequency nature of the tests [1–3]. Multiple linear regression was then performed with the torque (T), angular displacement (θ) and velocity ($\dot{\theta}$) values to extract elastic stiffness (K), viscous stiffness (B) and constant (e) values as shown in (1) below:

$$T = K|\theta| + B|\dot{\theta}| + e \tag{1}$$

VEPs were defined as the elastic stiffness and the viscous stiffness. Impedance (Z) was then calculated as the vectorial sum of the elastic and viscous stiffness [1–3] using (2):

$$Z = K + B\omega \tag{2}$$

where ω is the peak-to-peak frequency of the limb during an assessment. The average frequency for each assessment was used as the peak-to-peak frequency to calculate the impedance.

The RMS angular velocity from the gyroscopic data of the IMUs was utilised to quantify and model bradykinesia and the accelerometer data of the IMUs was used to identify and model tremor. Fast Fourier Transform (FFT) and peak power analysis were utilised to find instances of postural, kinetic and rest tremor.

2.5 Statistical Analysis

To evaluate if the sensor suite could identify a significant difference between the non-mimicked and mimicked data using the analysed parameters, an independent t-test was used. If $p \leq 0.15$, this was interpreted as a significant difference between the non-mimicked and mimicked data. All statistical analysis was undertaken in Matlab.

3 Results

3.1 Rigidity

Both impedance and VEPs were parameters used in this study to quantify rigidity. Force data were filtered using a fifth-order, low pass Butterworth filter with cut-off frequency at 20 Hz to remove any high frequency noise in the signal and to avoid aliasing when resampling. Force and IMU data were re-sampled to 125 Hz.

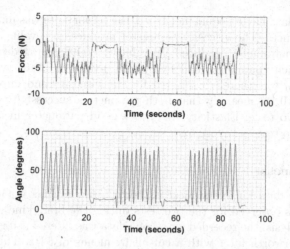

Fig. 2. Force and angle displacement for a non-mimicked test

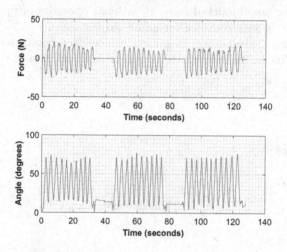

Fig. 3. Force and angle displacement for a mimicked test

Figure 2 shows the raw data for a test without any mimicked symptoms and Fig. 3 shows the raw data for a test with mimicked symptoms. The distinct differences in force magnitudes visually portrays the usefulness of the sensors in distinguishing between a non-mimicked, baseline reading (MDS-UPDRS = 0) and a mimicked reading (MDS-UPDRS ≠ 0). The impedance and VEPs were then calculated from the raw data to observe the presence of rigidity.

An independent t-test with unequal variances was conducted to assess the effectiveness of impedance and VEPs in detecting rigidity. The impedance and VEPs were the dependent variables for the test whilst the test type was the independent variable. The null hypothesis, H_0, is defined as both the non-mimicked and mimicked data having equal means, i.e. the sensor suite was unable to recognise any differences. The viscosity parameter produced a p-value of 0.0131 and the impedance parameter produced a p-value of 0.0092 averaged across all subjects. The elasticity parameter had a very high p-value of 0.8105 averaged across all subjects. H_0 was therefore rejected for the viscosity and impedance parameters as $p \leq 0.15$, meaning that both parameters successfully differentiated between the two tests. Elasticity was not a good parameter in distinguishing between the two tests.

3.2   Bradykinesia

The gyroscopic data for one subject is shown below in Figs. 4 and 5. Visually, a difference was observed for the two tests during the bradykinesia assessment. Before any analysis, the recorded gyroscope data was filtered using a fifth-order, low pass Butterworth filter with a cut-off frequency of 5 Hz. This was needed to remove any high frequency noise and to discard any movements related to tremor. Rotational velocities of up to 400°/s and an average velocity of 79°/s were recorded for non-mimicked tests. Rotational velocities of up to 50°/s and an average of 30°/s were recorded for mimicked tests.

An independent t-test with unequal variances was conducted to quantify any significant differences between the non-mimicked and mimicked bradykinesia

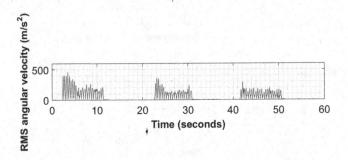

Fig. 4. RMS angular velocity of the wrist for a non-mimicked test

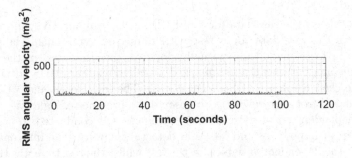

Fig. 5. RMS angular velocity of the wrist for a mimicked test

data. The average RMS angular velocity was used as the dependent variable and the test type was used as the independent variable. The null hypothesis, H_0, was defined as both types of test having equal means. Using the average RMS angular velocity parameter, a p-value of 0.0938 was produced. H_0 was hence rejected as $p \leq 0.15$. This verified that significant differences were observed between the non-mimicked and mimicked data.

3.3   Tremor

A fifth-order, band pass Butterworth filter with cut-off frequencies of 3–12 Hz was used because it has been recognised that Parkinsonian tremor occurs at this defined frequency band [9,10]. Tremor data were then Hamming windowed and zero-padded to isolate the relevant frequencies. Data from each tremor sub-assessment was analysed in an identical fashion.

The FFT of an accelerometer signal from the postural tremor assessment for one subject is shown below in Figs. 6 and 7. FFTs from all sub-assessments of tremor portrayed similar results. Significant tremors were produced during the mimicked tests and these were observed to be several magnitudes larger than their non-mimicked counterparts. This suggested that the peak magnitudes in the defined frequency band were an excellent indicator to detect the presence of tremor.

An independent t-test was conducted to statistically assess whether the peak magnitudes from FFTs of the accelerometer signal were able to distinguish differences between the non-mimicked and mimicked data. The peak magnitude for the FFT of the acceleration signal was used as the dependent variable and the test type was the independent variable. The dependent variable was restricted to the aforementioned tremor frequency band of 3–12 Hz. The null hypothesis, H_0, was defined as both test types having equal means. The postural tremor

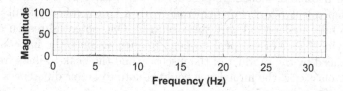

Fig. 6. FFT for a non-mimicked postural tremor test

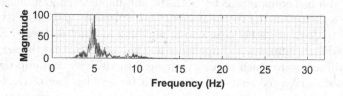

Fig. 7. FFT for a mimicked postural tremor test

t-test had a p-value of 0.0294, the kinetic tremor t-test had a p-value of 0.0028 and the rest tremor t-test had a p-value of 0.1045. All tremor sub-assessment t-tests had a p-value ≤ 0.15 which allowed the rejection of the null hypothesis.

4 Discussion

There were two reasons this study was conducted. Firstly, to record baseline results to compare with future tests on Parkinson's disease subjects. Baseline results are results from healthy control subjects during the non-mimicked tests essentially with a MDS-UPDRS score of zero. Secondly, the device needed validation by assessing working models to identify symptoms of Parkinson's disease. The results have shown that the device is capable in distinguishing between non-mimicked and mimicked results for all primary Parkinsonian symptoms. Following the outcome of this study, trials on PD subjects will next be conducted using the sensor suite.

4.1 Rigidity

Subjects were asked to stiffen their arm during the mimicked test. Out of the three primary symptoms, it is the most difficult to identify because of the external assistance needed. To assess whether rigidity is present in each subject, impedance and VEPs were analysed.

Mechanical impedance was calculated using (2). It has been highlighted that measures derived from torque, such as mechanical impedance, should depend significantly on the movement velocity and hence cannot properly represent the features of Parkinsonian rigidity [4]. However, throughout the rigidity assessment, constant velocity movements were used. Mechanical impedance was hence found to be an excellent differentiator between non-mimicked and mimicked tests with a p-value of 0.0092. Impedance had a smaller p-value than viscosity for the independent t-test. Further trials with a larger cohort of subjects are required to fully understand why this occurred. One potential reason could be the magnifying factor of the peak-to-peak frequency. Since the impedance was more influenced by the product of the viscosity and the peak-to-peak frequency, rather than elasticity, the impedance could equate to larger differences because of this magnifying factor.

The viscosity parameter was found to be slightly worse at differentiating between the two tests with a p-value of 0.0131. Elasticity did not reject the null hypothesis and had equal means for both the non-mimicked and mimicked tests. Impedance is defined as the vectorial sum of both viscosity and elasticity and yet it managed to distinguish between non-mimicked and mimicked tests. Elasticity therefore had little influence on impedance and this suggests that the magnitudes of elastic stiffness were much smaller compared to viscosity and impedance.

It has been suggested that viscosity was a better identifier of rigidity compared to mechanical impedance because it does not rely on movement velocity [4,11]. However, this study has shown that there is negligible difference between the viscosity and impedance when identifying rigidity between non-mimicked and mimicked tests. In addition, because of the torque model used to extract viscosity and elasticity values, viscosity actually relies on the movement velocity used in their studies. Larger cohort trials on PD subjects are required to distinguish the better performance index with this device.

The rigidity assessment in this study does have limitations. The device currently only identifies rigidity at the elbow. Rigidity can be found at other joints such as the wrist, knee and ankle joints. Future iterations of the device will incorporate functionality to be used on other joints. Moreover, further tests are required to detect rigidity at the higher resolution scale of the MDS-UPDRS. Higher resolution quantification can allow for more accurate and efficient treatments to symptoms.

4.2 Bradykinesia

Bradykinesia is another cardinal symptom of Parkinson's disease. One of the assessments for bradykinesia in the MDS-UPDRS protocol observes the speed at which the patient can pronate and supinate their wrist. In clinic, the patient is asked to do this as quickly as possible. As aforementioned, the first test was the non-mimicked test, whilst the second test was a symptom-mimicked test. In this bradykinesia assessment, subjects were asked to pronate and supinate their wrists slower in the mimicked test and IMUs were used to record the angular velocity of the wrists. RMS angular velocity has been shown to be a good performance index for bradykinesia in PD patients [5,6].

An independent t-test with unequal variances was conducted to test whether the sensor suite's algorithm can determine if there was a distinct difference between the non-mimicked and mimicked test. The p-value for the t-test was 0.0938 which indicates that the null hypothesis can be rejected, and that there was a significant enough difference between the RMS angular velocities of the two tests. Other assessments such as finger tapping would be considered for these trials [12]. If lower limb bradykinesia were to be explored, toe-tapping tests would also have to be considered [13,14]. Other parameters, such as peak power analysis and angular displacement will be contemplated for future tests on PD patients as an attempt to improve bradykinesia detection.

4.3 Tremor

Tremor in Parkinson's disease can be sub-categorised into postural, kinetic and rest tremor. As such, all three of these subcategories were tested in this study using the PDD. Postural tremor occurs when the patient is subject to any effects of gravity, i.e. holding their hand in the air, kinetic tremor occurs when the subject is asked to move their limb towards a target and rest tremor is observed when a patient is not doing anything.

Tremor has been detected in PD patients using accelerometers [7,8,10]. It has been suggested that peak power of accelerations was the best indicator of a tremor state [7,10]. Peak power analysis of the accelerations from the IMUs was used to identify tremor in this study.

The analysis for each type of tremor was identical. It was promising to observe that the peak power for every mimicked test occurred between the recommended 3–12 Hz range as suggested by the literature. This shows that healthy subjects were able to mimic tremor symptoms of PD patients very well in all tremor sub-assessments.

The FFT for non-mimicked and mimicked tests suggest that peak power analysis is an excellent detector for all types of tremor. The mimicked peak power magnitudes are at least a factor of 10 larger than their non-mimicked counterparts. Each of the subcategories of tremor have significantly different magnitudes between the non-mimicked and mimicked tests with p-values below 0.15. The sensor suite was able to distinguish successfully between mimicked and non-mimicked tests.

5   Conclusion

It was concluded that the system presented could distinguish between non-mimicked and mimicked tests for all symptom assessments, with very convincing differences. The sensor suite presented has become a platform to truly begin assessing all primary symptoms of Parkinson's disease in one session, rather than one at a time. Moreover, this study is a mandatory step towards closing the loop for DBS therapy using external sensors and improving rehabilitation with current rehabilitation robots with the potential to provide live feedback to the user. Future studies on PD subjects will include more detailed symptom detection algorithms. This is necessary because of the higher resolution scale of the MDS-UPDRS.

References

1. Dai, H., Otten, B., Mehrkens, J.H., D'Angelo, L.T., Lueth, T.C.: A novel glove monitoring system used to quantify neurological symptoms during deep-brain stimulation surgery. IEEE Sens. J. **13**(9), 3193–3202 (2013)
2. Prochazka, A., Bennett, D.J., Stephens, M.J., Patrick, S.K., Sears-Duru, R., Roberts, T., Jhamandas, J.H.: Measurement of rigidity in Parkinson's disease. Mov. Disord. **12**(1), 24–32 (1997)
3. Patrick, S.K., Denington, A.A., Gauthier, M.J.A., Gillard, D.M., Prochazka, A.: Quantification of the UPDRS rigidity scale. IEEE Trans. Neural Syst. Rehabil. Eng. **9**(1), 31–41 (2001)
4. Park, B.K., Kwon, Y., Kim, J.W., Lee, J.H., Eom, G.M., Koh, S.B., Jun, J.H., Hong, J.: Analysis of viscoelastic properties of wrist joint for quantification of Parkinsonian rigidity. IEEE Trans. Neural Syst. Rehabil. Eng. **19**(2), 167–176 (2011)
5. Salarian, A., Russmann, H., Wider, C., Burkhard, P.R., Vingerhoets, F.J.G., Aminian, K.: Quantification of tremor and bradykinesia in Parkinson's disease using a novel ambulatory monitoring system. IEEE Trans. Biomed. Eng. **54**(2), 313–322 (2007)

6. Jun, J.H., Kim, J.W., Kwon, Y., Eom, G.M., Koh, S.B., Lee, B., Kim, H.S., Yi, J.H., Tack, G.R.: Quantification of limb bradykinesia in patients with Parkinson's disease using a gyrosensor - improvement and validation. Int. J. Precis. Eng. Manuf. **12**(3), 557–563 (2011)

7. Dai, H., D'Angelo, L.: Quantitative assessment of tremor during deep brain stimulation using a wearable glove system. In: IEEE International Workshop of Internet-of-Things Networking and Control (IoT-NC), pp. 53–57 (2013)

8. Dai, H., Zhang, P., Lueth, T.C.: Quantitative assessment of Parkinsonian tremor based on an inertial measurement unit. Sensors **15**(10), 25055–25071 (2015)

9. Elble, R., Deuschl, G.: Milestones in tremor research. Mov. Disord. **26**(6), 1096–1105 (2011)

10. Harish, K., Venkateswara Rao, M., Borgohain, R., Sairam, A., Abhilash, P.: Tremor quantification and its measurements on Parkinsonian patients. In: 2009 International Conference on Biomedical and Pharmaceutical Engineering (ICBPE), pp. 1–3 (2009)

11. Kwon, Y., Park, S.H., Kim, J.W., Ho, Y., Jeon, H.M., Bang, M.J., Koh, S.B., Kim, J.H.H., Eom, G.M.: Quantitative evaluation of Parkinsonian rigidity during intra-operative deep brain stimulation. Bio-Med. Mater. Eng. **24**(6), 2273–2281 (2014)

12. Kim, J.W., Lee, J.H.H., Kwon, Y., Kim, C.S., Eom, G.M., Koh, S.B., Kwon, D.Y., Park, K.W.: Quantification of bradykinesia during clinical finger taps using a gyrosensor in patients with Parkinson's disease. Med. Biol. Eng. Comput. **49**(3), 365–371 (2011)

13. Kim, J.W., Kwon, Y., Kim, Y.M., Chung, H.Y., Eom, G.M., Jun, J.H., Lee, J.W., Koh, S.B., Park, B.K., Kwon, D.K.: Analysis of lower limb bradykinesia in Parkinson's disease patients. Geriatr. Gerontol. Int. **12**(2), 257–264 (2012)

14. Heldman, D.A., Filipkowski, D.E., Riley, D.E., Whitney, C.M., Walter, B.L., Gunzler, S.A., Giuffrida, J.P., Mera, T.O.: Automated motion sensor quantification of gait and lower extremity bradykinesia. In: Annual International Conference of the IEEE Engineering in Medicine and Biology Society, EMBS, pp. 1956–1959 (2012)

Systematic and Realistic Testing in Simulation of Control Code for Robots in Collaborative Human-Robot Interactions

Dejanira Araiza-Illan[1]([⊠]), David Western[1], Anthony G. Pipe[2], and Kerstin Eder[1]

[1] Department of Computer Science and Bristol Robotics Laboratory,
University of Bristol, Bristol, UK
{dejanira.araizaillan,david.western,kerstin.eder}@bristol.ac.uk
[2] Faculty of Engineering Technology and Bristol Robotics Laboratory,
University of the West of England, Bristol, UK
tony.pipe@brl.ac.uk

Abstract. Industries such as flexible manufacturing and home care will be transformed by the presence of robotic assistants. Assurance of safety and functional soundness for these robotic systems will require rigorous verification and validation. We propose testing in simulation using Coverage-Driven Verification (CDV) to guide the testing process in an automatic and systematic way. We use a two-tiered test generation approach, where abstract test sequences are computed first and then concretized (e.g., data and variables are instantiated), to reduce the complexity of the test generation problem. To demonstrate the effectiveness of our approach, we developed a testbench for robotic code, running in ROS-Gazebo, that implements an object handover as part of a human-robot interaction (HRI) task. Tests are generated to stimulate the robot's code in a realistic manner, through stimulating the human, environment, sensors, and actuators in simulation. We compare the merits of unconstrained, constrained and model-based test generation in achieving thorough exploration of the code under test, and interesting combinations of human-robot interactions. Our results show that CDV combined with systematic test generation achieves a very high degree of automation in simulation-based verification of control code for robots in HRI.

1 Introduction

Robotic assistants for industrial and domestic applications are designed to interact and collaborate directly with humans. These close interactions have ethical and legal implications. Consequently, the safety and functional soundness of such technologies needs to be demonstrated for them to become viable commercial products [6]. Currently, a physical separation between robots and humans is enforced for safety, besides restrictions of speed and force.[1] These restrictions

---

[1] Standards ISO 13482:2014 for robotic assistants and ISO 10218 (parts I and II) for industrial robotics.

© Springer International Publishing Switzerland 2016
L. Alboul et al. (Eds.): TAROS 2016, LNAI 9716, pp. 20–32, 2016.
DOI: 10.1007/978-3-319-40379-3_3

limit the scope of the applications for collaborative robots. To demonstrate that speed and force restrictions are being met, and thus to assure safety even without physical separation, the software that controls these robotic platforms must be subjected to rigorous verification and validation (V&V) processes. Software V&V needs to consider the robotic system as a whole entity, i.e. the software coupled with its hardware and electronics, as well as the reality and uncertainties of the target environments.

V&V of human-robot interactions (HRI) is challenging. The robot's environment is dynamic and uncertain (e.g., it includes people). Current V&V methods and tools are limited by computational resource bounds, restricting the degree of realism, detail, and exhaustiveness of exploration. Formal methods, e.g. model checking and theorem proving, are exhaustive and provide proof of requirement satisfaction, at the cost of employing highly abstracted models of the robotic systems and HRIs due to computational constraints, as in [23,26]. Testing in simulations allows realism and detail [19,20], at the cost of not being exhaustive with respect to the possibilities in the system under test (SUT), nor providing guarantees of requirement satisfaction.

Available verification methodologies from other domains, such as the microelectronics design industry, provide systematic and targeted approaches to maximize "coverage" (i.e., the extent to which a system's design has been explored) in testing. One of these methodologies is Coverage-Driven Verification (CDV), where various coverage models are used to assess exploration of the SUT and V&V completion [21]. Tests that maximize coverage –i.e., effective tests– are generated (mostly) automatically, coupled with feedback loops (automatic or manual) from automated coverage metrics collection, and automatic checks of (mostly) the SUT's response.

In test generation, constraints are commonly employed to bias testing towards rare events for coverage closure, after applying pseudo-random approaches to achieve exploration of the SUT [7,21]. Model-based test generation uses formal methods (e.g., model checking) or other techniques to explore models in order to bias or constrain tests [25]. Nonetheless, computing tests that stimulate robotic code in a realistic or human-like manner, as it would happen in a real-life HRI scenario, makes the test generation problem quite complex.

We manage complexity via a two-tiered test generation approach. Abstract test sequences are generated first, and then instantiated to obtain concrete tests that stimulate the robotic code indirectly –i.e., the tests stimulate the human, environment, sensors and actuators in simulation, these then stimulate the robot. For example, a test requires a human to send voice commands to activate the robot in a particular order, expressed as 'send voice command' actions in the abstract layer. Code that executes these 'human' actions is assembled according to the test action sequences. The concretization of these action sequences is the production of timed sequences from the human voice model in simulation, that will stimulate simulated voice sensors, and then will send their readings to the robot's code to stimulate it. This two-tiered process is employed in model-based testing [25]. In this paper we apply unconstrained, constrained, and model-based abstract test generation, coupled with test concretization via uniform sampling from classified ranges for variables and parameters. We demonstrate the complementary strengths of exploratory and targeted tests, particularly through model-based test generation, in achieving high levels of coverage for different coverage models, including code, cross-product, and assertions (requirements).

We tested the code for an object handover interaction between a humanoid torso and a person, envisaged for cooperative manufacture tasks, in a simulator developed in Robot Operating System[2] (ROS) and Gazebo[3], a 3D physics simulator. We employed a CDV testbench prototype developed for our simulator, fully compatible with ROS-Gazebo[4]. This paper extends our previous work in [2], with more requirements, coverage models, generated tests and simulation runs. Our testbench prototype is transferable and extendible to other robotic simulators based on ROS, and other collaborative and assistive applications.

The paper is structured as follows. We present the handover scenario in Sect. 2. The testbench components are presented in Sect. 3. A discussion of V&V and coverage results is presented in Sect. 4. Related work is presented in Sect. 5, and Sect. 6 concludes with an outlook on future work.

2 Case Study: Robot to Human Object Handover Task

The object handover case study was chosen because it is critical in many HRI tasks, such as cooperative manufacture, or home care. The robot platform, BERT2, is a humanoid torso with two arms [15]. A handover starts with voice activation from the person to the robot. The robot proceeds to pick up an object, holds it out to the human, and signals for the human to take it. The human indicates readiness to take the object through another voice command. Then, the robot will collect three sensor readings: "pressure," indicating whether the human is holding the object (applying force against the robot's hold of the object); "location," visually tracking that the person's hand is close to the object; and "gaze," visually tracking that the person's head is directed towards the object. Each sensor reading is classified into $G = P = L = \{\bar{1}, 1\}$, where 1 indicates the sensing was positive that the human is ready to receive the object, and $\bar{1}$ is any other sensing outcome, including null. After the sensing, the robot should decide to release the object if the human is ready, i.e. $GPL = (1, 1, 1)$ from the Cartesian product of the sensor readings (GPL for short), or it should decide not to release the object otherwise, i.e. $GPL \in \{(\bar{1}, *, *), (*, \bar{1}, *), (*, *, \bar{1})\}$, where $* \in \{1, \bar{1}\}$, within a time threshold. The person may disengage from the task before the robot makes a decision. The robot can time out whilst sensing, or while waiting for a signal.

A ROS 'node' contains the robot's action control code, comprising 212 statements in Python. The code was structured as a FSM using the SMACH modules [4], to facilitate computing a model of it for model-based test generation.

---

[2] http://www.ros.org/.

[3] http://gazebosim.org/.

[4] Available at: https://github.com/robosafe/testbench-v3.

2.1 Requirements List

The following safety and functional requirements need to be verified, derived from the standard ISO 13482:2014 and previous work on handover interaction protocols and their testing in [2,8]:

1. If the gaze, pressure and location are sensed as correct, then the object shall be released.
2. If the gaze, pressure or location are sensed as incorrect, then the object shall not be released.
3. The robot shall make a decision before a threshold of time.
4. The robot shall always either time out, decide to release the object, or decide not to release the object.
5. The robot shall not close the gripper when the human is too close.
6. The robot shall start in restricted speed.
7. The robot shall not collide with itself at high speeds.
8. The robot shall operate within allowable maximum values to avoid dangerous unintentional collisions with humans and other safety-related objects.

The last requirement was interpreted in four different quantifiable manners, considering a speed threshold of 250 mm/s based on standard ISO 120218-1:2011:

8a. The robot hand speed is always less than 250 mm/s.
8b. If the robot is within 10 cm of the human, the robot's hand speed is less than 250 mm/s.
8c. If the robot collides with anything, the robot's hand speed is less than 250 mm/s.
8d. If the robot collides with the human, the robot's hand speed is less than 250 mm/s.

2.2 Handover Simulator

A simulator of the handover scenario was developed in ROS-Gazebo. ROS is an open-source platform for the development and deployment of robotics code, using C++ and/or Python. Gazebo is a 3D physics simulator, compatible with ROS. BERT2, a cylindrical object, and the person's head and hand were modelled in Gazebo, as shown in Fig. 1. Models were developed in code for the sensors and the human action enactment.

3   A CDV Testbench for a ROS-GAZEBO Simulator

In the CDV methodology, a verification plan indicates the requirements to test, and the coverage models and metrics to use over the SUT [2,21]. A CDV testbench has four components: the **Test Generator**, the **Driver**, the **Checker** and the **Coverage Collector**. Figure 2 shows our testbench, considering the ROS-Gazebo simulator's components. The simulator's design ensures the access to internal parameters in the robot's code and data about the physical models from Gazebo, to facilitate checking and coverage collection. The dotted line indicates feedback to the test generation for coverage closure and verification completion that may require human input.

Fig. 1. The ROS-Gazebo simulation (LHS) and a real interaction from [8] (RHS).

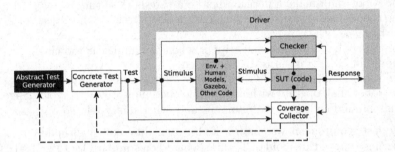

Fig. 2. Testbench and simulator elements in ROS-Gazebo. ROS nodes in gray; modules imported into ROS in white, and modules externally executed in black.

3.1   Test Generator

The aim of the test generation process is to trigger faults in the SUT (the robot's code), while exploring a wide range of scenarios. Guidance to produce effective tests comes from coverage and verification progress feedback. The generated tests must be valid and realistic, which makes the case for non-conventional software test generation approaches due to the complexity of "stimulating the robot's code in a human-like manner".

A test for the handover simulator is formed by an abstract test sequence for the human, environment, sensors and actuators (the environment surrounding the robotic code under test), which assembles code fragments to be executed concurrently by these simulator components. A concrete test is then computed, after parametrization, constraint solving and/or instantiation for all the individual parameters involved in the code fragments. We propose a two-tiered test generation approach to divide and simplify what would be a complex constraint solving, search or optimization problem. An abstract-to-concrete test construction is shown in Fig. 3.

We explored three options for the abstract test generation: pseudorandom, constrained and model-based. Pseudorandom (for repeatability purposes) is, in principle, unconstrained with respect to any assumptions about the HRI protocol. Thus, abstract test sequences are concatenated randomly, e.g., representing a person that disregards the handover protocol. To generate interesting tests, e.g., to verify a particular requirement, pseudorandom test generation can be biased using constraints. The implementation of these constraints requires significant

1	sendsignal	activateRobot	Send human voice A1 for 5 sec.
2	setparam	time = 40	Human waits 40 × 0.05 sec.
3	receivesignal	informHumanOfHandoverStart	Human waits for max. 60 sec.
4	sendsignal	humanIsReady	Send human voice A2 for 2 sec.
5	setparam	time = 10	Human waits 40 × 0.05 sec.
6	setparam	hgazeOk = true	Move human head in Gazebo to pose within ranges: offset [0.1, 0.2], distance [0.5, 0.6] and angle [15, 40)

Fig. 3. An abstract test sequence for the human to stimulate the robot's code (LHS), and its concretization from sampling from defined ranges (RHS).

manual input to be effective. Model-based test generation techniques [9, 14, 25] can target specific scenarios or requirements more effectively. In model-based test generation, a model of the system is explored or traversed in a systematic manner, e.g., through model checking for a requirement expressed as a temporal logic property [5]. A path through the model can be considered as a set of constraints [13, 17] for test generation.

For model-based test generation, the model captures both the ideal robot's code functionality and the human/environment's actions, assuming both follow the handover protocol. We chose probabilistic-timed automata (PTA) [10] models constructed manually in UPPAAL[5], to capture uncertain actions such as disengaging from the task, and the important aspect of human-like response timing in HRI. Requirements 1 to 4 (Sect. 2.1) were expressed as temporal logic properties, and model checked in UPPAAL. A witness trace (or path over the automata) is produced as a result of model checking, from which an abstract test sequence is extracted, disregarding the robot's actions in the trace.

3.2 Driver

The Driver distributes the resulting concrete tests into the simulator components, to be enacted to stimulate the robot indirectly. The Driver reacts to the responses of the SUT if necessary (a "reactive Driver").

3.3 Checker

The Checker monitors the response of the SUT during simulation, to detect requirement violations. Automata-based assertion monitors were implemented manually for all the requirements in Sect. 2.1, as in [2]. Events can be monitored at different abstraction levels, from "the robot received the correct command" (abstract), to "speed is less than the safe thresholds" (semi-continuous signals or variables). For example, the assertion monitor for Req. 5 is triggered every time the code executes the hand(close) function. The pose of the human hand is queried from the physical models in Gazebo. If the mass centre of the human hand is within a 0.05 m distance of the robot's hand, the monitor indicates Failed (requirement violation), or otherwise Passed (requirement satisfaction).

---

[5] http://www.uppaal.org/.

3.4   Coverage Collector

The Coverage Collector records the progress achieved by each test in exploring the SUT. We implemented three coverage models: requirements, cross-product and code. For requirements coverage, we assessed which assertion monitors were triggered by each test.

Cross-product coverage accounts for a complete set of conceivable scenarios. We computed the Cartesian product, $Human \times Robot$, focusing on tuples where the robot times out, and different GPL selections by the human element. The set of events to cover for the human comprised: failure to activate the robot at all, sending the first activation signal but not the second, setting any combination of GPL amongst the possible 8, and disengaging whilst the robot is sensing; i.e. $Human = \{NotActive, ActivSignal, GPL = (*, *, *), Disengaged\}$. The set of events to cover for the robot comprised: timing out whilst receiving any of the two signals (voice command) from the human or whilst sensing, releasing the object, and not releasing the object; i.e. $Robot = \{TimedOut, Released, NotReleased\}$. The total size of this cross-product is of 33 tuples, but 13 of them should not be reached if the code is functionally correct. Most of the tuples that should be reachable are meaningful for the handover, since to be covered in a test, at least part of the protocol was followed correctly by the human and the robot. The cross-product coverage was computed offline from the simulation reports. Cross-product coverage (*situation* coverage) has been proposed (independently) for the verification of autonomous robots [1], including combinations of environment events only.

For code coverage, we accumulate the number of executed code statements per test, through the 'coverage'[6] Python module.

4   Experiments and Results

We verified the robot's code for the handover, with respect to the requirements in Sect. 2.1. The simulator ran in ROS Hydro and Gazebo 1.9, on a PC with Intel i5-3230M 2.60 GHz CPU, 8 GB of RAM, and Ubuntu 13.03. We used UPPAAL 4.0.14 for model-based test generation.

4.1   Requirements Coverage

We first generated 100 unconstrained abstract tests from uniformly sampling the set of all possible abstract human actions and producing sequences of these. We concretized each abstract test by uniformly sampling from defined ranges of variables and parameters, as dictated by the abstract actions. The tests did not cover Reqs. 1 and 8d, and other assertions were triggered less frequently (e.g. Req. 5).

Subsequently, we generated 100 constrained abstract tests that enforced the activation of the robot, in an attempt to increase the coverage, concretized in

---

[6] http://nedbatchelder.com/code/coverage/.

Table 1. Requirements (assertion) coverage results

Req	Unconstrained			Constrained			Model-Based		
	C	P	F	C	P	F	C	P	F
1	0/100	0/100	0/100	0/100	0/100	0/100	2/4	2/4	0/4
2	30/100	30/100	0/100	94/100	94/100	0/100	2/4	2/4	0/4
3	30/100	30/100	0/100	94/100	94/100	0/100	4/4	4/4	0/4
4	100/100	100/100	0/100	100/100	100/100	0/100	4/4	4/4	0/4
5	46/100	44/100	2/100	100/100	100/100	0/100	4/4	4/4	0/4
6	100/100	0/100	100/100	100/100	0/100	100/100	4/4	0/4	4/4
7	14/100	14/100	0/100	22/100	22/100	0/100	2/4	2/4	0/4
8a	100/100	0/100	100/100	100/100	0/100	100/100	4/4	0/4	4/4
8b	98/100	0/100	98/100	100/100	0/100	100/100	4/4	0/4	4/4
8c	96/100	5/100	91/100	99/100	0/100	99/100	4/4	0/4	4/4
8d	0/100	0/100	0/100	0/100	0/100	0/100	1/4	0/4	1/4

the same manner as the unconstrained. We based our pseudorandom generators on the procedure described in [3] for software testing. Finally, we generated four model-based abstract tests targeting Reqs. 1 to 4, to target specifically Req. 1 (also concretized like the others). A test triggered the assertion for Req. 8d, as the robot collided with the human, an important safety violation. Overall, no assertion violations were found for Reqs. 1 to 4. These results are shown in Table 1. If the assertion monitors were Covered (C), either they Passed (P) or Failed (F). The colour code in the table helps to highlight the coverage level of each assertion monitor (green for high coverage, red for no coverage).

For requirements coverage, model-based test generation is most efficient, triggering all the monitors with just four tests. The checks for Reqs. 6 and 8a-d exposed some design flaws, as the robot violates the safety speed threshold of 250 mm/s at the start of the handover, and when picking the object. This could be improved by imposing speed constraints explicitly in the motion of the robot.

4.2 Cross-Product Coverage

To target the 20 reachable tuples in the cross-product coverage (Sect. 3.4), we began with a different set of 100 unconstrained abstract tests, concretized as for requirements coverage. Subsequently, we employed model-based test generation to target the uncovered tuples, formulating the reachability of each tuple as a temporal logic property and model checking it in UPPAAL. Each abstract test sequence was concretized with 20 different sampling instances (column "MB 1"). Finally, we added constraints in the concretization of these abstract tests, reducing the maximum length of timeout thresholds, to trigger the *TimedOut* event in the robot's code, and produced another set of 20 concrete tests for each abstract sequence (column "MB 2").

Table 2. Reachable cross-product coverage

Human × Robot	Unconstr.	MB 1	MB 2	TOTAL
$\langle NotActive, TimedOut \rangle$	55/100	0/160	0/180	55/440
$\langle ActivSignal, TimedOut \rangle$	11/100	0/160	0/180	11/440
$\langle GPL = (1,1,1), TimedOut \rangle$	0/100	3/160	18/180	21/440
$\langle GPL = (1,1,1), Released \rangle$	0/100	17/160	2/180	19/440
$\langle GPL = (\bar{1},\bar{1},\bar{1}), TimedOut \rangle$	1/100	0/160	19/180	20/440
$\langle GPL = (\bar{1},\bar{1},\bar{1}), NotReleased \rangle$	25/100	0/160	1/180	26/440
$\langle GPL = (\bar{1},\bar{1},1), TimedOut \rangle$	0/100	2/160	18/180	20/440
$\langle GPL = (\bar{1},\bar{1},1), NotReleased \rangle$	2/100	18/160	2/180	22/440
$\langle GPL = (\bar{1},1,\bar{1}), TimedOut \rangle$	0/100	0/160	16/180	16/440
$\langle GPL = (\bar{1},1,\bar{1}), NotReleased \rangle$	2/100	20/160	4/180	24/440
$\langle GPL = (\bar{1},1,1), TimedOut \rangle$	0/100	0/160	17/180	17/440
$\langle GPL = (\bar{1},1,1), NotReleased \rangle$	0/100	20/160	3/180	23/440
$\langle GPL = (1,\bar{1},\bar{1}), TimedOut \rangle$	0/100	2/160	18/180	20/440
$\langle GPL = (1,\bar{1},\bar{1}), NotReleased \rangle$	4/100	18/160	2/180	24/440
$\langle GPL = (1,\bar{1},1), TimedOut \rangle$	0/100	0/160	18/180	18/440
$\langle GPL = (1,\bar{1},1), NotReleased \rangle$	0/100	20/160	2/180	22/440
$\langle GPL = (1,1,\bar{1}), TimedOut \rangle$	0/100	0/160	19/180	19/440
$\langle GPL = (1,1,\bar{1}), NotReleased \rangle$	0/100	20/160	1/180	21/440
$\langle Disengaged, NotReleased \rangle$	0/100	20/160	3/180	23/440
$\langle Disengaged, TimedOut \rangle$	0/100	0/160	17/180	17/440

Table 2 shows the coverage results, with a column, "TOTAL", accumulating the coverage after all the tests. These results highlight the effectiveness of model-based test generation to target the possible functionalities of the robot's code and the expected critical human behaviours. For brevity, we omitted the cross-product tuples that were not reached (13/33 as mentioned in Sect. 3.4).

4.3 Code Coverage

The coverage of the code's 212 statements, shown in Fig. 4, was collected while running the tests for cross-product coverage. The code has been grouped using the SMACH FSM structure, and the percentages vary ±2% in inner decision branches. The block of code corresponding to the object's "release" was not covered by the unconstrained tests, but it was reached by the model-based tests.

In summary, while model-based test generation ensures that the requirements and the cross-product model are covered, unconstrained test generation can construct scenarios that the verification engineer has not foreseen, particularly from the environment stimulating a robot in the HRI domain.

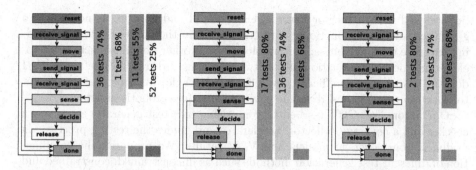

Fig. 4. Code coverage (percent values) from 100 unconstrained tests (LHS), 160 model-based (MB 1) tests (center), and 180 model-based (MB 2) tests (RHS).

5 Related Work

Although robotic code can be directly model checked, the focus of verification is on runtime errors, such as arrays out of bounds or unbounded loop executions, rather than functional requirements about the whole system interacting with its environment. Furthermore, formal tools are available only for selected sets of languages such as FRAMA-C or Ada-SPARK [24]. None of these tools are transferable to our robotic code in Python in a straight forward manner.

In generic software testing, research has focused on generating correct and valid data inputs, while exploring their state space through intelligent sampling [7], search [12], or constraint solving [16]. In robots for HRIs, however, the test generation problem goes beyond correct and valid data. The challenge is to include realistic, human-like, environment-like, timed streams of orchestrated stimuli, which interact concurrently with the robotic code. Robotic control code has been tested systematically in real-life experiments [16], in hybrid combinations of real-life and simulations [12], and in simulation [2]. Although hybrid systems methods might seem applicable, reducing our entire test generation problem to decidable hybrid automata for model checking, or hybrid models for search or sampling [11,22], is not straightforward.

Model-based test generation has been applied to software [25], either directly or modelled (e.g., timed automata in [18]). To be effective, such models must comprise enough details to be meaningful, yet must also be simple to traverse, modify and maintain [25]. Consequently, we propose to employ a two-tiered test generation approach, complementing model-based with unconstrained (pseudo-random) and constrained methods.

6 Conclusions

We presented an approach to verify and validate robotic code for HRI tasks in simulation-based testing, coupled with an automated CDV methodology to systematically explore the code under test, and reduce the likelihood that important scenarios will be overlooked. In simulation, a robot and its environment can be modelled with higher or lower levels of detail and realism, as necessary to guarantee safety and functional correctness, within the limits of testing regarding coverage exhaustiveness. Methodologies from other domains, such as microelectronics design verification and software testing, are transferable to the HRI domain, allowing more efficient and effective V&V for systems that are meant to work in uncertain and dynamic environments (e.g., robotic assistants).

Our automated CDV testbench, comprising of a test generator, a driver, a checker and a coverage collector, accelerates and guides the testing process, via feedback from coverage models and V&V results. We proposed the combination of different test generation methods such as unconstrained, constrained and model-based, towards coverage of the SUT from different angles, from respective coverage models. This reduces the need for hand-crafted directed tests. Additionally, a two-tiered test generation approach, from abstract to concrete, facilitates the efforts by dividing what otherwise would be a single complex constraint solving, search or optimization problem. Furthermore, we propose stimulating the robotic code through human, environment, sensor and actuator models –i.e., indirect stimulation–, to provide a greater level of realism in the V&V process.

Our approach is scalable not only in HRI, but for autonomous systems in general, as more complex systems can be verified using the same approach, for the actual system's code. The prototypes we have developed can be used for robot-in-the-loop and human-in-the-loop V&V, and can be adapted to work with other open-source or proprietary V&V software.

The handover example in this paper demonstrated the feasibility of implementing a systematic testing methodology, such as CDV, for a ROS-Gazebo based simulator. The experimental results demonstrate how feedback loops in the testbench can be exploited to seek coverage of the unexplored aspects of the code under test, or the environment's possibilities. Unconstrained test generation allows a degree of unpredictability in the human and/or environment, so that unexpected behaviours of the SUT may be exposed. Model-based test generation usefully complements the generation by systematically directing tests according to the requirements of the SUT, or towards combinations of simultaneous events in the environment and the robot.

In the future, we will apply systematic simulation-based testing to robots that learn, or that adapt to new situations. Additionally, we will explore different modelling formalisms for model-based test generation, seeking to include uncertainty, rationality and choice in different manners.

Acknowledgement. This work is part of the EPSRC-funded project "Trustworthy Robotic Assistants" (refs. EP/K006320/1 and EP/K006223/1).

References

1. Alexander, R., Hawkins, H., Rae, D.: Situation coverage - a coverage criterion for testing autonomous robots. Technical report, Department of Computer Science, University of York (2015)

2. Araiza-Illan, D., Western, D., Pipe, A., Eder, K.: Coverage-driven verification — an approach to verify code for robots that directly interact with humans. In: Piterman, N., et al. (eds.) HVC 2015. LNCS, vol. 9434, pp. 69–84. Springer, Heidelberg (2015). doi:10.1007/978-3-319-26287-1_5
3. Bird, D., Munoz, C.: Automatic generation of random self-checking test cases. IBM Syst. J. **22**(3), 229–245 (1983)
4. Boren, J., Cousins, S.: The SMACH high-level executive. IEEE Robot. Autom. Mag. **17**(4), 18–20 (2010)
5. Clarke, E.M., Grumberg, O., Peled, D.A.: Model Checking. MIT Press, Cambridge (1999)
6. Eder, K., Harper, C., Leonards, U.: Towards the safety of human-in-the-loop robotics: challenges and opportunities for safety assurance of robotic co-workers. In: Proceedings of ROMAN, pp. 660–665 (2014)
7. Gaudel, M.-C.: Counting for random testing. In: Wolff, B., Zaïdi, F. (eds.) ICTSS 2011. LNCS, vol. 7019, pp. 1–8. Springer, Heidelberg (2011)
8. Grigore, E.C., Eder, K., Lenz, A., Skachek, S., Pipe, A.G., Melhuish, C.: Towards safe human-robot interaction. In: Groß, R., Alboul, L., Melhuish, C., Witkowski, M., Prescott, T.J., Penders, J. (eds.) TAROS 2011. LNCS, vol. 6856, pp. 323–335. Springer, Heidelberg (2011)
9. Haedicke, F., Le, H., Grosse, D., Drechsler, R.: CRAVE: an advanced constrained random verification environment for SystemC. In: Proceedings of SoC, pp. 1–7 (2012)
10. Hartmanns, A., Hermanns, H.: A modest approach to checking probabilistic timed automata. In: Proceedings of QEST, pp. 187–196 (2009)
11. Julius, A.A., Fainekos, G.E., Anand, M., Lee, I., Pappas, G.J.: Robust test generation and coverage for hybrid systems. In: Bemporad, A., Bicchi, A., Buttazzo, G. (eds.) HSCC 2007. LNCS, vol. 4416, pp. 329–342. Springer, Heidelberg (2007)
12. Kim, J., Esposito, J.M., Kumar, R.: Sampling-based algorithm for testing and validating robot controllers. Int. J. Robot. Res. **25**(12), 1257–1272 (2006)
13. Lackner, H., Schlingloff, B.: Modeling for automated test generation a comparison. In: Proceedings of MBEES Workshop (2012)
14. Lakhotia, K., McMinn, P., Harman, M.: Automated test data generation for coverage: havent we solved this problem yet? In: Proceedings of TAIC (2009)
15. Lenz, A., Skachek, S., Hamann, K., Steinwender, J., Pipe, A., Melhuish, C.: The BERT2 infrastructure: an integrated system for the study of human-robot interaction. In: Proceedings of IEEE-RAS Humanoids, pp. 346–351 (2010)
16. Mossige, M., Gotlieb, A., Meling, H.: Testing robot controllers using constraint programming and continuous integration. Inf. Softw. Technol. **57**, 169–185 (2014)
17. Nielsen, B., Skou, A.: Automated test generation from timed automata. Int. J. Softw. Tools Technol. Transfer. **5**, 59–77 (2003)
18. Nielsen, B.: Towards a method for combined model-based testing and analysis. In: Proceedings of MODELSWARD, pp. 609–618 (2014)
19. Petters, S., Thomas, D., Friedmann, M., von Stryk, O.: Multilevel testing of control software for teams of autonomous mobile robots. In: Carpin, S., Noda, I., Pagello, E., Reggiani, M., von Stryk, O. (eds.) SIMPAR 2008. LNCS (LNAI), vol. 5325, pp. 183–194. Springer, Heidelberg (2008)
20. Pinho, T., Moreira, A.P., Boaventura-Cunha, J.: Framework using ROS and SimTwo simulator for realistic test of mobile robot controllers. In: Proceedings of CONTROLO, pp. 751–759 (2014)
21. Piziali, A.: Functional verification coverage measurement and analysis. Kluwer Academic (2004)

22. Sankaranarayanan, S., Fainekos, G.E.: Falsification of temporal properties of hybrid systems using the cross-entropy method. In: Proceedings of HSCC, pp. 125–134 (2012)
23. Stocker, R., Dennis, L., Dixon, C., Fisher, M.: Verifying brahms human-robot teamwork models. In: del Cerro, L.F., Herzig, A., Mengin, J. (eds.) JELIA 2012. LNCS, vol. 7519, pp. 385–397. Springer, Heidelberg (2012)
24. Trojanek, P., Eder, K.: Verification and testing of mobile robot navigation algorithms: a case study in SPARK. In: Proceedings of IROS, pp. 1489–1494 (2014)
25. Utting, M., Pretschner, A., Legeard, B.: A taxonomy of model-based testing approaches. Softw. Testi., Verification Reliab. **22**, 297–312 (2012)
26. Webster, M., Dixon, C., Fisher, M., Salem, M., Saunders, J., Koay, K.L., Dautenhahn, K.: Formal verification of an autonomous personal robotic assistant. In: Proceedings of AAAI FVHMS, pp. 74–79 (2014)

Congratulations, It's a Boy!
Bench-Marking Children's Perceptions
of the Robokind Zeno-R25

David Cameron[✉], Samuel Fernando, Abigail Millings, Michael Szollosy,
Emily Collins, Roger Moore, Amanda Sharkey, and Tony Prescott

Sheffield Robotics, University of Sheffield, Sheffield, UK
{d.s.cameron,s.fernando,a.millings,m.szollosy,e.c.collins,r.k.moore,
a.sharkey,t.j.prescott}@sheffield.ac.uk
http://easel.upf.edu/project

Abstract. This paper explores three fundamental attributes of the
Robokind Zeno-R25 (its status as person or machine, its 'gender', and
intensity of its simulated facial expressions) and their impact on children's
perceptions of the robot, using a one-sample study design. Results from a
sample of 37 children indicate that the robot is perceived as being a mix of
person and machine, but also strongly as a male figure. Children could label
emotions of the robot's simulated facial-expressions but perceived intensi-
ties of these expressions varied. The findings demonstrate the importance
of establishing fundamentals in user views towards social robots in supporting advanced arguments of social human-robot interaction.

Keywords: Human-robot interaction · Humanoid · Psychology

1 Introduction

The field of human-robot interaction (HRI) is rapidly developing [19]; while this
is extremely promising in addressing ground-breaking questions, it can mean that
fundamental assumptions are not critically assessed. In many cases, fundamental
assumptions based solely on good common-sense could lead to fruitful results,
but this comes at a risk of relying on the 'valuable but inherently dangerous
resource available' [8] of common-sense. Developing a solid empirical foundation
on which to support boundary-pushing research, serves to strengthen established
findings and generate new paths to explore.

This paper uses the Robokind, Zeno humanoid robot [9], Fig. 1, as a case-
study for exploration of three fundamentals in user perceptions of a robot[1]. The
Zeno series of robots are capable of generating life-like simulated facial expressions [9], developed with the aim of commercial release as interactive, educational
and play partners for children (as is the now retired-from-production Alice model
counterpart). The current production model of Zeno (Zeno-R25) is described as a

---

[1] These - or related - fundamentals could apply beyond the Zeno R-25 case-study to
other humanoids or even non-humanoid social-robots.

L. Alboul et al. (Eds.): TAROS 2016, LNAI 9716, pp. 33–39, 2016.
DOI: 10.1007/978-3-319-40379-3_4

Fig. 1. The Robokind Zeno R25 platform (humanoid figure approximately 60 cm tall)

humanoid boy robot in Robokind's promotional literature [15]. Further to this, various Zeno models are referred to as humanoid-boys, or variations thereof, across multiple papers [4,5,10,14,17,18].

To date, it seems there is no bench-marking of Zeno's 'gender' in the research literature. Appropriate confirmation of this common-sense assumption about a fundamental perception: Zeno's 'gender' is important. Recent literature suggests differences in boys' and girls' interactions with, and responses towards, Zeno could be due to their perceptions of the robot as being a boy (who just happens to be mechanical) [5].

A further fundamental assumption made across the literature is the life-like or human-like appearance and nature of the robot [1,2,5,14]. At present, one study indicates that while children perceive Zeno to be marginally more like a person than a machine; however, this is influenced by child's age [6]. Again, fundamental assumptions made about the nature of the robot could impact findings upstream; the expectancies children have for Zeno's social behaviour could be impacted by their particular beliefs surrounding its life-like (or otherwise) appearance.

One fundamental that the literature *does* address is the validity of the robot's simulated facial expressions, particularly people's accuracy in decoding these. Zeno's simulated facial expressions have been accurately decoded by adults [18] and children [7,16]. While research explores people's decoding of different expressive states, it currently does not yet consider children's interpretations of the *intensity* of emotion communicated by the facial expressions.

2   Method

2.1   Design

The bench-marking study primarily used a one-sample design; participant responses were compared against predefined means; predefined means for this study comprised of the midpoints of each scale (for scale midpoints, see Sect. 2.3). Significant differences between participants' scores and the predefined means indicate that participants had, on average, made a choice tending towards the scales' endpoints.

2.2   Participants

The study took place as part of a single-day exhibit at a museum in the UK. Children visiting the exhibit were invited to participant in the study by playing a game with Zeno titled 'Guess the robot's expressions'. Forty-Three children took part in the study in total, although six children (each aged three or below) did not complete all measures and are not included in analysis. Of the remaining participants, 15 were female and 22 were male (M age = 6.73, SD = 2.16).

2.3   Measures

Our first measure was children's perceptions of Zeno as being more like a person or machine. To assess this, children completed the statement 'I think Zeno is' with one of the following five responses on a Likert scale: 'Lots more like a machine', 'A bit more like a machine', 'An even mix of machine and person', 'A bit more like a person', 'Lots more like a person' (scale endpoints of -2: 'a lot like a machine' and 2: 'a lot like a person').

Children were given a follow-up question (regardless of their first response) concerning their perceptions of the robot's gender. This followed same format as the prior question with the terms 'machine' and 'person' replaced with 'boy' and 'girl', respectively. Children were also given the option for 'not either'.

Two further measures were taken to assess children's perceptions of Zeno's 'Happy' and 'Sad' facial expressions. Children pointed to one of the five pictures on the Self-Assessment Manikin for valence that they thought best matched Zeno's expression for each (SAM; valence scale endpoints: -2 to 2 [3]).

2.4   Procedure

The experiment took place in a publicly accessible lab. Interactions lasted approximately five minutes. Brief information about the experiment was provided to parents/carers and informed consent for participation was obtained prior to participation. If children were hesitant in the study, parents were available to offer reassurance or help clarify questions asked all parents were made aware of the children's right to withdraw at any point in the study. Ethical approval for this study was obtained prior to any data collection.

The children were told that they were going to play a quick game of guessing the expression made by Zeno. The experimenter pressed one of two buttons on the front of the Zeno-R25 to generate the Happiness and Sadness simulated facial expressions (children were given no indication as to which expression the button corresponded to). Children responded by pointing to their chosen answers on the SAM on a sheet of paper.

The children then answered the two questions about the robot's status (person/machine, boy/girl, see Sect. 2.3), by pointing to the answers that best matched their opinion. The order of questions presented was alternated (i.e. questions of the robot's status, before or after expressions) to counterbalance any effect the expressions may have.

After all interactions, children and their parents were given opportunity to ask questions and interact with the robot. At the close of the interaction, the experimenter pressed the button for Zeno to wave goodbye.

3   Results

3.1   Robot Animacy

A one-sample T-test indicated that children's classification of the robot as machine or person was not significantly different from reporting the robot as being an even mix of machine and person, $t(36) = 1.34$ $p = .19$ (M = -.30 SD = 1.35). Responses were distributed across all five points on the scale (see figure), with a plurality of responses (27.03 %) of 'an even mix of machine and person'.

3.2   Robot Gender

A one-sample T-test indicated that children's classification of the robot's gender were significantly different from reporting the robot as being an even mix of boy and girl, $t(36) = 18.12$ $p < .01$ (M = -1.64, SD = .54). The majority of responses (66.67 %) given were for Zeno being 'a lot like a boy', 30.56 % of responses for 'a bit like a boy' and only one response (2.78 %) for 'an even mix of boy and girl'. While not formally assessed, children often volunteered their reasoning, with modal reason given regarding Zeno's 'short hair, like a boy's'.

There was no difference in children's ratings based on their own gender ($F(1,32) = .65$, $p = .43$) nor their age (Median split of groups aged six-and-under or over-six; $F(1,32) = .92$, $p = .34$). Across all groups, children consistently report the Zeno robot as being more like a boy. Only one child (age 12) responded that robots are neither boys nor girls.

3.3   Robot Expressions

Children accurately classified the robot's simulated expressions. For simulated happiness expression (M = 1.32, SD = .71), a one-sample T-test indicated that children's responses were significantly different from reporting the robot as having a neutral expression $t(36) = 11.36$, p < .01. For simulated sadness expression (M = -1.14, SD = .79), a one-sample T-test indicated that children's responses were significantly different from reporting the robot as having a neutral expression $t(36) = -8.77$, p < .01. Participant ratings are presented in Fig. 2.

4   Discussion

Our findings indicate that assumptions made in the research literature about the Zeno R-25 have good grounding. Children view the robot as male, albeit a mix of person and machine. This outcome supports existing research [6] and suggests

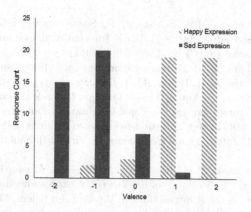

Fig. 2. Frequency count of participant ratings for the robot's 'Happy' and 'Sad' expressions

that exploring which factors children use to identify a humanoid-robot's status as person or machine could benefit future robot and research designs.

Children perceiving the robot as male has implications for studies concerning social HRI (e.g., [5]) because children typically show gender differences in their preference of, and behaviour towards, play-partners [13]. In particular, younger children may view humanoid robots as persons to play with rather than objects to interact with [6]. Seemingly arbitrary design choices may discourage individuals, in this case girls, from interaction with humanoid robots and the learning opportunities this may afford.

Children successfully identify simulated facial expressions on the Zeno-R25 but vary in their interpretation of the affect intensity. Further work could establish if this recognition is a cognitive process (e.g., primary channel processing [11]) or if this is supported by processes such as mimicry and emotion contagion [12]. Establishing the mechanisms that children use to determine the robot's expressed state could further improve robot and HRI-scenario development.

These findings indicate the importance of critically assessing the fundamental assumptions that can be made in HRI. Addressing these creates steps towards a robust and empirically-based taxonomy of robots to support boundary-pushing HRI research, strengthen established findings, and generate new paths to explore.

Acknowledgments. This work was supported by European Union Seventh Framework Programme (FP7-ICT-2013-10) under grant agreement no. 611971.

References

1. Admoni, H., Bank, C., Tan, J., Toneva, M., Scassellati, B.: Robot gaze does not reflexively cue human attention. In: Proceedings of the 33rd Annual Conference of the Cognitive Science Society, Boston, MA, USA pp. 1983–1988 (2011)

2. Bethel, C.L., Stevenson, M.R., Scassellati, B.: Secret-sharing: interactions between a child, robot, and adult. In: 2011 IEEE International Conference on Systems, man, and cybernetics (SMC) pp. 2489–2494 (2011)
3. Bradley, M.M., Lang, P.J.: Measuring emotion: the self-assessment manikin and the semantic differential. J. Behav. Ther. Exp. Psychiatry **25**, 49–59 (1994)
4. Bugnariu, N., Garver, C., Young, C., Ranatunga, I., Rockenbach, K., Beltran, M., Patterson, R.M., Torres-Arenas, N., Popa, D.: Human-robot interaction as a tool to evaluate and quantify motor imitation behavior in children with autism spectrum disorders. In: 2013 International Conference on Virtual Rehabilitation (ICVR) pp. 57–62 (2013)
5. Cameron, D., Fernando, S., Collins, E.C., Millings, A., Moore, R.K., Sharkey, A., Evers, V., Prescott, T.: Presence of life-like robot expressions influences children's enjoyment of human-robot interactions in the field. In: Salem, M., Weiss, A., Baxter, P., Dautenhahn, K., (eds.) 4th International Symposium on New Frontiers in Human-Robot Interaction, pp. 36–41 (2015)
6. Cameron, D., Fernando, S., Millings, A., Moore, R., Sharkey, A., Prescott, T.: Children's Age Influences Their Perceptions of a Humanoid Robot as Being Like a Person or Machine. In: Wilson, S.P., Verschure, P.F.M.J., Mura, A., Prescott, T.J. (eds.) Living Machines 2015. LNCS, vol. 9222, pp. 348–353. Springer, Heidelberg (2015)
7. Costa, S., Soares, F., Santos, C.: Facial expressions and gestures to convey emotions with a humanoid robot. In: Herrmann, G., Pearson, M.J., Lenz, A., Bremner, P., Spiers, A., Leonards, U. (eds.) ICSR 2013. LNCS, vol. 8239, pp. 542–551. Springer, Heidelberg (2013)
8. Fletcher, G.J.: Psychology and common sense. Am. Psychol. **39**, 203 (1984)
9. Hanson, D., Baurmann, S., Riccio, T., Margolin, R., Dockins, T., Tavares, M., Carpenter, K.: Zeno: a cognitive character. In: AI Magazine, and special Proceedings of AAAI National Conference, Chicago (2009)
10. Hanson, D., Mazzei, D., Garver, C., Ahluwalia, A., De Rossi, D., Stevenson, M., Reynolds, K.: Realistic humanlike robots for treatment of ASD, social training, and research; shown to appeal to youths with ASD, cause physiological arousal, and increase human-to-human social engagement. Environment. In: PETRA (PErvasive Technologies Related to Assistive) (2012)
11. Hareli, S., Rafaelli, A.: Emotion cycles: on the social influence of emotion in organizations. Res. Organ. Behav. **28**, 35–99 (2008)
12. Hatfield, E., Cacioppo, J.T., Rapson, R.L.: Emotional Contagion. Cambridge University Press, New York (1994)
13. Martin, C.L., Fabes, R.A.: The stability and consequences of young children's same-sex peer interactions. Dev. Psychol. **3**, 431–446 (2001)
14. Ranatunga, I., Rajruangrabin, J., Popa, D.O., Makedon, F.: Enhanced therapeutic interactivity using social robot Zeno. In: Proceedings of the 4th International Conference on PErvasive Technologies Related to Assistive Environments pp. 57–62 (2011)
15. http://www.robokindrobots.com/robots4autism-home/
16. Salvador, M.J., Silver, S., Mahoor, M.H.: An emotion recognition comparative study of autistic and typically-developing children using the Zeno robot. In: 2015 IEEE International Conference on Robotics and Automation (ICRA) pp. 6128–6133 (2015)
17. Sanders, D.: Progress in machine intelligence. Ind. Robot Int. J. **35**, 485–487 (2008)

18. Si, M., McDaniel, D.J.: Creating genuine smiles for digital and robotic characters: an empirical study. In: Games Entertainment Media Conference (GEM) pp. 1–6 (2015)
19. Yanco, H.A., Drury, J.L.: A taxonomy for human-robot interaction. In: Proceedings of the AAAI Fall Symposium on Human-Robot Interaction pp. 111–119 (2002)

A Bioinspired Approach to Vision

Daniel Camilleri[(✉)], Luke Boorman, Uriel Martinez,
Andreas Damianou, and Tony Prescott

Psychology Department, University of Sheffield,
Western Bank, Sheffield, UK
d.camilleri@sheffield.ac.uk
http://www.sheffield.ac.uk

Abstract. This paper describes the design of a computational vision framework inspired by the cortices of the brain. The proposed framework carries out visual saliency and provides pathways through which object segmentation, learning and recognition skills can be learned and acquired through experience.

Keywords: Human visual system · Bioinspired computing · Computational model · Vision model

1 Introduction

Vision processing is the major signal processing pipeline of the human brain since more than 70 % of outside information is assimilated and understood through our visual sense [1] thus making it the richest source of information on the immediate surroundings. It is no wonder therefore that this sensory system started very early on in the history of evolution. Fast forwarding to today we find that this visual system is now present in most cognitive creatures not least of all in humans who have a very complex and powerful visual system.

Thus, given that the visual system is so important to the understanding of everyday life, much work has gone into identifying first of all the roles of the different visual cortices in the human brain and secondly the processing occurring within each cortex in the hopes of implementing a computational model of the Human Vision System (HVS). The implementation of such a model would be very beneficial in areas such as mobile robotic navigation and machine vision to name a few.

The layout of this paper is thus as follows. Section 2 contains the investigation of the eye and the various cortices of the human brain together with their individual function and how they are connected together. Section 3 builds on the information in Sect. 2 by identifying computational equivalents to create the final computational model. Section 4 details the work that has been carried out so far and finally Sect. 5 concludes with a recap followed by a description of upcoming work.

© Springer International Publishing Switzerland 2016
L. Alboul et al. (Eds.): TAROS 2016, LNAI 9716, pp. 40–52, 2016.
DOI: 10.1007/978-3-319-40379-3_5

2 Visual Processing in the Human Brain

The research being carried out on the visual cortices of the brain in literature takes the form of two distinct but overlapping streams. On one side psychological, neurological and physiological research aims to understand the exact methods by which the brain carries out its day to day visual processing with carefully structured studies based on human, monkey and rat subjects. This research contributes to the understanding of the brain and has been going on since the mid-19th Century.

The other stream is that of Computational Vision Modelling which started in the late 20th Century and whose research focuses on the implementation of visual models which are either derived from the understanding obtained in the other stream, or a simplification thereof.

This computational implementation together with validation against respective datasets constitutes a way of not only improving the understanding derived in the psychological studies but also a way of validating them and their underlying assumptions of operation. This contributes to further refining the initial research and/or point out differences and cases which are irreconcilable with the current model. In this section an overview is given of some of the visual cortices in the brain and the starting point of this overview is the eye itself.

2.1 Mechanical Structure of the Eye

The eye is made up of three pairs of opposing extra-ocular muscles that provide a one to one mapping of muscle configuration to eye gaze [2]. The eye is capable of performing five main types of movements. The first two, related to involuntary reflexes, called the vestibular-ocular and optokinetic reflexes, keep the point of fixation constant in the presence of bodily motion [3].

Another two types of movement, relate to the pursuit and the vergence systems [4] of the eye that track moving objects both in terms of its planar position (eyes moving in parallel to each other) as well as in depth (eyes converging or diverging to focus on an object at a particular depth). The fifth and final eye movement is the saccade which directs attention to specific areas within the surrounding environment.

One can already see a central theme emerging from the movements of the eyes demonstrating that the motion of the eyes is completely dedicated towards focusing attention in a directional manner on separate objects at a time. Moreover as described in [5] the direction depends on the informational and behavioural value of the particular point in space at that point in time.

2.2 Visual Processing in the Retina

The next step in the visual hierarchy is the retina and this is where the processing of the raw input information starts to occur. As shown by [6] the retina is responsible for the low-level image processing within the HVS and is made up of photoreceptors, bipolar cells, and ganglion cells amongst others which are loosely grouped to form two cell layers called the Outer Plexiform Layer (OPL) and the Inner Plexiform Layer (IPL).

Starting at the OPL, this layer performs log luminance equalisation of the incoming scene and applies a non-separable spatio-temporal filter to the input. This both increases the dynamic range of the input and also highlights areas within the image that contain high frequencies either in the spatial or in the temporal domain. This information is then passed to the IPL which applies two different operations. The first operation described as the parvocellular pathway further enhances the textures and contours of the scene. The second operation which occurs in parallel to the first enhances the motion detection in the scene and is carried out in the magnocellular pathway.

Once this processing has been carried out the information is then passed to the rest of the brain via the optic nerve which branches and enters both the Lateral Geniculate Nucleus and the Superior Colliculus [7].

2.3   The Lateral Geniculate Nucleus (LGN) and the Superior Colliculus (SC)

The LGN and the SC are both situated in the mid-brain and each cortex executes a different processing function. The LGN on one hand is responsible for extracting features such as colour and contrast [8] from the information extracted by the parvocellular pathway and it is thought to do so through the extraction of colour-antagonists [9]. This compresses the information from the 3 colours detected by 3 separate cone types [10] in the retina to 2 streams of antagonistic colours, namely red-green and blue-yellow.

The SC on the other hand is subdivided into 7 cellular layers which are divided into the superficial and the deep layers. The latter receives sensory input from multiple senses including vision and sound but the former exclusively receives visual input directly from the retina. This retinal input has been shown by Sabes et al. [11] to form retinotopic maps on the surface of this cortex which preserve the link to the spatial location of each input region.

This input to the superficial layers is then processed to extract visually transient information such as flicker (temporal changes in light intensity) as well as motion stimuli [12]. The output of the SC has been demonstrated in primates to stimulate the generation of spatially averaged saccades through the combination of multi-sensory information in the deep layers [13] but it also performs saccadic suppression [14] resulting in the inhibition of saccades towards uninteresting regions of the visual field.

2.4   Visual Processing Streams in Biology

From the LGN and SC onwards, the information flow splits into two main streams which are the Dorsal Stream and the Ventral Stream. These streams have been advocated and refined since their initial proposition by Mishkin et al. [15] in 1983 and can be seen in Fig. 1. The dorsal and ventral streams are also colloquially called the 'where' and 'what' streams respectively. As described by

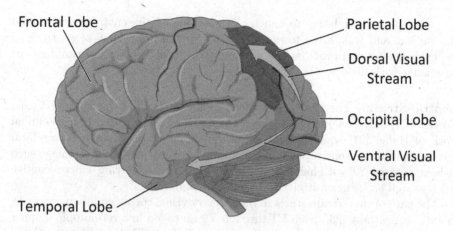

Fig. 1. Diagram of the position of visual cortices in the brain [17].

[16], the dorsal stream is concerned with processing the spatial features of an image and is the main driver of saccadic movements and thus visual attention.

On the other hand the ventral stream is tasked with identifying objects in the scene through the high level representation of said objects in the subject's memory. This stream drives saccades in a more indirect manner and often only in the availability of task specific demands. The following paragraphs will describe in more detail the functionality and role of the cortices for the dorsal and ventral stream which are relevant to the scope of this project.

Dorsal Stream. As stated above, the Dorsal stream is the 'where' stream and thus its main role is in the redirection of gaze as part of the oculomotor system. The structure of this stream developed in [18] depicts the connections between the different cortical regions in the human brain for the dorsal stream, of which the Frontal Eye Fields (FEF) and Supplementary Eye Fields (SEF) are of particular interest with relevance to visual saliency, which although not the sole function of these cortices, is a principal component.

Frontal Eye Fields (FEF). The FEF are described by literature as the "principle saccadic decision structure together with the SC" [16] and just like the SC, this cortex keeps a retinotopic map of the field of vision. Furthermore the work in [19] has shown that this cortex combines not only the information from both the dorsal and ventral stream to decide on a saccade target but has also been shown to accept input biases from areas in the Pre-Frontal Cortex (PFC) cortex that modify the selectiveness of certain properties providing a path where a high-level cortical process can tweak the functionality or priority of lower-level cortical operation such as saliency.

Supplementary Eye Fields (SEF). The SEF are linked heavily with the FEF but while the latter generates saccadic targets, the former's task is in keeping a craniotopic [20] (relative to the head) mapping of the environment around the subject and thus hints towards a region of the brain that keeps track of the environment and provides the required data to fill in gaps in instantaneous knowledge with historic data.

Ventral Stream. The Ventral or 'what' stream starts off at the Primary Visual Cortex (V1) and was one of the first cortices to be investigated in the seminal work of Hubel [21] which describes the functionality of V1 as a hierarchical arrangement of neurons capable of extracting local features from an image such as bars or edges [22] with higher levels of the hierarchy displaying wider receptive fields as well as an insensitivity to orientation and scale.

The path of this stream starts in the primary visual cortex and then proceeds, mainly sequentially [23], from V1 through V2 up to V5 in a retinotopic manner [24,25]. Finally, it passes through the Posterior Inferior Temporal Cortex (PIT) followed by the Anterior Inferior Temporal Cortex (AIT) which is where this visual stream ends. From here on, it branches into the medial temporal lobe and PFC cortex whose feature biasing role has been described initially. Two interesting features of the Inferior Temporal Cortex (IT) include highly specialised cells that respond to very specific stimulus along with a response that is invariant to the number of objects present called Cardinality Blindness which is an effect of the trend for wider receptive fields and rotation invariance displayed by the ventral stream.

As such, instead of having a retinotopic map like the dorsal stream, the ventral stream at the level of the IT is described as having a sparse representation of all current recognisable and behaviourally relevant objects in the visual field with more of an emphasis on the classification rather than the location of said object.

In view of this basic introduction to some of the cortical regions relevant to the HVS, the next section presents a computational framework that encompasses the functionality of all the different regions.

3   Human Visual System Computational Framework

The consolidation of facts from the previous section points towards the 6 low-level features of importance to the HVS: Intensity, Red-Green Antagonist and Blue-Yellow Antagonist attributed to the LGN, Flicker and Motion attributed to the SC and Orientation of edges attributed to the Primary Visual Cortex V1. These 6 features form the basis upon which the HVS carries out the rest of its functions. As such the proposed Computational Framework looks at the extraction of these features followed by their possible application specifically to visual saliency, object learning and object recognition from a robotic viewpoint.

3.1   Visual Saliency

According to [16], visual attention is defined as the "process of enhancing the responses of neurons that relate a subset of the visual field with the purpose of overcoming the computational limitations of the visual system" and its importance as mentioned before lies in the reduction of the computational burden on the human brain.

Building on this, the book *Selective Visual Attention : Computational Models and Applications* [1], through the use of various experiments and observations, identifies two important aspects of visual attention. The first aspect is that visual attention is divided temporally into pre-attention followed by attention. The second aspect is that the attention phase has two operating modes which are Top-Down and Bottom-Up.

Pre-Attention and Attention. This aspect of visual attention consisting of two sequential processes is described by [26–28]. The first process, pre-attention, performs the extraction of the low level features. This is performed very quickly, in parallel and on the whole visual field imitating the functionality of the LGN, V1 and part of the SC. Subsequently, once all the features have been explicitly extracted, the second process, attention, takes over. In this stage, the low level features are combined together through a process called feature integration which is proposed in [28].

However according to [28], feature integration in the HVS is only carried out through visual attention. Therefore, this means that although the visual system could perceive the presence of multiple objects with a particular combination of features, it does not explicitly know their location due to the cardinality of the ventral stream. Thus the HVS has to resort to a serial search over the visual field in order to locate objects with a combination of features.

Top-Down and Bottom-Up Drives. Secondly, according to [29–31], a subject's gaze is drawn to a particular point due to a combination of bottom-up stimuli that arise solely from changes within the scene and top-down stimuli that modulate the attention depending on the task at hand. This separation of function provides a powerful scheme for object recognition mimicking the feedback loop of the FEF and PFC. Furthermore it has led to the development of two classes of saliency algorithms in literature which are either unguided (bottom-up) or guided (top-down effects).

In the unguided scenario, also known as 'free-roaming', the weights for all feature maps are identical and the final saliehcy map is generated through the summation of all equally weighted feature maps like Itti et al's Baseline Saliency Model [32]. On the other hand the guided scenario modifies the feature weightings independently depending on the task at hand, examples of which are [29, 30] which demonstrate limited top-down effects due to the complexity of factors contributing to this mode.

Of important note is that humans start out life with unguided saliency [33] and later start paying guided attention to specific objects signifying an accumulation of knowledge before the application of top-down effects: knowledge which arises from object learning.

3.2   Proposed Model and Its Implementation

Figure 2 provides a visualisation of the information flow within the model incorporating all the processes which have been mentioned so far while extending the application to object recognition and including a path for top-down modulation of saliency. The main contribution of this model is the use of a memory model that is very similar in functionality to that of the a human memory system and the use of this memory model for the purpose of object learning, recognition and searching. The process starts with the eyes where image capture is followed by the extraction of the corresponding parvo and magno images using the algorithm developed in [6].

Subsequently, the overlapping area of both parvo images is extracted in order to be processed for R-G, B-Y and Orientation while both images are stitched together for the extraction of Intensity and Flicker. This is analogous to human vision where the central region of focus is processed to identify details and has a high presence of cones attributed to colour perception while changes in light intensity are computed over the whole field of view due to the uniform density of rods in the retina. [34]. Similairly, motion is described as the temporal change of light intensity on an array of detectors by [35] thus the extraction of motion is also carried out over the whole image which is why magno images are stitched together.

From this point, base feature map extraction at the capture resolution for the LGN and V1 block follows the process of visual saliency extraction by Itti et al. [32]. In the case of the SC block, flicker base map is computed as described in [36] while the extraction of motion requires no further computation to extract the base feature map. Subsequently a Gaussian pyramid with 8 levels is created for each base feature image imitating the effect of center-on surround-off effects of the retina. The 8 levels are then normalised and collapsed to a level of choice into a conspicuity map. The lower the level, the higher the spatial resolution of the saliency map but the lower the noise to signal ratio.

From here on, the conspicuity maps are combined and averaged depending on the weights from the PFC block which are initially set as equal. The rest of the blocks then carry out object learning and recognition based on the extracted features.

In this framework, object learning is focused and directed by saliency much like infants. [37]. Thus once a salient point has been identified, a feature vector describing the point is created which, in turn, is used to find all areas within the image that correspond to this vector thus creating blobs of common regions. The blob which contains the salient point is subsequently chosen and a super-vector is created from the feature vectors of all the pixels in the blob. This feature vector is then passed to the PFC which carries out statistical learning on the super-vector with the use of SAM [38]. Thus given multiple inputs over the course of multiple frames, SAM creates a latent feature model of the input data which condenses the data into its most important features and also allows for the labelling of blobs. This then allows for one of two processes to take place.

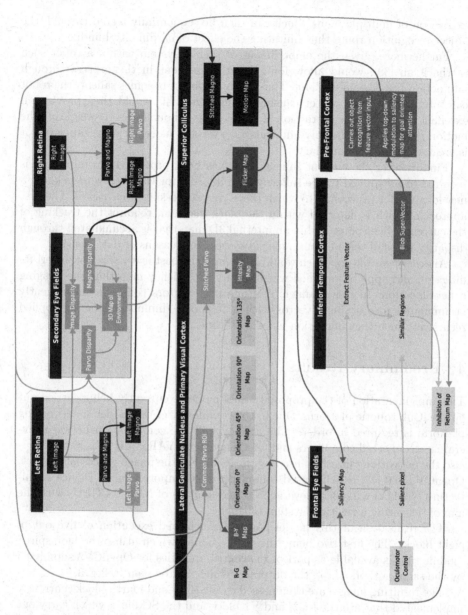

Fig. 2. Proposed Block Diagram for the Computational Framework (Color figure online)

. The first, given an input blob, one could carry out object recognition and return the label of the currently salient object. The second, given a label to look for, SAM returns a feature vector best describing the object with a mean and variance, which is applied as weighting to the saliency map. This results in the extraction of multiple blobs which can then be sequentially tested through the object recognition route thus imitating the effect of cardinality blindness.

Furthermore, given the embodiment of such a system within a robot such as the iCub [39], would allow for a great refinement in the process through interaction just like infants [40]. Thus when an object becomes salient, the robot can interact with the object causing its motion. This would then provide an excellent trigger for the extraction of precise object boundaries by weighting the motion saliency more heavily and thus a better and more accurate super-vector is created given this boundary.

Finally, after the current salient point has been processed, an inhibition is required to be applied to the saliency map in order for the observer to choose a new location for processing to which there are two possible approaches. The first approach, which is depicted within the block diagram, requires the tracking of the current salient point within an internal 3D map that is accumulated through disparity maps obtained from multiple sources for reasons of data density.

Another possible approach would be to keep the past feature vector, invert its direction and apply a modified weighting to the saliency map thus encouraging a new location. In this manner, multiple past vectors could be subsequently accumulated into a single vector with a temporal diminishing factor applied recursively such that inhibition fades with time.

4   Preliminary Results

The implementation of the proposed visual model is currently being carried out on the iCub robotic platform. Due to the complexity of the model, a networking platform is required in order to be able to split the computation between several interconnected computers and for this purpose YARP [41] is used since it is also the networking interface used by the iCub. For the image processing aspect, OpenCV 3.0.0 [42] is being used compiled with Compute Unified Device Architecture (CUDA) 6.5 [43] to leverage the acceleration of NVIDIA GPUs with the aim of delivering a realtime system (30 fps).

The realisation of this model has so far achieved execution of five out of eight blocks. The first two being the retina models which utilise the bioinspired module that is available as part of the contrib modules for OpenCV3 submitted by the same authors of [6], the output of which can be seen in Fig. 3.

The resulting images are then passed to the third and fourth blocks currently implemented that are the LGN and V1 block and the SC block which carry out Itti's Baseline Saliency Model [32] for the generation of a saliency map as shown in Fig. 4 at a resolution of 40×32 for input images at 640×480 resolution.

From this Saliency Map, the most salient pixel is chosen and passed to the oculomotor controller that directs the gaze of the iCub towards that location. Furthermore, a primitive inhibition of return has also been applied as a substitute for the inhibition of return that will be implemented after object learning has been achieved. The current method retains a 40×32 map in memory with saliency inhibition values applied to the respective pixel that is currently the

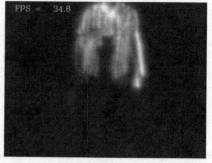

(a) Enhanced Texture and Contours   (b) Motion map

Fig. 3. Parvo and Magno output for the Retina Model with their corresponding frame rates for CPU computation at an input resolution of 640×480

Fig. 4. Saliency Map at a resolution of 40×32

focus of attention. This primitive implementation does not take into consideration that the iCub head changes orientation as it looks towards a salient point but still, the result, which can be seen in [44], displays promising behaviour.

Furthermore there are two main avenues for future improvement. The first is the implementation of a computationally fast stitching process that takes advantage of the known camera orientations which would allow for the application of separate regions for colour processing and motion processing because so far, all stages beyond the retina use the overlapped region thus mimicking a retina that has a constant distribution of both cones and rods.

The second improvement deals with the effect of camera motion on the magno and parvo images. Currently, the embodiment of the system results in blurry magno and parvo images whenever a saccade or head motion is executed. It just happens that the HVS also encounters this problem which it solves through the application of a process called saccadic masking [45]. Thus blur will be mitigated in the future through a communication loop that blocks image input to the model whenever an oculomotor command is sent.

The model's computation is divided between 2 computers. The first carries out both retina models on a Xenon 6 Core CPU at 2.8 GHz which is a purely CPU based implementation running at an average of 25 fps. The second carries out the saliency computation and robot control on a Core i7-3630QM CPU running at 2.4 GHz with NVIDIA GTX 675MX GPU having 4 GB of VRAM at an average rate of 15 fps. Thus assuming negligible transfer delays due to low resolution images being transferred on high bandwidth local networks, the whole system runs at a throughput of 15 fps with an end-to-end delay of approximately a 100 ms.

5   Conclusion

In conclusion, [28] states that "without attention, general purpose vision is not possible" thus this work has reviewed the different cortices in the human brain that process the visual input of the eyes and identified the key functionalities of each. Subsequently, these key functions were summed into an extraction of six important low-level features and their role within visual saliency.

Following this, bioinspired computational implementations for visual saliency were investigated to establish the computation of these features as well as that of the saliency map. A method for object learning, object recognition and possible interactive object segmentation are then proposed, together with two possibilities for the application of inhibition of return which leads to dynamic behaviour. Finally a description of the work that has been carried out is provided with performance results for the currently implemented blocks as well as some pitfalls that have been encountered along the way.

As one can see the implications of the model are very powerful and allow for a range of possibilities in behaviour. Currently the implementation of this model is in development and future work will look into documenting the efficacy of learning using the process state above as well as the level of fidelity with datasets of human visual saliency.

Acknowledgments. I would like to thank all those who helped and inspired me especially Tony Prescott, Luke Boorman, Uriel Martinez and Andreas Damianou.

References

1. Zhang, L., Lin, W.: Selective Visual Attention: Computational Models and Applications. John Wiley & Sons, Hoboken (2013)
2. Subramanian, P.S.: Active vision: the psychology of looking and seeing. J. Neuro Ophthalmol. **26**(1), 69–70 (2006)
3. Schweigart, G., Mergner, T., Evdokimidis, I., Morand, S., Becker, W.: Gaze stabilization by optokinetic reflex (okr) and vestibulo-ocular reflex (vor) during active head rotation in man. Vis. Res. **37**(12), 1643–1652 (1997)

4. Robinson, D.A.: Eye movement control in primates. Science **161**(3847), 1219–1224 (1968)
5. Henderson, J.M.: Human gaze control during real-world scene perception. Trends Cogn. Sci. **7**(11), 498–504 (2003)
6. Benoit, A., Caplier, A., Durette, B., Hérault, J.: Using human visual system modeling for bio-inspired low level image processing. Comput. Vis. Image Underst. **114**(7), 758–773 (2010)
7. Schiller, P.H.: Central connections of the retinal on and off pathways. Nature **297**, 580–583 (1982)
8. Gao, D., Mahadevan, V., Vasconcelos, N.: The discriminant center-surround hypothesis for bottom-up saliency. In: Advances in neural information processing systems. pp. 497–504 (2008)
9. Rodieck, R.: Which cells code for color? In: Valberg, A., Lee, B.B. (eds.) Advances in Understanding Visual Processes. NATO ASI Series, vol. 203, pp. 83–93. Springer, New York (1991)
10. Mullen, K.T.: The contrast sensitivity of human colour vision to red-green and blue-yellow chromatic gratings. J. Physiol. **359**(1), 381–400 (1985)
11. Sabes, P.N., Breznen, B., Andersen, R.A.: Parietal representation of object-based saccades. J. Neurophysiol. **88**(4), 1815–1829 (2002)
12. Wurtz, R.H., Albano, J.E.: Visual-motor function of the primate superior colliculus. Ann. Rev. Neurosci. **3**(1), 189–226 (1980)
13. Glimcher, P.W., Sparks, D.L.: Representation of averaging saccades in the superior colliculus of the monkey. Exp. Brain Res. **95**(3), 429–435 (1993)
14. Sommer, M.A., Wurtz, R.H.: Visual perception and corollary discharge. Perception **37**(3), 408 (2008)
15. Mishkin, M., Ungerleider, L.G., Macko, K.A.: Object vision and spatial vision: two cortical pathways. Trends in Neurosci. **6**, 414–417 (1983)
16. Cope, A.J.: The role of object recognition in active vision : a computational study. Ph.d, September 2011
17. Illustration from anatomy & physiology, connexions web site. https://commons. wikimedia.org/wiki/File:1424_Visual_Streams.jpg. Anatomy & Physiology, Connexions Web site. http://cnx.org/content/col11496/1.6/, 19, June 2013. Accessed 1 Feb 2016
18. Hikosaka, O., Takikawa, Y., Kawagoe, R.: Role of the basal ganglia in the control of purposive saccadic eye movements. Physiol. Rev. **80**(3), 953–978 (2000)
19. Bichot, N.P., Thompson, K.G., Rao, S.C., Schall, J.D.: Reliability of macaque frontal eye field neurons signaling saccade targets during visual search. J. Neurosci. **21**(2), 713–725 (2001)
20. Schall, J.D., Morel, A., Kaas, J.H.: Topography of supplementary eye field afferents to frontal eye field in macaque: implications for mapping between saccade coordinate systems. Vis. Neurosci. **10**(02), 385–393 (1993)
21. Hubel, D.H., Wiesel, T.N.: Receptive fields, binocular interaction and functional architecture in the cat's visual cortex. J. Physiol. **160**(1), 106 (1962)
22. Lee, T.S., Mumford, D., Romero, R., Lamme, V.A.: The role of the primary visual cortex in higher level vision. Vis. Res. **38**(15), 2429–2454 (1998)
23. Felleman, D.J., Van Essen, D.C.: Distributed hierarchical processing in the primate cerebral cortex. Cereb. Cortex **1**(1), 1–47 (1991)
24. Tanaka, K.: Inferotemporal cortex and object vision. Ann. Rev. Neurosci. **19**(1), 109–139 (1996)
25. Gross, C.G.: Single neuron studies of inferior temporal cortex. Neuropsychologia **46**(3), 841–852 (2008)

26. Neisser, U.: Cognitive Psychology, Classic edn. Psychology Press, New York (2014)
27. Hoffman, J.E.: Hierarchical stages in the processing of visual information. Percept. Psychophys. **18**(5), 348–354 (1975)
28. Koch, C., Ullman, S.: Shifts in selective visual attention: towards the underlying neural circuitry. In: Vaina, L.M. (ed.) Matters of Intelligence. Synthese Library, vol. 188, pp. 115–141. Springer, Dordrecht (1987)
29. Wolfe, J.M.: Guided search 2.0 a revised model of visual search. Psychon. Bull. Rev. **1**(2), 202–238 (1994)
30. Wolfe, J.M., Cave, K.R., Franzel, S.L.: Guided search: an alternative to the feature integration model for visual search. J. Exp. Psychol. Hum. Percept. Perform. **15**(3), 419 (1989)
31. Navalpakkam, V., Itti, L.: Top-down attention selection is fine grained. J. Vis. **6**(11), 4 (2006)
32. Itti, L., Koch, C., Niebur, E.: A model of saliency-based visual attention for rapid scene analysis. IEEE Trans. Pattern Anal. Mach. Intell. **11**, 1254–1259 (1998)
33. Galazka, M., Nyström, P.: Visual attention to dynamic spatial relations in infants and adults. Infancy **21**(1), 90–103 (2016)
34. Saleh, M., Debellemanière, G., Meillat, M., Tumahai, P., Garnier, M.B., Flores, M., Schwartz, C., Delbosc, B.: Quantification of cone loss after surgery for retinal detachment involving the macula using adaptive optics. Br. J. Ophthalmol. **98**, 1343–1438 (2014). bjophthalmol-2013
35. Borst, A., Egelhaaf, M.: Principles of visual motion detection. Trends in Neurosci. **12**(8), 297–306 (1989)
36. Itti, L., Baldi, P.: A principled approach to detecting surprising events in video. In: 2005 IEEE Computer Society Conference on Computer Vision and Pattern Recognition, CVPR 2005. vol. 1, pp. 631–637. IEEE (2005)
37. Pruden, S.M., Hirsh-Pasek, K., Golinkoff, R.M., Hennon, E.A.: The birth of words: ten-month-olds learn words through perceptual salience. Child Dev. **77**(2), 266–280 (2006)
38. Damianou, A., Boorman, L., Lawrence, N.D., Prescott, T.J.: A top-down approach for a synthetic autobiographical memory system. In: Wilson, S.P., Verschure, P.F.M.J., Mura, A., Prescott, T.J. (eds.) Living Machines 2015. LNCS, vol. 9222, pp. 280–292. Springer, Heidelberg (2015)
39. IIT: icub: an open source cognitive humanoid robotic platform. http://www.icub.org/. Accessed 1 Feb 2016
40. Kellman, P.J., Spelke, E.S., Short, K.R.: Infant perception of object unity from translatory motion in depth and vertical translation. Child Dev. **57**(1), 72–86 (1986)
41. YARP: Yet another robot platform. http://wiki.icub.org/yarpdoc/. Accessed 10 Feb 2016
42. Itseez: Opencv 3.0.0. http://opencv.org/opencv-3-0.html. Accessed 10 Feb 2016
43. NVIDIA: Cuda developer website. https://developer.nvidia.com/cuda-toolkit. Accessed 10 Feb 2016
44. Camilleri, D.: icub visual saliency. https://www.youtube.com/watch?v=_OgBuLZHCh8. Accessed 10 Feb 2016
45. Burr, D.C., Morrone, M.C., Ross, J., et al.: Selective suppression of the magnocellular visual pathway during saccadic eye movements. Nature **371**(6497), 511–513 (1994)

An Overview of the Ongoing Humanoid Robot Project LARMbot

Marco Ceccarelli[1], Daniele Cafolla[1], Mingfeng Wang[1,2(✉)], and Giuseppe Carbone[1,3]

[1] LARM: Laboratory of Robotics and Mechatronics, DICeM-University of Cassino and South Latium, Via Di Biasio 43, 03043 Cassino, FR, Italy
wang@unicas.it
[2] School of Mechanical and Electrical Engineering, Central South University, No. 932, Lushan South Road, Changsha 410083, Hunan, China
[3] Department of Engineering and Mathematics, Sheffield Hallam University, Sheffield S1 1WB, UK

Abstract. LARMbot project aims to develop a humanoid robot with biomimetic inspiration from human anatomy by using parallel mechanisms. Previous related work is presented particularly referring to torso and leg modules. A specific design of LARMbot is proposed by using proper parallel mechanisms in torso and leg designs. A CAD model is elaborated in SolidWorks® environment and the corresponding prototype is fabricated with low-cost user-oriented features by using commercial components and parts manufactured using 3D printing. Preliminary results of experiment tests are also reported for operation evaluation and architecture design characterization.

Keywords: Humanoid robots · Mechanism design · Parallel mechanisms

1 Introduction

Humanoid robots are designed to possess high mobility and operability as expected to interacting with human beings in daily life [12]. Thanks to the fast improvements in the field of sensors, computers and actuators, the development of humanoid robots has increased rapidly in the past two decades and a lot of prototypes have been built in laboratories or companies, such as ASIMO of Honda [10], HRP of AIST/KAWADA [11], HUBO of KAIST [24], LOLA of Technical University Munich [1], WALK-MAN of IIT [19], and ATLAS of Boston Dynamics [13].

Mechanism design plays a key role for designing a humanoid robot, since it determines its motion capability and operation performance. Most of existing humanoid robots are based on serial architectures by using serial mechanisms From the design point of view, serial mechanisms are easy to be implemented and capable of relatively large workspace and dexterous maneuverability, which can guarantee humanoid robots' capability of imitating human-like movements.

© Springer International Publishing Switzerland 2016
L. Alboul et al. (Eds.): TAROS 2016, LNAI 9716, pp. 53–64, 2016.
DOI: 10.1007/978-3-319-40379-3_6

However, they suffer from several drawbacks such as low payload to own weight ratio, additive joint errors, poor dynamic performance and relatively poor system stiffness, due to the serial nature of actuation and transmission [23]. Compared with serial mechanisms, parallel mechanisms have several advantages in terms of higher payload to weight ratio, higher stiffness, higher dexterity, higher dynamic response, and compact size. Therefore, parallel mechanisms can be considered as excellent candidates for designing components of new humanoid robots with the aim of improving their operation performance. Additionally, it is to note that human body has several muscles and tendons in each body part that act in parallel to each other similarly to parallel architectures [18]. With such a biomimetic inspiration a humanoid robot can be designed and operated with a kinematic structure made of several parallel mechanisms with different design solutions as depending of the body parts [5].

At Laboratory of Robotics and Mechatronics (LARM) in Cassino, a research line has been undergoing with the aim to develop a humanoid robot with features of low-cost and easy-operation in the past decade [6–9, 14–17]. This paper describes the ongoing LARMbot humanoid prosect at LARM by focusing in particular on torso and leg modules.

2   Previous Related Work

At first stage, for the torso module, experiments with basic movement have been conducted to characterize the behavior of the human torso by using four different IMU sensors on the body of different subjects [4]. A scheme for a suitable analysis procedure of human torso is shown in Fig. 1 and a system layout of the IMUs sensor system is shown in Fig. 2, respectively. For an example, an experimental test is performed with slowly turning right and left for three times

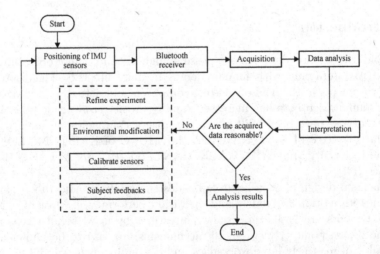

Fig. 1. A scheme for the analysis process of human torso motion with IMU sensors

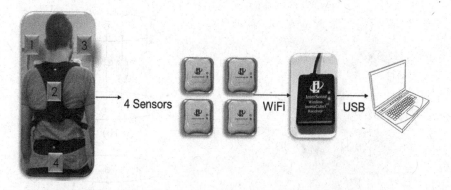

Fig. 2. A system layout of the sensing for human torso motion

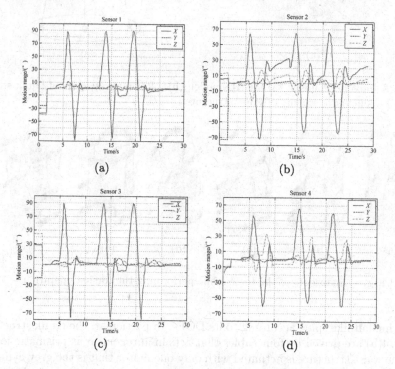

Fig. 3. Angles measured with IMUs sensor system during an example test

while maintaining the torso in a standstill position and the acquired data from the IMU sensors is presented with measured angles, as show in Fig. 3.

A torso mechanism has been proposed as based on waist and trunk designs of parallel mechanisms through which is a human-like spine mechanism [2]. The kinematic schemes of trunk and waist structures are shown in Fig. 4(b) and (c), respectively, where cylindrical joints have been used to schematize the movement

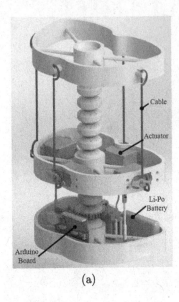

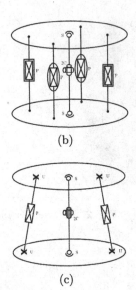

(b)

(c)

(a)

Fig. 4. A torso mechanism: (a) a CAD model; (b) a kinematic scheme of trunk structure; (c) a kinematic scheme of waist structure

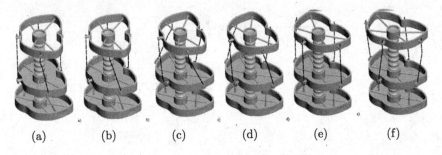

(a)   (b)   (c)   (d)   (e)   (f)

Fig. 5. Snapshots of the torso while performing the torsion operation

of the human-like spine according to its DOFs. In particular, the trunk structure in Fig. 4(b) are driven by four cables that acts in antagonism as prismatic joints and the waist structure is actuated with only one motor that is the grey cylinder shown in Fig. 4(c). A CAD design of the trunk-waist torso is shown in Fig. 4(a) with the trunk and waist modules connected to each other through a common platform. A Dynamic simulation has been carried out in ADAMS® environment to replicate the pure torsion of a human torso. Figure 5 shows the computed simulation sequence of the torso operation in six snapshots, where the waist starts to rotate from Fig. 5(a) to (c), and with the waist standstill the torso starts to perform the torsion from Fig. 5(d) until (f).

For the leg module, a LARM tripod leg mechanism has been firstly proposed [20] as based on a topology search [22] and its preliminary prototype and

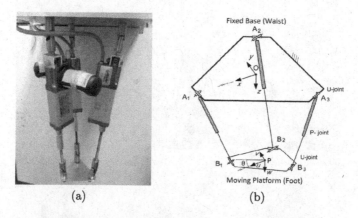

(a)   (b)

Fig. 6. A LARM tripod leg mechanism: (a) a prototype; (b) a kinematic scheme

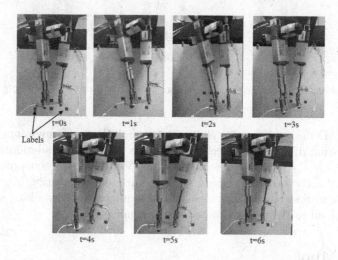

Fig. 7. A sequence of snapshots of the LARM tripod leg during a test

kinematic scheme of a 3-UPU parallel manipulator are shown in Fig. 6(a) and (b). The tripod leg mechanism is a parallel manipulator consisting of a waist plate, a moving foot plate, and three identical limbs of linear actuators with U-joints at each end. An experimental test has been performed with the prototype to follow a prescribed step movement of the foot platform. A sequence of leg configurations during an experimental test is shown in Fig. 7, where experimental test time are indicated for each snapshot and six reference points of the prescribed trajectory are marked by labels.

A biped locomotor have been proposed consisting of two identical LARM tripod leg mechanisms and a waist, on which the two leg mechanisms are installed [21]. In order to achieve static equilibrium walking, a planning of biped walking gait is performed by coordinating the motions of the two leg mechanisms and

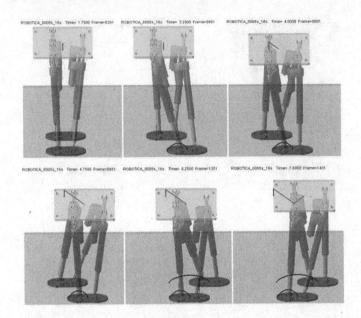

Fig. 8. Snapshots of the biped locomotor while performing the walking gait

waist. A CAD design has been elaborated in SolidWorks® environment for a characterization of a feasible mechanical design. Dynamic simulation is carried out in ADAMS® environment with the aim to evaluate the dynamic walking performance of the proposed design with its fairly easy-operation high-payload features. Figure 8 shows the snapshots of the simulated biped walking sequences that correspond to the planned biped walking gait.

3   LARMbot

The LARMbot design has been elaborated in SolidWorks® environment, as shown in Fig. 9(a), consisting of a torso system, a locomotion system, two arms and a neck mechanism. A first prototype of humanoid robot LARMbot has been built following the CAD design solutions in Fig. 9(a) by using commercial components and 3D printed parts, as shown in Fig. 9(b).

In particular, the torso system CAssino hUmanoid Torso (CAUTO) [3] consists of 4 disc bodies that replicate the function of the vertebrae in the human spine and an actuating system of cables in a parallel architecture, as shown in Fig. 10. The vertebrae are interconnected with each other by means of flexible couplings that behave as spherical spring joints. The trunk is fixed on the spine through a vertebral disc and the spine is also connected to the abdomen using another vertebral disk. The pelvis is connected to the abdomen and houses 5 actuators 4 of which operate the 3 DOFs of the humanoid spine. The four cables are fixed to the platform through the trunk and then pass through the holes of the

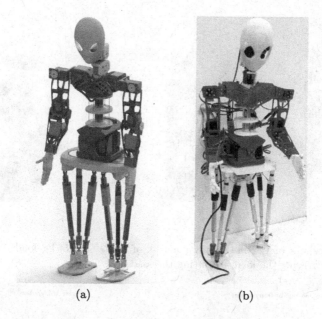

Fig. 9. Humanoid robot LARMbot: (a) a CAD design; (b) a prototype

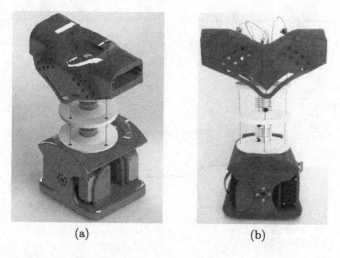

Fig. 10. CAUTO: (a) a CAD design; (b) a prototype

other vertebral discs reaching the actuators on which they are fixed. The cables are pulled by the actuators and act antagonistically as a cable parallel manipulator in order to perform left-right and forward-backward bending motions of the humanoid spine, as well as a circular motion thanks to the combination of the above two movements. An experimental test of CAUTO is carried out for lifting a load of 0.56 kg with a motion of 20.40 s and snapshots of motion sequences

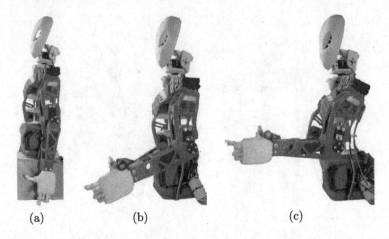

(a)   (b)   (c)

Fig. 11. Snapshots of CAUTO during a test of lifting a 0.56 kg load: (a) starting position; (b) carrying the load; (c) lifting the load

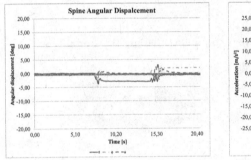

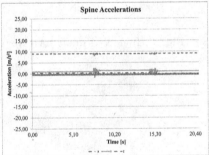

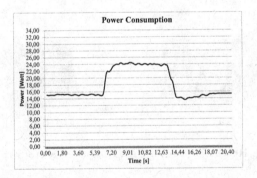

Fig. 12. Results of the experimental test in Fig. 11

are shown in Fig. 11. Test results are reported in terms of spine angular displacement, acceleration and power consumption, as shown in Fig. 12, where the load is lifted at 7.6 s and maintained for about 8.44 s. In Fig. 12(a), the acquired angle ranges of θ, ϕ and ψ are respectively $-3.76°$ to $0.23°$, $-2.44°$ to $2.37°$, and

$-2.07°$ to $3.68°$. When the load is applied on the arms it causes the bending forward of the torso but it balances the additional weight going backward maintaining a constant position without falling down as shown in the first spike of θ in Fig. 12(a). In Fig. 12(b), the linear acceleration ranges along X-axis, Y-axis and Z-axis are respectively $-2.24\,\mathrm{m/s^2}$ to $2.20\,\mathrm{m/s^2}$, $-0.24\,\mathrm{m/s^2}$ to $2.55\,\mathrm{m/s^2}$, and $8.16\,\mathrm{m/s^2}$ to $9.81\,\mathrm{m/s^2}$ while performing the motion. There are spikes only when lifting up or down operation are performed due to the vibration that matches the inertia of the load. In Fig. 12(c), the power consumption is constant up to the lifting operation, then it absorbs more current that allows the arms to maintain the additional mass and lifts it up. When the load is lifted down, the power consumption decrease. The power consumption has a range between $13.32\,\mathrm{W}$ and $24.09\,\mathrm{W}$ and a mean value of $18.07\,\mathrm{W}$. This experiment shows that the proposed torso can satisfactory lift a $0.56\,\mathrm{kg}$ load, that is about $35\,\%$ of its weight, with two arms.

The locomotion system Cassino Biped Locomotor, as shown in Fig. 13(a), is composed of a waist platform, a turning mechanism, two 3-DOF leg mechanisms, and two foot platforms. The turning mechanism is installed on the waist platform engaged with the driven gears, which can perform turning motion while walking. The two foot platforms are connected to the lower platforms of two leg mechanisms as end-effector, which can perform forward-backward, left-right and up-down movements. A preliminary prototype is built without turning mechanism, as shown in Fig. 13(b). An experimental test of the prototype is implemented to perform a straight walking on a horizontal plane with a $3.50\,\mathrm{kg}$ payload. The payload is fixed on a passive two-wheeled chariot. Snapshots of motion sequences are shown in Fig. 14, where experimental test time has been indicated for each snapshot. Experiment results in terms of ground contact forces and power consumption are measured, as show in Fig. 15. In Fig. 15(a), the maximum value of contact forces between feet and ground is computed as about $17.6\,\mathrm{N}$. There are

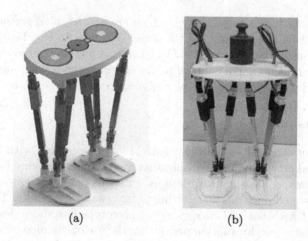

(a)   (b)

Fig. 13. Cassino Biped Locomotor: (a) a CAD design; (b) a prototype

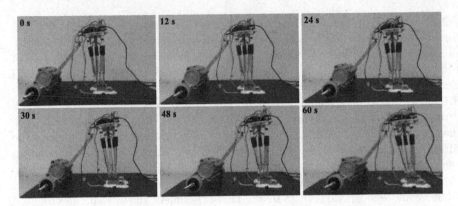

Fig. 14. Snapshots of Cassino Biped Locomotor during a test of straight walking with a 3.50 kg payload

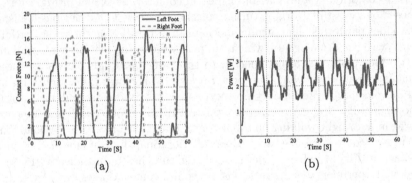

Fig. 15. Results of the experimental test in Fig. 14: (a) contact force between feet and ground; (b) power consumption

one or two minor jumps in each step, which are due to the corresponding foot landing impact on the ground and the waist swinging. In Fig. 15(b), the minimum power consumption is about 0.4 W when it is powered on but without any operation and the maximum value of power consumption is computed as 3.6 W, which shows that the built prototype requires very low power in operation.

4 Conclusion

An ongoing humanoid robot project LARMbot is presented with its peculiar design as based on a biomimetic inspiration from human anatomy by using parallel mechanisms in torso and leg designs. Previous related work on LARMbot project is presented particularly referring to torso and leg modules. The mechanical design of LARMbot and corresponding prototype have been developed in a compact and light weight solution permitting the use of commercial components and 3D printed parts. Experimental tests have been carried out for basic motions to confirm the design features in terms of low-cost and user friendliness.

References

1. Buschmann, T., Schwienbacher, M., Favot, V., Ewald, A., Ulbrich, H.: The biped walking robot lola. J. Robot. Soc. Jpn. **30**(4), 363–366 (2012)
2. Cafolla, D., Ceccarelli, M.: Design and fem analysis of a novel humanoid torso. In: Multibody Mechatronic Systems, pp. 477–488. Springer (2015)
3. Cafolla, D., Ceccarelli, M.: Design and simulation of a cable-driven vertebra-based humanoid torso. Int. J. Humanoid Robot. (2015)
4. Cafolla, D., Chen, I.M., Ceccarelli, M.: An experimental characterization of human torso motion. Front. Mech. Eng. **10**(4), 311–325 (2015)
5. Carbone, G., Liang, C., Ceccarelli, M.: Using parallel architectures for humanoid robots. In: Kolloquium Getriebetechnik, Aachen 2009, pp. 177–188 (2009)
6. Ceccarelli, M.: LARM PKM solutions for torso design in humanoid robots. Front. Mech. Eng. **9**(4), 308–316 (2014)
7. Ceccarelli, M.: Kinematic design problems for low-cost easy-operation humanoid robots. In: Interdisciplinary Applications of Kinematics, pp. 91–99. Springer (2015)
8. Copilusi, C., Ceccarelli, M., Carbone, G.: Design and numerical characterization of a new leg exoskeleton for motion assistance. Robotica **33**(05), 1147–1162 (2015)
9. Gu, H., Ceccarelli, M.: A multiobjective optimal path planning for a 1-dof clutched ARM. Mech. Based Des. Struct. Mach. **40**(1), 109–121 (2012)
10. Hirose, M., Ogawa, K.: Honda humanoid robots development. Philos. Trans. R. Soc. Lond. A: Math. Phys. Eng. Sci. **365**(1850), 11–19 (2007)
11. Kaneko, K., Kanehiro, F., Morisawa, M., Miura, K., Nakaoka, S., Kajita, S.: Cybernetic human HRP-4C. In: 9th IEEE-RAS International Conference on Humanoid Robots, 2009, Humanoids 2009, pp. 7–14. IEEE (2009)
12. Kemp, C.C., Fitzpatrick, P., Hirukawa, H., Yokoi, K., Harada, K., Matsumoto, Y.: Humanoids. In: Siciliano, B., Khatib, O. (eds.) Springer Handbook of Robotics, pp. 1307–1333. Springer, New York (2008)
13. Kuindersma, S., Deits, R., Fallon, M., Valenzuela, A., Dai, H., Permenter, F., Koolen, T., Marion, P., Tedrake, R.: Optimization-based locomotion planning, estimation, and control design for the atlas humanoid robot. In: Autonomous Robots, pp. 1–27 (2015)
14. Li, T., Ceccarelli, M.: Design and simulated characteristics of a new biped mechanism. Robotica **33**(07), 1568–1588 (2015)
15. Liang, C., Ceccarelli, M.: Design and simulation of a waist-trunk system for a humanoid robot. Mech. Mach. Theory **53**, 50–65 (2012)
16. Liang, C., Gu, H., Ceccarelli, M., Carbone, G.: Design and operation of a tripod walking robot via dynamics simulation. Robotica **29**(05), 733–743 (2011)
17. Rodriguez, N.E.N., Carbone, G., Ceccarelli, M.: Simulation results for design and operation of CALUMA, a new low-cost humanoid robot. Robotica **26**(5), 601–618 (2008)
18. Saladin, K.S.: Human Anatomy, 2nd edn. McGraw Hill Higher Education, New York (2008)
19. WALK-MAN: Whole-body adaptive locomotion and manipulation, European Community's 7th Framework Programme: FP7-ICT 611832, Cognitive Systems and Robotics: FP7-ICT-2013-10. http://www.walk-man.eu (2013–2017)
20. Wang, M., Ceccarelli, M., Carbone, G.: Experimental tests on operation performance of a LARM leg mechanism with 3-DOF parallel architecture. Mech. Sci. **6**(1), 1 (2015)

21. Wang, M., Ceccarelli, M.: Design and simulation of walking operation of a Cassino biped locomotor. In: New Trends in Mechanism and Machine Science, pp. 613–621. Springer (2015)
22. Wang, M., Ceccarelli, M.: Topology search of 3-DOF translational parallel manipulators with three identical limbs for leg mechanisms. Chin. J. Mech. Eng. **28**(4), 666–675 (2015)
23. Zhang, D.: Parallel Robotic Machine Tools. Springer Science & Business Media, New York (2009)
24. Zucker, M., Joo, S., Grey, M.X., Rasmussen, C., Huang, E., Stilman, M., Bobick, A.: A general-purpose system for teleoperation of the DRC-HUBO humanoid robot. J. Field Robot. **32**(3), 336–351 (2015)

Force and Topography Reconstruction Using GP and MOR for the TACTIP Soft Sensor System

G. de Boer[1(✉)], H. Wang[2], M. Ghajari[1], A. Alazmani[2], R. Hewson[1], and P. Culmer[2]

[1] Department of Aeronautics, Imperial College London, London, SW7 2AZ, UK
{g.de-boer,m.ghajari,r.hewson}@imperial.ac.uk
[2] School of Mechanical Engineering, University of Leeds, Leeds, LS2 9JT, UK
{h.wang1,a.alazmani,p.r.culmer}@leeds.ac.uk

Abstract. Sensors take measurements and provide feedback to the user via a calibrated system, in soft sensing the development of such systems is complicated by the presence of nonlinearities, e.g. contact, material properties and complex geometries. When designing soft-sensors it is desirable for them to be inexpensive and capable of providing high resolution output. Often these constraints limit the complexity of the sensing components and their low resolution data capture, this means that the usefulness of the sensor relies heavily upon the system design. This work delivers a force and topography sensing framework for a soft sensor. A system was designed to allow the data corresponding to the deformation of the sensor to be related to outputs of force and topography. This system utilised Genetic Programming (GP) and Model Order Reduction (MOR) methods to generate the required relationships. Using a range of 3D printed samples it was demonstrated that the system is capable of reconstructing the outputs within an error of one order of magnitude.

Keywords: Soft-sensing · Genetic Programming · Model Order Reduction

1 Introduction

Tactile sensors are an essential sense in robotics to safely explore the external world and to precisely manipulate objects by providing force and contact information. Soft forms of tactile sensors offer improved interaction with complex environments since they can inherently conform to complex surfaces and deform to avoid damage. A number of soft tactile sensor systems have been developed, using a range of sensing technologies, with notable examples including TakkTile [1], GelForce [2], BioTac [3, 4], and TACTIP [5]. However, the inherent nonlinearities in soft sensing systems (e.g. contact forces, material properties and complex geometries) make it difficult to process and relate their output to the real world.

The biologically inspired TACTIP system, which features a deformable 'finger-tip' membrane upon which traceable markers are placed [6, 7], is a robust and economic soft sensor. The TACTIP system has previously been used for shape recognition [8], edge detection analysis [9] and determining surface texture [10]. However, obtaining quantitative force and topography information from TACTIP is non-trivial and complicated

© Springer International Publishing Switzerland 2016
L. Alboul et al. (Eds.): TAROS 2016, LNAI 9716, pp. 65–74, 2016.
DOI: 10.1007/978-3-319-40379-3_7

by the presence of nonlinear material behaviour, larger deformations, and complex geometries.

Computational optimisation techniques provide an efficient way to address these challenges. Genetic Programming (GP) is a biologically inspired evolutionary based algorithm for defining an equation which gives the best evaluation of an output based on a set of inputs [11]. GP has been used to design sensors associated with autonomous robotics [12], vision [13], and locomotion [14]. GP has also been successfully applied to soft sensors associated with biochemical applications [15, 16]. Other methods have also been used in the design of soft sensor systems such as Artificial Neural Networks (ANN) [17] and Response Surface Methods (RSM) [18]. In conjunction, Singular Value Decomposition (SVD) provides a means to decompose a set of discrete data into a lower order model which maintains the highest possible level of accuracy [19]. This is a useful approach because it efficiently and accurately provides a method, known as Model Order Reduction (MOR), for describing a large amount of data with a much smaller subset. MOR has been used in the design of piezoelectric [20], magnetic resonance [21] and soft sensing applications [18].

Here we describe how a combination of GP and MOR techniques can be used for complex force and topography reconstruction in soft tactile sensors, using the TACTIP sensor system as an example. The method developed is applicable to a wide range of applications beyond soft sensors, the fidelity of the responses generated using the method will depend upon the level of training of the sensor system and the intended sensing purpose.

2   Materials and Methods

2.1   TACTIP Sensor

The TACTIP sensor is a biologically inspired soft tactile sensor designed by the Bristol Robotics Laboratory [6], it uses a camera to track the movement of markers on a compliant skin. As shown in Fig. 1(a), TACTIP consists of a compliant skin with markers on the inner surface, a soft body covered by the compliant skin is filled with clear Gel, an IR LED is the illumination source, a clear Acrylic sheet separates the Gel inside the

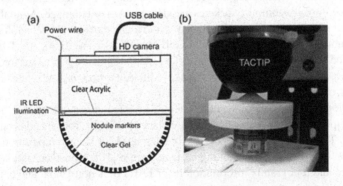

Fig. 1. TACTIP sensor (a) cross-section schematic and (b) test bed.

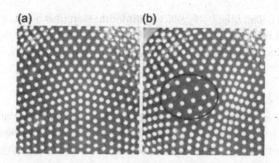

Fig. 2. Captured image from TACTIP camera (a) unloaded (b) loaded (red circled region) (Color figure online).

soft body with the camera system, and a USB HD camera captures the image of the inner surface of the skin. Details of the marked skin design and fabrication are described in [7]. A photograph of the TACTIP indenting a surface is presented in Fig. 1(b), and the images captured by the internal USB HD camera are given in Fig. 2. In order to recognize the white markers (pins) and track their movement, a real-time image processing programme was implemented in LabVIEW (National Instruments, USA).

2.2 Indentation Test Apparatus

A test platform was built to repeatedly probe the sensor system (Fig. 3) and includes a micropositioning linear stage (T-LSR75B, Zaber Technologies Inc., Canada), the TACTIP sensor with USB camera, a 6-axis load cell (Nano 17-E, ATI Industrial Automation, USA), and a computer based data acquisition system (myRio, National

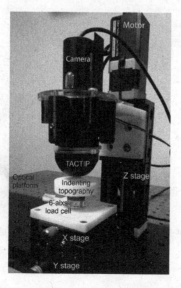

Fig. 3. Photograph of the indentation test apparatus.

Instruments, USA). The linear stage has a minimum step of 0.5 µm, a travel range of 75 mm, and repeatability of 2.5 µm. The load cell was capable of measuring a range of ±35 N in the Z axis, with a resolution of 6.25 mN.

2.3 Topography

In order to investigate a range of topographies a selection of samples with axisymmetric features were fabricated. Figure 4(a) illustrates the cross section of the topography, the maximum radius of the samples was 21 mm. The height of topography, h, is described by Eq. (1),

$$h = A \exp \left(\frac{-r^2}{2c^2} \right) \tag{1}$$

where r is the sample radius. A and c are parameters which differ for the m = 12 samples, A represents the maximum height and c the rate of decay with increasing radius. parameterizing the topography as according to Eq. (1) means that a lower order model can be used to accurately reconstruct the range of shapes, since the modal decomposition of the parametrised topography will have similar properties (see Sect. 3.2). The values of the parameters relating to topography for the samples used are given in Table 1. Each sample was manufactured by 3D printer (Objet 1000, Stratasys Ltd., USA). The material of the manufactured samples is rigid in comparison to the surface of the TACTIP sensor, therefore during indentation only the surface of the probe deforms and the topography remains unchanged. Two example 3D printed samples with topography are shown in Fig. 4(b).

Table 1. Topography parameters for the 3D printed samples.

Sample	#1	#2	#3	#4	#5	#6	#7	#8	#9	#10	#11	#12
A [mm]	5	5	5	3	3	1	−5	−5	−5	−3	−3	−1
c [mm]	8	4	2	6	3	2	8	4	2	6	3	2

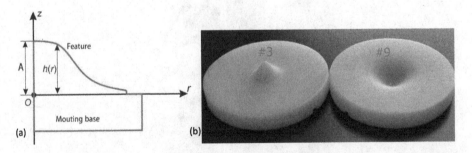

Fig. 4. (a) Cross-section of the parameterised topography. (b) photography of 3D printed samples #3 and #9.

3 Theory

3.1 Force Reconstruction

The normal forces, $\mathbf{F}$, were recorded by the force sensor over the duration of indentation and range of samples. This is defined by Eq. (2),

$$\mathbf{F} = \begin{bmatrix} F_z^{1,1} & \cdots & F_z^{1,n} & \cdots & F_z^{m,1} & \cdots & F_z^{m,n} \end{bmatrix} \tag{2}$$

where $F_z^{i,j}$ is the normal force for the i'th sample at the j'th time step. As there are m samples and n time steps, the size of $\mathbf{F}$ is $[1 \times mn]$. $\mathbf{D}$ is the TACTIP pin deformations which correspond to the same time steps and samples used to construct the normal force vector as described by Eq. (3), where $D_{x,k}^{i,j}$; $D_{y,k}^{i,j}$ are the k'th pin deformations for the i'th sample at the j'th time step. As there are p samples and the size of $\mathbf{D}$ is $[2p \times mn]$.

$$\mathbf{D} = \begin{bmatrix} D_{x,1}^{1,1} & \cdots & D_{x,1}^{1,n} & \cdots & D_{x,1}^{m,1} & \cdots & D_{x,1}^{m,n} \\ \vdots & \vdots & \vdots & \vdots & \vdots & \vdots & \vdots \\ D_{x,p}^{1,1} & \cdots & D_{x,p}^{1,n} & \cdots & D_{x,p}^{m,1} & \cdots & D_{x,p}^{m,n} \\ D_{y,1}^{1,1} & \cdots & D_{y,1}^{1,n} & \cdots & D_{y,1}^{m,1} & \cdots & D_{y,1}^{m,n} \\ \vdots & \vdots & \vdots & \vdots & \vdots & \vdots & \vdots \\ D_{y,p}^{1,1} & \cdots & D_{y,p}^{1,n} & \cdots & D_{y,p}^{m,1} & \cdots & D_{y,p}^{m,n} \end{bmatrix} \tag{3}$$

In order to correlate the force as a function of time and sample selection to the pin deformations, $\mathbf{F}$ is related to the matrix $\mathbf{D}$ using GP. GP was used to create an equation linking the pin deformations to normal force by generating a range of possible algebraic descriptions from combinations of the input variables. These descriptions can contain any set of prescribed expressions and as such can describe complex non-linear trends which are not obtained through simple data fitting analyses. The general statement of the expression obtained from GP in this case is given by Eq. (4),

$$\mathbf{F} = f(D_{x,q}, D_{y,q}) q \in \mathbb{Z}_p^+ \tag{4}$$

This equation does not necessarily contain all input variables as their usefulness is evaluated in determining the output, hence q describes a subset of all p pins. Running GP multiple times produces a different result because of the complexity associated with the number of possible combinations of expressions and input variables which are associated with determining the output relationship. Running the solver over more iterations improves the likelihood that the fit achieved is more accurate. The best fit is determined by an evolutionary algorithm which learns by assessment of a fitness function the best selection and combination of input variables in minimising the error in the output [11].

3.2 Topography Reconstruction

The topography heights for the samples are arranged into a matrix $\mathbf{A}$ which is defined by Eq. (5),

$$\mathbf{A} = \begin{bmatrix} h_1^1 & \cdots & h_1^m \\ \vdots & \vdots & \vdots \\ h_s^1 & \cdots & h_s^m \end{bmatrix} \tag{5}$$

where h_j^i is the j'th location for the i'th sample. In total there are s heights per sample and the size of $\mathbf{A}$ is [s × m]. Importantly the definition of topography is discrete such that any numerical description of topography can be included and does not rely on the analytical description of Eq. (1) for the 3D printed topography. The SVD of $\mathbf{A}$ allows the matrix to be written as the product of three component matrices $\mathbf{U}$, $\mathbf{\Sigma}$, and $\mathbf{V}^T$. The SVD of $\mathbf{A}$ can be truncated by defining a rank K which determines the amount of information kept by the approximation. This leads to Eq. (6) which gives the MOD of $\mathbf{A}$,

$$\mathbf{A} \cong \mathbf{U}_K \mathbf{\Sigma}_K \mathbf{V}_K^T, K \leq \min(m, s) \tag{6}$$

where $\mathbf{U}_K$ is the first K columns of $\mathbf{U}$[s × K], $\mathbf{\Sigma}_K$ is the first K columns and rows of $\mathbf{\Sigma}$[K × K], and $\mathbf{V}_K$ is the first K columns of $\mathbf{V}$[m × K]. The matrix $\mathbf{V}_K$ are known as the modes of the SVD of $\mathbf{A}$. $\mathbf{d}$ is defined as a matrix of pin deformations at a specific instance in time. In order to correlate the modes of topography to the pin deformations each component in $\mathbf{V}_K^T$[K × m] were related to the matrix $\mathbf{d}$[2p × m] by using GP in a similar way to that described in Sect. 3.1. The relationships which are generated describe the correlation between the pin deformations and modes of the reduced order model for topography as given by Eq. (7),

$$\mathbf{V}_l^T = f(d_{x,q}, d_{y,q}) l = 1, \ldots, K \ q \in \mathbb{Z}_p^+ \tag{7}$$

4 Results and Discussion

4.1 Force Reconstruction

Indentation was undertaken at the centre of the samples and data was recorded for n = 61 time steps over a period of 16 s. The depth was linearly increased over time to the maximum 6 mm at the halfway point and then back to zero, in total p = 134 pins were recorded during the indentation. Analysing the data produced using the multi-gene GP toolbox in Matlab *gptips* [22] produced expressions for the normal force as a function of a subset of pin deformations. The GP solver was run 10 times and the result which produced the lowest root-mean-squared-error over the complete set was selected as the overall best fit. The number of generations used was 500, the population size was 300, the number of genes was 6, and the number of terms each gene could have was 12. The total time to compute was ~ 120 min using a 2.8 GHz 4-core CPU running with 3 GB of RAM for the process, the minimum RMS error over all samples and time steps achieved was 53.2 mN with a mean of 34.4 mN and variance of 9.8 mN.

The equation generated by GP indicates how the normal force can be reconstructed from the pin deformations, not all of the pins are included in the terms and as such only those pins with a significant influence are used. Figure 5 shows the normal force reconstruction for two example samples, in these plots blue represents the reconstructed and red represents the recorded data. The accuracy of the reconstructed points compared to the recorded is reasonable for each of the forces investigated, with the error found to be an order of magnitude smaller than the recorded forces themselves. Generally the shape of the force responses is well represented and the peak value is obtained to within an order of magnitude. The low resolution of the pin deformations can be seen to influence the types of responses generated by using them, whereby a higher resolution result is generated but is still subjected to certain regions of pixelation. Sub-pixel tracking of pin deformations would allow a continuous expression to be generated in this way.

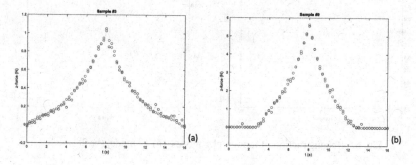

Fig. 5. Normal force reconstruction for (a) sample #3, (b) sample #9.

Further investigation the GP solver tolerances and number of terms in the resulting equation would be explored to potentially improve the force reconstruction. Another point to consider is the types of expressions which can be used to create the GP solution, which can be any set of mathematical expressions. Changing the types of expressions which the GP explores will change the types of response which can be generated and may improve the resultant fit for a given data set.

4.2 Topography Reconstruction

Topography coordinates were generated for the samples and arranged into the matrix **A** as outlined in Sect. 3.2, the SVD of **A** was undertaken using Matlab (TheMathsWorks Inc., USA) and the reduced order model for topography was then chosen by setting the rank $K = 3$, this represents 25 % of the total number of modes. Using the same procedure as described in Sect. 4.1 the modes of topography were correlated to the maximum pin deformations using GP. The equations generated indicate how the modes of the topography can be reconstructed from the pin deformations at the maximum indentation as a function of the sample selection.

Figure 6 shows two of the three mode reconstructions using the equations generated by GP. These are of a higher level of accuracy than the force reconstruction, this is because m = 12 points need to be considered for topography reconstruction in comparison to mn = 732 for the force reconstruction. Increasing the number of samples tested increases the likelihood that the reconstruction will be of a lower accuracy. It is interesting to note that each of the modes has a very different type of response and that GP is able to find a relationship that accurately correlates them all to the pin deformations, which themselves have similar trends. Using the modes determined from GP the topography was subsequently reconstructed. Figure 7 shows the topography reconstructions for two samples #3 and #9, chosen as an example. In this figure blue represents the reconstructed and red represents the recorded data, the reconstructed topography can be seen to be accurate to within an order of magnitude using MOR and GP.

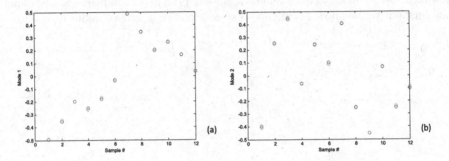

Fig. 6. Reconstruction of topography modes, (a) 1st mode, (b) 2nd mode.

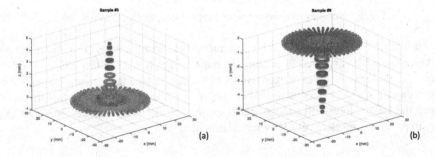

Fig. 7. Topography reconstruction for (a) sample #3 (b) sample #9 (Color figure online).

The minimum RMS error in the topography reconstruction was 0.0512 mm with a mean of 0.0813 mm and variance of 0.00578 mm. The error in the reconstructed points is at least two orders of magnitude smaller than the range of recorded topography data. The rank of K = 3 was chosen to demonstrate that the topographies can be accurately reconstructed from a limited number of modes. As the number of modes is increased the accuracy of the reconstruction increases however so does the computational expense.

5   Conclusion

A method for soft-sensor force and topography reconstruction using the TACTIP sensor as an example is presented. Physical testing was undertaken to evaluate a novel method in which GP derived equations were obtained to link the sensor pin deformations with a corresponding force/topography. In the case of topography MOR was used to decompose the response into modes which simplified the reconstruction process. Both force and topography were reconstructed to within an order of magnitude of the known values. It was shown for force reconstruction that low resolution pin deformations can be used to give a high resolution result via the GP procedure and that inaccuracies in the resulting relationships could be improved by sub-pixel resolution imaging. While this work focuses on the TACTIP soft sensor, the method provides a more general approach to reconstructing physical quantities with high fidelity from nonlinear low fidelity inputs – a process which is non-trivial or impossible with analytical approaches. The method is a promising approach to be further explored in soft sensing applications such as grasping and edge detection. For real-time sensing ANN or RSM provide good alterative opportunities to develop and evaluate the required relationships.

Acknowledgements.   We would like to thank The Leverhulme Trust (Grant number: RPG-2014-381) for funding this work.

References

1. Tenzer, Y., Jentoft, L.P., Howe, R.D.: Inexpensive and easily customized tactile array sensors using mems barometers chips. IEEE Robot. Autom. Mag. (2014)
2. Sato, K., et al.: Finger-shaped gelforce: sensor for measuring surface traction fields for robotic hand. IEEE Trans. Haptics 3(1), 37–47 (2010)
3. Xu, D., Loeb, G.E., Fishel, J.A.: Tactile identification of objects using Bayesian exploration. In: 2013 IEEE International Conference on Robotics and Automation (ICRA), pp. 3056–3061. IEEE (2013)
4. Wettles, N., Santos, V.J., Johansson, R.S., Loeb, G.E.: Biomimetic tactile sensor array. Adv. Robot. 22, 829–849 (2008)
5. Assaf, T., et al.: Seeing by touch: evaluation of a soft biologically-inspired artificial fingertip in real-time active touch. Sensors 14(2), 2561–2577 (2014)
6. Chorley, C., Melhuish, C., Pipe, T., Rossiter, J.: Development of a tactile sensor based on biologically inspired edge encoding. In: International Conference on Advanced Robotics, ICAR 2009, 22 Jun 2009, pp. 1–6. IEEE (2009)
7. Winstone, B., Griffiths, G., Melhuish, C., Pipe, T., Rossiter, J.: TACTIP – tactile fingertip device, challenges in reduction of size to ready for robot hand integration. In: Proceedings of the 2012 IEEE, International Conference on Robotics and Biomimetics, Guangzhou, China, 11–14 December 2012
8. Assaf, T., Chorley, C., Rossiter, J., Pipe, T., Stefanini, C., Melhuish, C.: Realtime processing of a biologically inspired tactile sensor for edge following and shape recognition. In: Towards Autonomous Robotic Systems (TAROS) Conference, Plymouth, UK (2010)

9. Roke, C., Melhuish, C., Pipe, T., Drury, D., Chorley, C.: Deformation-based tactile feedback using a biologically-inspired sensor and a modified display. In: Groß, R., Alboul, L., Melhuish, C., Witkowski, M., Prescott, T.J., Penders, J. (eds.) TAROS 2011. LNCS, vol. 6856, pp. 114–124. Springer, Heidelberg (2011)

10. Winstone, B., Griffiths, G., Pipe, T., Melhuish, C., Rossiter, J.: TACTIP - tactile fingertip device, texture analysis through optical tracking of skin features. In: Lepora, N.F., Mura, A., Krapp, H.G., Verschure, P.F., Prescott, T.J. (eds.) Living Machines 2013. LNCS, vol. 8064, pp. 323–334. Springer, Heidelberg (2013)

11. Koza, J.: Genetic Programming: on the Programming of Computers by Means of Natural Selection, vol. 1. MIT Press, Cambridge (1992)

12. Terence, S., Heckendorn, R.: A practical platform for on-line genetic programming for robotics. In: Riolo, R., Vladislavleva, E., Ritchie, M.D., Moore, J.H. (eds.) Genetic Programming Theory and Practice X, pp. 15–29. Springer, New York (2013)

13. Chih-Hung, W., et al.: Target position estimation by genetic expression programming for mobile robots with vision sensors. IEEE Trans. Instrum. Meas. 62(12), 3218–3230 (2013)

14. Pedro, S., et al.: Automatic generation of biped locomotion controllers using genetic programming. Robot. Auton. Syst. 62(10), 1531–1548 (2014)

15. Kordon, A., Smits, G., Jordaan, E., Rightor, E.: Robust soft sensors based on integration of genetic programming, analytical neural networks, and support vector machines. In: Proceedings of the 2002 Congress on Evolutionary Computation, vol. 1, 12–17 May 2002, pp. 896–901 (2002)

16. Suraj, S., Tambe, S.: Soft-sensor development for biochemical systems using genetic programming. Biochem. Eng. J. 85, 89–100 (2014)

17. Alexandridis, A.: Evolving RBF neural networks for adaptive soft-sensor design. Int. J. Neural Syst. 23(06), 1350029 (2013)

18. Shi, J., Xing-Gao, L.: Product quality prediction by a neural soft-sensor based on MSA and PCA. Int. J. Autom. Comput. 3(1), 17–22 (2006)

19. Buljak, V.: Inverse Analyses with Model Reduction: Proper Orthogonal Decomposition in Structural Mechanics. Computational Fluid and Solid Mechanics. Springer, Heidelberg (2012)

20. Zu-Qing, Q.: An efficient modelling method for laminated composite plates with piezoelectric sensors and actuators. Smart Mater. Struct. 10(4), 807–818 (2001)

21. Kudryavtsev, M., et al.: A compact parametric model of magnetic resonance micro sensor. In: 2015 16th International Conference on Thermal, Mechanical and Multi-physics Simulation and Experiments in Microelectronics and Microsystems (EuroSimE). IEEE (2015)

22. Searson, D.: GPTIPS. https://sites.google.com/site/gptips4matlab/. Accessed 15 Feb 2016

Agent-Based Autonomous Systems and Abstraction Engines: Theory Meets Practice

Louise A. Dennis[1]($\boxtimes$), Jonathan M. Aitken[2], Joe Collenette[1], Elisa Cucco[1],
Maryam Kamali[1], Owen McAree[2], Affan Shaukat[3], Katie Atkinson[1],
Yang Gao[3], Sandor M. Veres[2], and Michael Fisher[1]

[1] Department of Computer Science, University of Liverpool, Liverpool, UK
L.A.Dennis@liverpool.ac.uk
[2] Department of Autonomous Systems and Control,
University of Sheffield, Sheffield, UK
[3] Surrey Space Centre, University of Surrey, Guildford, UK

Abstract. We report on experiences in the development of hybrid autonomous systems where high-level decisions are made by a rational agent. This rational agent interacts with other sub-systems via an *abstraction engine*. We describe three systems we have developed using the EASS BDI agent programming language and framework which supports this architecture. As a result of these experiences we recommend changes to the theoretical operational semantics that underpins the EASS framework and present a fourth implementation using the new semantics.

1 Introduction

Translating continuous sensor data into abstractions suitable for use in Beliefs-Desires-Intentions (BDI) style agent programming languages is an area of active study and research [6,7,17]. Work in [7] provides an architecture for autonomous systems which explicitly includes an *abstraction engine* responsible for translating continuous data into discrete agent beliefs and for reifying actions from the agent into commands for the underlying control system. The architecture includes an operational semantics that specifies the interactions between the various sub-systems.

This paper reports on three autonomous systems based on this architecture, recommends changes to the semantics and then presents a fourth system using the new architecture. In particular we recommend moving away from a view based on a fixed number of sub-systems, to one based on a variable number of communication channels; and abandoning the idea that the abstraction engine works with data present in a logical form.

2 Agents with Abstraction Engines

Hybrid models of control are of increasing popularity in the design and implementation of autonomous systems. In particular there has been interest in systems in which a software agent takes high-level decisions but then invokes lower

© Springer International Publishing Switzerland 2016
L. Alboul et al. (Eds.): TAROS 2016, LNAI 9716, pp. 75–86, 2016.
DOI: 10.1007/978-3-319-40379-3_8

level controllers to enact those decisions [14, 15, 20]. In many applications the ability of a *rational* agent to capture the "reasons" for making decisions is important [11]. As the key programming paradigm for rational agents, the BDI model (*Beliefs-Desires-Intention* [18]) is of obvious interest when designing and programming such systems.

A key problem in integrating BDI programs with control systems is that the data generated by sensors is generally continuous in nature where BDI programming languages are generally based on the logic-programming paradigm and prefer to manage information presented in the format of discrete first order logic predicates. A second problem is that the data delivered from sensors often arrives faster than a BDI system is able to process it, particularly when attempting to execute more complex reasoning tasks at the same time.

Work in [7] proposes the architecture shown in Fig. 1 in which an *abstraction engine* is inserted between the rational agent (called the *reasoning engine*) and the rest of the system. The abstraction engine is able to rapidly process incoming sensor data and forward only events of interest to the reasoning engine for decision-making purposes. The abstraction engine mediates between the reasoning engine and a *physical engine* (representing the software control systems and sensors of the underlying autonomous system) and a *continuous engine* which can perform calculation and simulation, specifically, in the original architecture, with a view to path planning and spatial prediction.

The reasoning engine reasons with discrete information. The abstraction engine is responsible, therefore, for abstracting the data from sensors (e.g., real number values, representing distances) to predicates (e.g., *too_close*, etc.). In [7] it is assumed that the abstraction engine works with sensor readings represented in logical form – for instance of the form *distance*(2.5). A function, *fof*, is assumed which transforms data from sensors into this representation. The reasoning engine makes decisions about actions, but may also request calculations

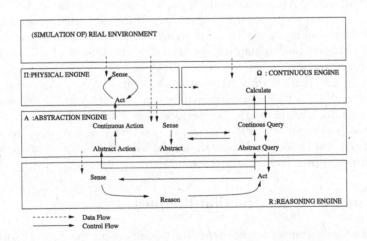

Fig. 1. Abstract architecture for agent based autonomous systems

(for instance estimates of whether any collisions are anticipated in the next time window). The abstraction engine is responsible for translating these actions, which it is anticipated will be expressed in a high level fashion, into appropriate commands for either the physical engine or the continuous engine.

A variant of the GWENDOLEN BDI programming language [8], named EASS, implements this architecture. Specifically it provides BDI-based programming structures for both abstraction and reasoning engines (so both are implemented as rational agents). It also provides Java classes for building middleware *environments* for the agents which implement the operational semantics of interaction from [7]. These environments can communicate with external systems by a range of methods (e.g., over sockets or using the Robot Operating System (ROS)[1] [16]) and provide support for transforming data from these systems into first order predicate structures (i.e., providing the *fof* function). We refer to the EASS language and associated environment support as the EASS framework. The EASS framework was used in an extensive case study for the architecture involving satellite and space craft systems [10,13].

3   Practical Systems

We now discuss three systems built using the EASS framework. The first is a demonstration system in which a robot arm performs *sort and segregate* tasks. The second is a simulation of a convoy of road vehicles. The third is a public engagement activity involving LEGO robots.

3.1   An Autonomous Robotic Arm

The autonomous robotic arm system performs sort and segregate tasks such as waste recycling or nuclear waste management[2] [1,2]. The system is required to view a set of items on a tray and identify those items. It must determine what should be done with each one (e.g. composted, used for paper recycling or glass recycling, etc.) and then move each item to a suitable location.

The system integrates computer vision, a robot arm and agent-based decision making. It is implemented in the Robot Operating System (ROS) [16]. Computer vision identifies items on a tray [19]. These identities and locations are published to a *ROS topic* – a communication channel. The abstraction engine subscribes to this topic and abstracts away the location information informing the reasoning engine what types of object can be seen. The reasoning engine makes decisions about what should be done with each object. These decisions involve, for instance, sending: anything that is plant matter to be composted; paper for recycling; and bricks for landfill. These decisions are published to a different topic

---

[1] www.ros.org.

[2] The robotic arm system involves proprietary software developed jointly by the universities of Liverpool, Sheffield and Surrey and National Nuclear Labs. Requests for access to the code or experimental data should be made to Profs Fisher, Veres or Gao.

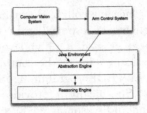

Fig. 2. Architecture for the Robot Arm

by the abstraction engine (adding back the information about object location) to which the robot arm control system subscribes. The control system publishes information about what it is doing which the reasoning engine uses to make sure new instructions are not sent until previous ones have completed.

The architecture is shown in Fig. 2. The abstraction and reasoning engine operate as described in [7]. We also show the Java Environment that supports interaction. There is no equivalent to the continuous engine. The physical engine is a combination of the arm control system and the computer vision system. The computer vision and robotic arm sub-systems communicate with each other for coordination when moving objects.

This system has subsequently been extended to deal with the *sort and disrupt* problem [1]. A canister is presented, which must be lifted and placed in a set of v-shaped grooves, before it is opened using a pneumatic cutting tool to inspect the contents which then undergo sort and segregate. This demonstrator consists of a KUKA IIWA arm, with a payload of 7 kg. The location of disruption is indicated on a canister via a laser-pen to prevent physical destruction of the test pieces. As well as handling the sort and disrupt task the agent can reason about faulty equipment (simulated using laser-pen failure). The reasoning engine uses a reconfiguration strategy [9] to instruct the arm to use a different laser-pen to complete the task. Similarly the agent reasons about the accuracy and reliability of the information received from the computer vision system.

3.2   Vehicle Convoying

We view an autonomous vehicle convoy as a queue of vehicles in which the first is controlled by a human driver, but subsequent vehicles are controlled autonomously. The autonomously controlled "follower" vehicles maintain a safe distance from the vehicle in front. When a human driving a vehicle wishes to join a convoy they signal their intent to the convoy lead vehicle, together with the position in the convoy they wish to join. Autonomous systems in the lead vehicle then instructs the vehicle that will be behind the new one to drop back, creating a gap for it to move into. When the gap is large enough, the human driver is informed that they may change lane. Once this is achieved, autonomous systems take control and move all the vehicles to the minimum safe convoying distance. Similar protocols are followed when a driver wishes to leave the convoy.

Maintenance of minimum safe distances between vehicles is handled by two low level control systems. When the convoy is in formation, control is managed using distance sensors and wireless messages from the lead vehicle. These messages inform the convoy when the lead vehicle is braking or accelerating and so allow smooth responses from the whole convoy to these events. This reduces the safe minimum distance to one where fuel efficiency gains are possible. In some situations control uses sensors alone (e.g., during leaving and joining). In these situations the minimum safe distance is larger.

The agent system manages the messaging protocols for leaving and joining, and switches between the control systems for distance maintenance. For instance, if a communication break-down is detected, the agent switches to safe distance control based on sensors alone. The abstraction engine, therefore, is involved primarily in monitoring distance sensors and communication pings.

The system was developed in order to investigate issues in the verification and validation of these convoying systems [12][3]. A simulation of the vehicle control systems was created in MATLab and connected to the the TORCS[4] racing car simulator. The architecture for this system is close to that in [7] with a single physical engine, but still no continuous engine. Several agents connect to the simulator, via the Java environment. The agents use the Java environment for messaging between agents. This architecture is shown in Fig. 3.

Following verification and validation phases based on simulation, we are in the process of transferring the system to Jaguar 4×4 wheeled robots[5] for hardware testing in outdoor situations.

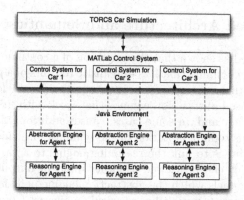

Fig. 3. Architecture for a simulated convoying system

---

[3] Software available from https://github.com/VerifiableAutonomy.

[4] torcs.sourceforge.net.

[5] jaguar.drrobot.com.

3.3 LEGO Rovers

The LEGO Rovers system was developed to introduce the concepts of abstraction and rational agent programming to school children[6]. It is used in science clubs by volunteer members of the STEM Ambassador scheme, and has also been used in larger scale events and demonstrations. The activity introduces the user to a teleoperated LEGO robot and asks them to imagine it is a planetary rover. The robot's sensors are explained, the user is shown how the incoming data is abstracted into beliefs such as *obstacle* or *path* using simple thresholds and can then create simple rules, using a GUI, which dictate how the robot should react to the appearance and disappearance of obstacles, etc.

This activity has been through two versions. In the first, the EASS framework was used off-the-shelf with LEGO NXT robots, once again with no continuous engine. The GUI ran on laptops. The system used the leJOS Java-based operating system for Mindstorms robots [4,5] and classes from this were used for direct communication with the robot. Sensors needed to be polled for data in contrast two our other applications where sensors continuously published data to a stream. While the activity worked well, some issues were observed, particularly the robot's response to rules sometimes lagged more than could be accounted for simply by delays in Bluetooth communication. Investigation suggested that, even when concerned only with processing sensor data, a rational agent was more heavy-weight technology than was required for abstraction. The rational agent used logical reasoning to match plans to events which were then executed to produce abstractions. We discuss an amended version of the system in Sect. 5.

4 An Improved Architecture and Semantics

Practical experience has shown that the provision of a continuous engine is not a fundamental requirement and that it is unrealistic to think of a physical engine as a monolithic entity that encompasses a single input and a single output channel. The use of agent-based abstraction engines that work with first order formulae also seems unnecessary and, in some cases, causes additional inefficiency in the system. This leads us to an adaptation of the semantics in which the purpose of the abstraction engine is to link the reasoning engine to a variety of communication channels which can be viewed as either input or output channels. Input channels include channels which the abstraction engine polls for data so long as it does not execute further until it has received an answer. Communications, like those handled by the continuous engine, in which a request is made for a calculation and, some time later, an answer is received can be handled by placing a request on an output channel and then receiving an answer on an input channel. The abstraction engine (and reasoning engine) can be responsible for matching received answers to requests in an appropriate fashion.

We group these channels into two sets. Π are output channels where the abstraction engine writes information or sends commands to be interpreted by

---

[6] www.csc.liv.ac.uk/~lad/legorovers.

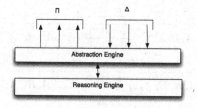

Fig. 4. Refined architecture for agent-based hybrid systems with abstractions

other parts of the system. Δ are input channels. Input channels may operate on a request-reply basis but the abstraction engine does nothing between request and reply. Figure 4 shows this new architecture.

In [7], Π is the physical engine, Ω the continuous engine and Δ is a set of sensor inputs. The semantics also references A (the abstraction engine), R (the reasoning engine) and Σ, Γ and Q, sets of predicates which are used in communication between the abstraction and reasoning engines. Σ is a set of *shared beliefs* which stores abstractions, Γ is a set of commands the reasoning engine has requested for execution and Q is a set of queries from the reasoning engine. The whole system is represented as a tuple. In the modified version of the operational semantics, we no longer consider Ω and Q. Δ and Π are now sets of channels rather than explicit sub-systems. Therefore, we represent the system as a tuple $\langle \Delta, \Pi, A, R, \Sigma, \Gamma \rangle$. The operational semantics specifies a labelled transition system on this tuple. For clarity in discussion, we will sometimes replace parts of this tuple with ellipsis (...) if they are unchanged by a transition.

Most of the rules governing interaction between the abstraction and reasoning engine are largely unchanged. We show these rules in Fig. 5 for completeness but do not discuss them further here. We refer the reader to [7] for further discussion and motivation. Three rules involving "queries" (the set Q) intended for the continuous engine have been removed.

We turn our attention to the remaining rules in the semantics. It should be noted that all subsystems may take internal transitions which change their state.

Only one rule involving interaction between abstraction and reasoning engine needs modification. This semantic rule governs transitions the system takes after the reasoning engine has placed a command in the set Γ for the abstraction engine to reify. The original semantics views this reification process as a combination of the transitions taken by the abstraction engine in order to transform the command *and* any subsequent changes to the physical engine. We simplify our view of this so we consider only the change in A when it reads in the command $A \xrightarrow{read_C \gamma} A'$. The abstraction engine may make subsequent internal transitions as it processes the command. The changes to the system when it passes the command on to the physical engine are shown in semantic rule (9). The new version of this rule is shown in (8). In this rule when A performs a read on Γ the whole system makes a transition in which A is transformed to A' and γ, the

$$\frac{A \xrightarrow{per(\Sigma)} A'}{\langle \ldots, A, R, \Sigma, \ldots \rangle \xrightarrow{per_A(\Sigma)} \langle \ldots, A', R, \Sigma, \ldots \rangle} \tag{1}$$

$$\frac{R \xrightarrow{per(\Sigma)} R'}{\langle \ldots, A, R, \Sigma, \ldots, \rangle \xrightarrow{per_R(\Sigma)} \langle \ldots, A, R', \Sigma, \ldots \rangle} \tag{2}$$

$$\frac{A \xrightarrow{+\Sigma^b} A'}{\langle \ldots, A, R, \Sigma, \ldots \rangle \xrightarrow{+\Sigma, A^b} \langle \ldots, A', R, \Sigma \cup \{b\}, \ldots \rangle} \tag{3}$$

$$\frac{A \xrightarrow{-\Sigma^b} A'}{\langle \ldots, A, R, \Sigma, \ldots \rangle \xrightarrow{-\Sigma, A^b} \langle \ldots, A', R, \Sigma \backslash \{b\}, \ldots \rangle} \tag{4}$$

$$\frac{R \xrightarrow{+\Sigma^b} R'}{\langle \ldots, R, \ldots \rangle \xrightarrow{+\Sigma, R^b} \langle \ldots, R', \Sigma \cup \{b\}, \ldots \rangle} \tag{5}$$

$$\frac{R \xrightarrow{-\Sigma^b} R'}{\langle \ldots, R, \Sigma, \ldots \rangle \xrightarrow{-\Sigma, R^b} \langle \ldots, R', \Sigma \backslash \{b\}, \ldots \rangle} \tag{6}$$

$$\frac{R \xrightarrow{do(\gamma)} R'}{\langle \ldots, R, \ldots, \Gamma \rangle \xrightarrow{do_R(\gamma)} \langle \ldots, R', \ldots, \{\gamma\} \cup \Gamma \rangle} \tag{7}$$

Fig. 5. Unchanged semantic rules

command A has read, is removed from Γ.

$$\frac{\gamma \in \Gamma \qquad A \xrightarrow{read_C(\gamma)} A'}{\langle \ldots, A, \ldots, \Gamma \rangle \xrightarrow{do_A(\gamma)} \langle \ldots \ldots, A', \ldots, \Gamma \backslash \{\gamma\} \rangle} \tag{8}$$

Example. Assume a simple BDI agent, A, represented as a tuple, $\langle \mathcal{B}, \mathcal{P}, \mathcal{I} \rangle$ of a set $\mathcal{B}$ of beliefs, a set $\mathcal{P}$ of plans and an intention stack $\mathcal{I}$ of commands to be executed. Assume that when A reads a formula, $\gamma \in \Gamma$, it places it in $\mathcal{B}$ as $do(\gamma)$. Consider the Lego Rover example and a request from the reasoning engine for the robot to *turn_right* when A has an empty belief set. Before execution of rule (8), $A = \langle \emptyset, \mathcal{P} \rangle$. After the execution of Eq. 8, $A = \langle \{do(turn\_right)\}, \mathcal{P}, \mathcal{I} \rangle$.

Our view of the interaction of the abstraction engine with the rest of the system is considerably simplified by the removal of an explicit continuous engine. Two complex semantic rules are removed and we need only consider what happens when the abstraction engine publishes a command to an output channel and when it reads in data from an input channel.

A, the abstraction engine, can place data, γ (assumed to be the processed form of a command issued by the reasoning engine – though this is not enforced[7]), on

---

[7] Particularly since a single command from the reasoning engine can be transformed into a sequence of commands by the abstraction engine.

some output channel, π from Π. It does this using the transition $A \xrightarrow{run(\gamma,\pi)} A'$ and we represent the change caused to π when a value is published on it as $\pi \xrightarrow{pub(\gamma)} \pi'$. $\Pi\{\pi/\pi'\}$ is the set Π with π replaced by π'. Rule (9) shows the semantic rule for A publishing a request to π.

$$\frac{\pi \in \Pi \qquad A \xrightarrow{run(\gamma,\pi)} A' \qquad \pi \xrightarrow{pub(\gamma)} \pi'}{\langle \ldots, \Pi, A, \ldots \rangle \xrightarrow{run(\gamma,\pi)} \langle \ldots, \Pi\{\pi/\pi'\}, A', \ldots \rangle} \qquad (9)$$

This rule states that if A makes a transition where it publishes some internally generated command, γ, to π, $A \xrightarrow{run(\gamma,\pi)} A'$, then assuming π is one of the output channels and then effect of publishing γ to π is π' then the whole system makes a transition in which A is replaced by A' and π by π'.

Example. Returning to our example above, assume that A has plans that reify *turn_right* into a sequence of two commands, one for each engine controlling a wheel on the robot. Internal execution of these plans places these commands on the intention stack[8].

$$A = \langle \{\mathcal{B}, \mathcal{P}, pilot(right.back()) : pilot(left.forward()) : \mathcal{I} \rangle$$

We assume Π consists of two output channels *pilot* (which transmits commands over bluetooth to the leJOS `Pilot` class) and *gui* which sends status information to the GUI controlling the robot. The agent executes the command $pilot(\gamma)$ by placing γ on the *pilot* channel and removing the command from the intention stack. Assume that the *pilot* channel is empty and the GUI channel contains the notification of an obstacle $gui = \{obstacle\}$. So before execution of rule 9, $\Pi = \{\emptyset, \{obstacle\}\}$. After one execution, $\Pi = \{\{right.back()\}, \{obstacle\}\}$ and $A = \langle \{\mathcal{B}, \mathcal{P}, pilot(left.forward()) : \mathcal{I} \rangle$.

Similarly, we indicate the process of abstraction engine, A, reading a value from a channel d by $A \xrightarrow{read(d)} A'$. We represent any change in state on the channel if a value is read from it as $d \xrightarrow{read} d'$. This allows us to simply define a semantics for perception in (10).

$$\frac{d \in \Delta \qquad A \xrightarrow{read(d)} A' \qquad d \xrightarrow{read} d'}{\langle \Delta, \ldots A, \ldots \rangle \xrightarrow{read(d)} \langle \Delta\{d/d'\}, \ldots, A', \ldots \rangle} \qquad (10)$$

Example. Returning to the example, assume that $A = \langle \emptyset, \mathcal{P}, \mathcal{I} \rangle$ reads from two input channels: a distance sensor and an RGB colour sensor. Let $\Delta = \{\{50\}, \{\langle 1, 5, 6 \rangle\}\}$ (i.e., there is an object 50 centimetres from the robot and the color sensor is looking at a very dark (almost black) surface). Suppose, when A reads data, d, from either channel it removes the reading from the channel and turns it into a belief $distance(d)$ or $rgb(d)$ respectively. If rule (10) is executed

---

[8] We use : to indicate concatenation of an element to the top of a stack.

and A reads from the distance channel. Then Δ becomes $\{\emptyset, \{\langle 1, 5, 6 \rangle\}\}$ and $A = \langle \{distance(50)\}, \mathcal{P}, \mathcal{I} \rangle$.

The main advantage of this new semantics is the simplification of the system by the removal of semantic rules involving calculations and queries and the removal of related components from the state tuple. We also believe it has the pragmatic virtue of representing more accurately the architectures of systems that people actually build using abstraction engines and reasoning agents.

5   Implementation (LEGO Rovers v.2)

The new operational semantics required few changes to the EASS framework, many of which had been implemented incrementally during our experience building the systems in Sect. 3. Support for continuous engines was dropped, simplifying both the language and the Java Environment classes. Support for agent-based abstraction engines was kept, but their use became optional. The implementation had never enforced the use of single input and output channels and the nature of Java meant these were trivial to include as an application required.

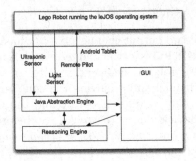

Fig. 6. Architecture for the LEGO rover system

A second version of the LEGO Rover activity was then developed for LEGO EV3 robots with a GUI running on Android Tablets. This version used the new semantics with the EASS based abstraction engine being replaced by an efficient Java class that handled abstraction of data and reification of commands.

The architecture for the second version is shown in Fig. 6. Both abstraction and reasoning engines interact with the GUI which displays the sensor data and the abstractions, and allows the user to define the rules in the reasoning engine. The abstraction engine polls two sensors (ultrasonic and light) for data and publishes commands to the robot. Feedback from demonstrators and teachers suggests that the changes to the activity, including the removal of lag caused by the agent-based abstraction engine, provide a better experience for children.

6 Further Work and Conclusions

In this paper we amended the operational semantics for communication between rational agents, abstraction engines and the rest of a hybrid autonomous system that was presented in [7]. This results in a simplification of the semantics, particularly the removal of an explicit continuous engine and the insistence that abstraction engines handle data in a logical form. The changes are based on extensive practical experience in developing such systems, three of which we have outlined in order both to illustrate the development of agent-based autonomous systems and to motivate the changes to the semantics.

In future, we intend to provide more software support for the implemented framework, particularly for the development of abstraction engines. For instance we aim to expand our abstraction engines, which are currently based almost exclusively on thresholding data, to richer formalisms that would allow abstractions to be formed based on observed complex events and to explore the use of external systems and frameworks to perform the necessary stream processing. In particular it seems desirable to consider abstraction engines either developed in, or incorporating, efficient stream processing tools such as Esper[9] or ETALIS [3].

We now have a refined and simplified a framework for agent-based hybrid systems that use abstraction engines. The framework is well-supported by practical experience in the construction of such systems and represents a strong practical theoretical basis for agent-based hybrid systems.

Acknowledgments. The work in this paper was funded by EPSRC grants Reconfigurable Autonomy (EP/J011770/1, EP/J011843/1, EP/J011916/1) and Verifiable Autonomy (EP/L024845/1, EP/L024942/1, EP/L024861/1) and STFC Grant LEGO Rovers Evolution (ST/M002225/1).

References

1. Aitken, J.M., Shaukat, A., Cucco, E., Dennis, L.A., Veres, S.M., Gao, Y., Fisher, M., Kuo, J.A., Robinson, T., Mort, P.E.: Autonomous nuclear waste management. Robotics and Automation (2016, under Review)
2. Aitken, J.M., Veres, S.M., Judge, M.: Adaptation of system configuration under the robot operating system. In: Proceedings of the 19th world congress of the international federation of automatic control (IFAC) (2014)
3. Anicic, D., Rudolph, S., Fodor, P., Stojanovic, N.: Stream reasoning and complex event processing in etalis. Semant. web **3**(4), 397–407 (2012)
4. Bagnall, B.: Maximum LEGO NXT: Building Robots with Java Brains. Variant Press, Winnipeg (2013)
5. Bagnall, B.: Maximum LEGO EV3: Building Robots with Java Brains. Variant Press, Winnipeg (2014)
6. Cranefield, S., Ranathunga, S.: Handling agent perception in heterogeneous distributed systems: a policy-based approach. In: Holvoet, T., Viroli, M. (eds.) Coordination Models and Languages. LNCS, vol. 9037, pp. 169–185. Springer, Heidelberg (2015)

---

[9] www.espertech.com.

7. Dennis, L.A., Fisher, M., Lincoln, N.K., Lisitsa, A., Veres, S.M.: Declarative abstractions for agent based hybrid control systems. In: Omicini, A., Sardina, S., Vasconcelos, W. (eds.) DALT 2010. LNCS, vol. 6619, pp. 96–111. Springer, Heidelberg (2011)

8. Dennis, L.A., Farwer, B.: Gwendolen: a bdi language for verifiable agents. In: Löwe, B. (ed.) Logic and the Simulation of Interaction and Reasoning. AISB, Aberdeen, AISB 2008 Workshop (2008)

9. Dennis, L.A., Fisher, M., Aitken, J.M., Veres, S.M., Gao, Y., Shaukat, A., Burroughes, G.: Reconfigurable autonomy. KI-Künstliche Intelligenz **28**(3), 199–207 (2014)

10. Dennis, L.A., Fisher, M., Lincoln, N.K., Lisitsa, A., Veres, S.M.: Practical verification of decision-making in agent-based autonomous systems. Autom. Softw. Eng., pp. 1–55 (2014)

11. Fisher, M., Dennis, L.A., Webster, M.P.: Verifying autonomous systems. Commun. ACM **56**(9), 84–93 (2013)

12. Kamali, M., Dennis, L.A., McAree, O., Fisher, M., Veres, S.M.: Formal Verification of Autonomous Vehicle Platooning. ArXiv e-prints, February 2016. under Review

13. Lincoln, N.K., Veres, S.M., Dennis, L.A., Fisher, M., Lisitsa, A.: Autonomous asteroid exploration by rational agents. IEEE Comput. Intell. Mag. **8**(4), 25–38 (2013)

14. Muscettola, N., Nayak, P.P., Pell, B., Williams, B.C.: Remote agent: to boldly go where no ai system has gone before. Artif. Intell. **103**(1–2), 5–47 (1998)

15. Patchett, C., Ansell, D.: The development of an advanced autonomous integrated mission system for uninhabited air systems to meet uk airspace requirements. In: 2010 International Conference on Intelligent Systems, Modelling and Simulation (ISMS), pp. 60–64, January 2010

16. Quigley, M., Conley, K., Gerkey, B.P., Faust, J., Foote, T., Leibs, J., Wheeler, R., Ng, A.Y.: ROS: an open-source robot operating system. In: Proceedings of the ICRA Workshop on Open Source Software (2009)

17. Ranathunga, S., Cranefield, S., Purvis, M.: Identifying events taking place in second life virtual environments. Appl. Artif. Intell. **26**(1–2), 137–181 (2012)

18. Rao, A.S., Georgeff, M.P.: An abstract architecture for rational agents. In: Proceedings of the 3rd International Conference on Principles of Knowledge Representation and Reasoning (KR), pp. 439–449 (1992)

19. Shaukat, A., Gao, Y., Kuo, J.A., Bowen, B.A., Mort, P.E.: Visual classification of waste material for nuclear decommissioning. Rob. Auton. Syst. **75**, 365–378 (2016). Part B

20. Wei, C., Hindriks, K.V.: An agent-based cognitive robot architecture. In: Dastani, M., Hübner, J.F., Logan, B. (eds.) Programming Multi-Agent Systems: 10th International Workshop, ProMAS 2012, Valencia, Spain, June 5, 2012, Revised Selected Papers, pp. 54–71. Springer, Berlin Heidelberg, Berlin, Heidelberg (2013)

Experimental Evaluation of a Multi-modal User Interface for a Robotic Service

Alessandro Di Nuovo[1]([✉]), Ning Wang[2], Frank Broz[3], Tony Belpaeme[2],
Ray Jones[4], and Angelo Cangelosi[2]

[1] Department of Computing, Sheffield Hallam University, Sheffield, UK
a.dinuovo@shu.ac.uk
[2] Centre for Robotics and Neural Systems, Plymouth University, Plymouth, UK
[3] School of Mathematical and Computer Sciences,
Heriot-Watt University, Edinburgh, UK
[4] School of Nursing and Midwifery, Plymouth University, Plymouth, UK

Abstract. This paper reports the experimental evaluation of a Multi-Modal User Interface (MMUI) designed to enhance the user experience in terms of service usability and to increase acceptability of assistive robot systems by elderly users. The MMUI system offers users two main modalities to send commands: they are a GUI, usually running on the tablet attached to the robot, and a SUI, with a wearable microphone on the user. The study involved fifteen participants, aged between 70 and 89 years old, who were invited to interact with a robotic platform customized for providing every-day care and services to the elderly. The experimental task for the participants was to order a meal from three different menus using any interaction modality they liked. Quantitative and qualitative data analyses demonstrate a positive evaluation by users and show that the multi-modal means of interaction can help to make elderly-robot interaction more flexible and natural.

Keywords: Human-robot interaction and interfaces · Service robotics · Socially Assistive Robotics

1 Introduction

Recent technological developments have made Socially Assistive Robotic (SAR) platforms a realistic option for providing services and care to our ageing population [22]. As an example they are able to provide assistance inside the apartment or serving in shared facilities of the building or accompanying people outdoors [11].

In this application domain, one of the current challenges is to identify the modalities of interaction that can be feasible and acceptable by the specific user population, also to overcome physical limitations and the digital divide. To this end, SAR interfaces share the same design principles and guidelines that are derived from Human Computer Interaction [21]. Examples are mobile robotic telepresence systems that incorporate video conferencing devices onto mobile robots which can be steered from remote locations [18]. General principles of

© Springer International Publishing Switzerland 2016
L. Alboul et al. (Eds.): TAROS 2016, LNAI 9716, pp. 87–98, 2016.
DOI: 10.1007/978-3-319-40379-3_9

recent interfaces for SAR are presented and discussed in [19], that present five experimental projects focused on a single platform that was operating inside the user's home. The majority of them use a wheel drive platform with a fixed touch screen as the main Graphic User Interface (GUI). The limitation of the fixed touch screen is that a short distance is needed to interact with the robot. In the KSERA project the GUI is projected on a wall with a pico projector [20]. This solution has the advantage to provide a bigger screen and to allow interaction from a distance, but the main drawback is that a suitable surface is always required. Speech and gesture interfaces may help human-robot interaction when a GUI is not available, but there are services that require the use of a graphic interface (e.g. shopping, video calling, food ordering, etc.).

In this area multi-modal interaction has developed considerably in the past fifteen years [15], thanks to the wide availability of devices and sensors that can support this approach. Multi-Modal User Interfaces (MMUI) convincingly mimic social interaction between a human and robot by means of speech, gestures or other modalities, that may be preferred over unimodal interfaces by elderly users [16]. Multimodal interfaces have been demonstrated to offer better flexibility and reliability than other human/machine interaction means [14]. However, it is clear that further research in this area is needed to prove the added value and soundness for care provision [5]. Particular emphasis should be given to the types and quality of feedback that the robot interface provides in order to make the system more intuitive for elderly users [9] and how often this feedback is given [8].

In this paper, we present the user evaluation of a MMUI, which was designed to enhance the user experience in terms of usability and to increase acceptability of assistive robot systems by elderly users [13] as part of the activities of the EU FP7 Robot-Era project. The MMUI is described in Sect. 2.1, which gives detail of the graphic and the speech user interfaces. The description of the robotic platform and of the service tested is in Sect. 2.2. Participants, setting and instruments used for the evaluation are presented in Sect. 2.3. The experimental results are discussed in Sect. 3. Finally, our conclusion is given in Sect. 4.

2   Material and Methods

In this section, we describe the MMUI tested in our experiment. The MMUI was developed as part of the EU FP7 Robot-Era project (2012–2015), which integrated three robotic platforms in the Ambient Assisted Living (AAL) paradigm [1]. The project developed, implemented and demonstrated the general feasibility, scientific/technical effectiveness and social/legal plausibility and acceptability by end-users of a plurality of complete advanced robotic services. Integrated services demonstrated their effectiveness in real conditions and cooperate with real people to provide favourable independent living, improving the quality of life and the efficiency of care for elderly people.

2.1   Multi-modal Interface for Elderly-Robot Interaction

The MMUI system offers users two main modalities to send commands: they are a GUI, usually running on the tablet attached to the robot, and a SUI, with a wearable microphone on the user. Two main modules implement the MMUI: the *Web Interface System* that includes the graphic user interface (GUI) and the text-to-speech (TTS) software; the *Dialog Manager* that implements the Speech User Interface (SUI) with the Automatic Speech Recognition (ASR) software. The MMUI software is also responsible for providing feedback to users via the same modalities that they use to command the robot. This was implemented by making both robotic platforms and web graphic interface able to produce sounds, including speech. As visual feedback the tablet can show specific text messages (e.g. yes/no buttons for confirmation) and robots can also change colour of their LEDs (e.g. they change colour when the robot is waiting for the user action).

The Web-Based Graphic User Interface (GUI). The scientific literature includes many contributions that provide guidelines and design recommendations about web and tablet interfaces for the elderly. Web design guidelines to address the web experiences of older users can be found in [3,25], while a survey on touch-based mobile phone interface for the elderly can be found in [2]. In our implementation we used a tablet because it has been found that elderly users are very receptive to these devices [24].

The GUI uses web technology to support the widest range of devices that can be connected to the system network. A main menu index page allows the user to navigate between the different service pages that compose the GUI. There are additional web pages, not accessible by the user, that are available from the web server to allow testing and to support other actors of the Robot-Era system (grocer, caregiver, etc.). A settings page, that is not accessible from the main page, is also present to modify the language, items in the shopping page and to switch test site apartment map. Pages are implemented with HTML5 and Javascript so that they can ideally be available on any browser running on any device. On the web server side, there are scripts that do the reasoning and communicate via the PEIS middleware with the other sensors and modules that compose the ambient intelligence.

The graphic interface for the Robot-Era services is developed as a web based server-client architecture in order to allow remote control through mobile devices like tablets and smart-phones. Information from the robot or the ambient environment is also made available to the user via notifications and warnings. The interface is complementary to speech control of the robot. The two modalities are usually interchangeable, except for shopping and communication services where the tablet is required for the shopping list and the video call. For video calling, Skype was integrated with the interface using its web API. As guideline for the graphic design we mainly followed the recommendations provided by the World Wide Web Consortium (W3C) about Web Accessibility for Older Users [12]. The graphics interface is intended to run on any platform (e.g. PC, tablet, smart-phones) with any web browser, but at the same time the design of the graphic interface aims to maximize the integration with the host device, in order to give the impression that it can be a real product and, moreover, to provide people that have previous knowledge of the device with the basic commands that they already know.

More details and a preliminary evaluation of the Robot-Era user interfaces can be found in [13].

The Speech User Interaction (SUI) Module, Multi-language Support and Feedback. Two significantly different versions of the SUI were implemented during the Robot-Era project. First, a basic speech recognition system was implemented to simply allow the user to call the robot and order the robotic services by using easy spoken commands. The first version was designed to be simple in order to perform a first test with elderly participants and, then, decide to increase the complexity if that was successful. Indeed, after the success of the general experimentation and the positive user feedback, a more complex dialogue manager and a refined speech recognition system were implemented for the final version that was tested in the focused study.

Both versions shared a TTS module implemented as a ROS module on robots. The Acapela *Voice As A Service*(VAAS) is used for TTS. This is a web service that receives any text with voice parameters and responds with an audio file containing the speech. The Acapela web service is a module integrated into the web interface and on the robot platforms. Acapela web VAAS was preferred because of its easy integration with the web architecture of the interface. Voices were selected based on an informal survey of native speakers among the choices available for each language (many options for English but only a couple for each other language). The domestic robot has a female voice, as requested by the elderly in preliminary study, while condominium and outdoor have male voices to distinguish them and their service roles from the domestic robot.

Speech recognition is implemented using the Nuance SDK and is based on a set of restricted grammars. Nuance was preferred because it supports all the languages used in our experimentation (Italian, Swedish and English, which was used also for debugging and demonstration). The program runs as a single-threaded application whose flow is controlled by the audio input stream. Audio events trigger callbacks to handle the processing of sound input. The recognition grammars are loaded dynamically to change what input the system is *listening for* based on the context and stage of the verbal interaction.

The speech recognition is done out-of-the-box, i.e. there was no training session. Users begin verbal interaction with the robot by calling the robot by name using their wearable microphone. The robot's name is defined as a *wake-up* word which must be recognised before a service request interaction is initiated by the speech interface. This prevents service requests from being issued based on false positives from the speech recognition (which could otherwise occur in situations where the user is speaking to another person present rather than the robot). The keywords used to identify each service are specified in the grammars and may be uttered alone or as part of a longer natural language phrase.

During a service request interaction, the user may request any service. The following interaction will be determined by which service was selected. After the user has called the robot, the dialogue proceeds in a system-initiative manner. The speech interface is designed to produce short, simple, command-oriented dialogues with the user. In the case of services which require complex or extended user input, (such as creating a shopping list or entering an appointment for a reminder), the SUI directs the user to use the GUI for input and hands the interaction over to the GUI. This design choice was made to avoid the need for numerous confirmation or error-recovery dialogues which could frustrate the user in the event of low speech recognition accuracy.

Considering that using context aware models can help improve recognition accuracy and system efficacy, the original speech interaction architecture was upgraded by incorporating a more flexible and efficient dialogue flow control mechanism, as well as a more powerful dialogue manager. The dialogue manager was based on the open-source Olympus dialogue management architecture. The Ravenclaw dialogue manager, part of Olympus, simplifies the authoring of complex dialogues, has general support handling speech recognition errors, and can be extended to support multi-modal input and output [6]. The main task in achieving a context-dependent spoken dialogue system was to design dialogue task specifications according to user expectations and service requirements. We did this by following three steps: User expectation exploitation; Service-specific grammar design; Context-aware grammar flow switch.

More details and a preliminary evaluation of the SUI can be found in [23].

2.2 Robotic Platform and Description of the Service: The Food Delivery Service

The robotic platform used in our experiments is shown in Fig. 1. It is a customized G5 platform produced within the Robot-Era project. The appearance of the platforms was also studied with elderly and specifically designed for the project [10]. The robot is equipped with a tablet that is mounted on a magnetic frame that can be detached at will. A wearable microphone was provided to the users and used as input for the speech user interface. The microphone was placed on the user near the neck in order to maximize the quality of the speech recording.

The MMUI is designed to allow the user to browse three different options: meat, fish and vegetarian (veggie). Each option has 3 courses and the total calories are shown. A price is also shown to be more realistic. The user can also use the SUI to navigate among the menus and read the items aloud. After deciding which option they prefer, users can select it using the SUI or press the *order* button to proceed with the food order. Figure 1 presents the GUI and the dialogue flow for the SUI.

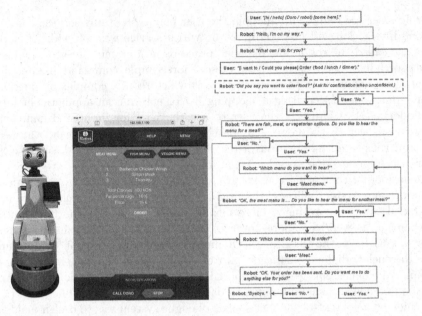

Fig. 1. (Left) The customized robotic platform used in our experiment. (Center) The food delivery page on the GUI. (Right) Dialogue flow (SUI) for the food delivery service.

2.3   User Recruitment, Setting and Evaluation Instruments

To evaluate the SUI and GUI we developed a specific use case and we carried out a series of HRI experiments at Plymouth University. To recruit participants we held a workshop at a sheltered housing facility for the accommodation of retired people in Plymouth, United Kingdom. During the workshop, we introduced the latest elderly-tailored user interface developed under the Robot-Era Project and gave demonstrations of the interaction capabilities of an humanoid robotic platform. Around 30 residents in the apartment building attended the workshop. At the end of the workshop, the attendants were invited to participate in the follow-up study on elderly-robot interaction at Plymouth University. A total of 15 subjects (3 males, 12 females, all native British English speakers) from Wesley Court have participated in the experiments. Participants' age was in the range 70–89, average 80.53, they were all well educated, with higher or further education degrees, including 4 engineers. Almost all use (6) or know how to use (8) a computer, majority (10) never used a smartphone but Eight regularly use a tablet and Three know how to use it.

The food delivery service was chosen to be the main task; subjects could use speech, tablet, or both of them to complete the task. Figure 2 shows the experimental setting: the elderly participant is sitting in front of the robot while the tablet is detached and placed on the table easily accessible by the user. The researcher is behind the user and he did not interact with the participant unless

Fig. 2. Focused experiment in a controlled setting. The elderly participant is sitting in front of the robot, while the tablet is detached and placed on the table easy accessible by the user.

not explicitly prompted by them. The test for each subject lasted for 45 min. During the experiments, in addition to questionnaires and open-question interview opinions, frontal and profile video recordings were collected. Participants always filled all the questionnaires on their own with no conversation to avoid influencing them.

Participants completed the System Usability Scale (SUS). This is a reliable, lightweight usability scale that can be used for global assessments of technological systems usability. SUS was developed by Brooke in 1996 [7], it is a simple, ten-item, five point attitude Likert scale giving a global view of subjective assessments of usability. SUS yields a single score on a scale of 0–100, this is obtained converting the range of possible values from 0 to 100 instead of from 10 to 50. The SUS has been widely used in the evaluation of a range of systems. Bangor, Kortum and Miller [4] have used the scale extensively over a ten-year period and have produced normative data that allow SUS ratings to be positioned relative to other systems. According to them, products which are at least passable have SUS scores above 70, with better products scoring in the high 70s to upper 80s. Truly superior products score better than 90.

In addition to the standard instrument, an ad-hoc questionnaire was administered to evaluate some aspects of the human robot interaction for each service. The ad-hoc questionnaire has three constructs (HRI, GUI, SUI) and it has been presented in [13]. Participants could indicate their level of agreement to the statements on a five point Likert scale including verbal anchors: totally disagree (1) disagree (2) neither agree nor disagree (3) agree (4) totally agree (5). Here we refer to GUI, SUI and HRI as the final score obtained by averaging all the corresponding scores for each service experienced by the participant.

As a preliminary requisite for the analysis, we tested the reliability of the ad-hoc questionnaire constructs by means of Cronbach's Alpha (α) analysis. Cronbach's alpha is a measure of internal consistency, that is, how closely related a set of items are as a group, for details, see [17]. According to Cronbach's alphas all constructs of the ad-hoc questionnaire are solid enough as Alpha is at least 0.7: $\alpha_{GUI} = 0.958$; $\alpha_{SUI} = 0.893$; $\alpha_{HRI} = 0.747$.

3   Experimental Results and Discussion

This section presents the analyses of the data gathered from the experiments with elderly participants. Descriptive statistics reported are: Average (Avg), Median (Med), and Standard Deviation (Stdev).

Table 1 presents descriptive statistics of the scores of the usability questionnaires (ad-hoc and SUS) given by participants of the experiment. For all the dimensions considered the results are positive, showing a good usability and acceptability of the MMUI for the service tested.

To further analyse the results we observed and annotated the video of the experiment in order to identify the participants' gaze direction. Three different directions were evaluated: the robot, the tablet, other objects in the room (including the researcher). Results of the video analysis are reported in Fig. 3 for all the fifteen participants to the focused experiment. From the gaze analysis, we see that almost all (14 out of 15) participants demonstrated to focus their attention on the technological devices. In particular, 12 participants (80%) spent the majority of the time (>50%) looking at the robot, while just two (13%) participants preferred to look at the tablet. From the discussion with them, we know that it is a personal preference not influenced by their physical conditions or previous experience. Indeed, ones never used a tablet before, while ones was using it often, and none of them has hearing loss. One participant never looked

Table 1. Usability analysis and comparison. Descriptive statistics reported are: Average (Avg), Median (Med), and Standard Deviation (Stdev).

	avg	med	stdev
HRI	3.99	4	0.58
GUI	4.13	4	0.45
SUI	3.90	4	0.65
SUS	80.67	80	9.04

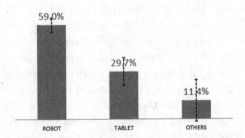

Fig. 3. Participants' gaze direction during the entire experiment. The graph presents the average percentage of time spent by participants looking at: the robot, the tablet, other objects in the room (including the researcher). Confidence interval bars are also shown.

at the tablet because her physical condition, indeed she has a severe visual impairment and she didn't have her magnifying glasses during the experiment. However, she enjoyed to speak with the robot and she scored 5 (fully agree) to the question "I'm confident that I can use only speech (no tablet) to complete the food delivery service".

The result obtained with the gaze analysis are directly related to the actual mean of interaction chosen by the participants. Indeed, those that spent the majority of the time looking at the robot also preferred to use the SUI to interact with it, while the others mixed the two modalities. This confirms the preference given by the participants of the general experimentation.

Furthermore, to better analyse the behaviour of participants, we identified the following 3 steps related to the dialogue phases:

1. *Waking up-Which meal (to hear?)*. The participant calls the robot, which then asks "how can I help?" and starts the Food delivery service. Then, the robot offers to read aloud the items of the three available menus. The user can ask to read menus one by one or all of them together.
2. *Reading menu*. The robot reads the menu(s) aloud. Note that when the robot is reading a menu, the system will automatically show the items on the tablet screen.
3. *Ordering-Ending*. The participant selects and confirms the meal to order among the three available choices. The robot asks if it can do anything else, but the participant closes the interaction by saying "no". The robot closes the interaction with a "Goodbye".

The breakdown of the activities during the three steps is reported in Fig. 4, where it is shown the proportion (percentage) of the time spent by the robot and user interacting and waiting each other.

Figure 4 [Left] presents the average percentage of the total experimental time spent by participants looking at the tablet, broken down for each step of the experiment. It can be easily seen that participants prefer to read the menu items on the tablet while the robot is reading them aloud. Indeed, 12 participants out of 15, i.e. 80 %, looked at the tablet for more than 50 % of the time in the *Reading menu* step of the experiment. This is a clear advantage of the multi-modality of the interface that allows elderly users to select the modality they like to interact with the robot according to the different situations.

In fact, even if we see in the video a preference of the SUI as mean of interaction, the majority of the participants (8: 61 %) gave the same score to the questions "I prefer to use the [tablet — speech] rather than [speech — tablet] for the food delivery service". Moreover, as an additional confirmation of their preference of the multi-modal interaction capability of the system, all participants scored at least 4, with an average score of 4.14, "I like the idea of using speech and tablet together to complete the food delivery service".

As final remark, all the participants indicated that they enjoyed the experience during the experiments with the robot and many offered their availability for further studies.

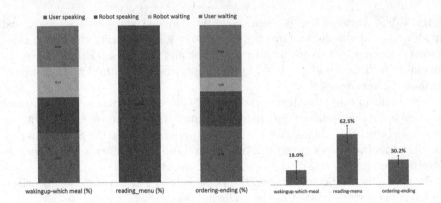

Fig. 4. [Right] Breakdown of the activities during each step of the experiment. *User waiting* indicates the time spent by the user thinking (e.g. what to answer). *Robot waiting* indicates the time needed by software system to elaborate (e.g. speech recognition). [Left] Average percentages of the time spent by participants looking at the tablet during each step of the experiment.

4   Conclusion

This paper presented the experimental evaluation of a software interface designed to enhance the user experience in terms of service usability and to increase acceptability of an assistive robot system by elderly users. The system offered two main modalities to interact with the robot: a GUI on a tablet detachable from the robot, and a SUI, which decoded speech commands recorded via wearable microphone on the user. Fifteen participants, aged between 70 and 89 years old, were invited to interact with a robotic platform using any modality they liked. The task was to order a meal among three different set menus.

The study identified a common behaviour of the elderly participants to switch their attention between the different interaction modalities according to the current situation. Indeed, all the participants were able to successfully perform the task using the SUI, which they stated to be preferable as mean of interaction. However, when the robot was reading aloud the items of the menus, all the users focused their attention on the GUI to follow the list and perform their selection. This behaviour suggests that a MMUI facilitates a more personalised and flexible interaction, which is clearly preferred by the elderly participants of our experiment. Indeed, all participants liked the idea of using speech and tablet together to complete the service.

All the participants indicated that they enjoyed the experience during the experiments with the robot and many offered their availability for further studies.

Finally, we can conclude that multi-modality can add value to the entire system and be a further step towards more usable and widely accepted robot as companions in the every-day elderly care.

References

1. Robot-Era project: Implementation and integration of advanced robotic systems and intelligent environments in real scenarios for the ageing population, FP7 - ICT - Challenge 5: ICT for Health, Ageing Well, Inclusion and Governance. Grant agreement number 288899. www.robot-era.eu
2. Al-Razgan, M.S., Al-Khalifa, H.S., Al-Shahrani, M.D., AlAjmi, H.H.: Touch-Based mobile phone interface guidelines and design recommendations for elderly people: a survey of the literature. In: Huang, T., Zeng, Z., Li, C., Leung, C.S. (eds.) ICONIP 2012, Part IV. LNCS, vol. 7666, pp. 568–574. Springer, Heidelberg (2012)
3. Arch, A., Abou-Zahra, S., Henry, S.L.: Older users online: WAI guidelines address the web experiences of older users. User Experience Mag. 8(1), 18–19 (2009)
4. Bangor, A., Kortum, P.T., Miller, J.T.: An empirical evaluation of the system usability scale. Int. J. Hum. Comput. Interact. 24(6), 574–594 (2008)
5. Bemelmans, R., Gelderblom, G.J., Jonker, P., de Witte, L.: Socially assistive robots in elderly care: a systematic review into effects and effectiveness. J. Am. Med. Directors Assoc. 13(2), 114–120 (2012)
6. Bohus, D., Rudnicky, A.I.: The ravenclaw dialog management framework: architecture and systems. Comput. Speech Lang. 23(3), 332–361 (2009)
7. Brooke, J.: Sus-a quick and dirty usability scale. Usability Eval. Ind. 189, 194 (1996)
8. Broz, F., Di Nuovo, A., Belpaeme, T., Cangelosi, A.: Talking about task progress: towards integrating task planning and dialog for assistive robotic services. Paladyn, J. Behav. Robot. 6(1), 111–118 (2015)
9. Broz, F., Nuovo, A.D., Belpaeme, T., Cangelosi, A.: Multimodal robot feedback for eldercare. In: Workshop on Robot Feedback in Human-Robot Interaction: How to Make a Robot Readable for a Human Interaction Partner at Ro-MAN 2012 (2012)
10. Casiddu, N., Cavallo, F., Divano, A., Mannari, I., Micheli, E., Porfirione, C., Zallio, M., Aquilano, M., Dario, P.: Robot interface design of domestic and condominium robot for ageing population. In: ForITAAL, October 2013
11. Cavallo, F., Limosani, R., Manzi, A., Bonaccorsi, M., Esposito, R., Rocco, M., Pecora, F., Teti, G., Saffiotti, A., Dario, P.: Development of a socially believable multi-robot solution from town to home. Cogn. Comput. 6(4), 954–967 (2014)
12. Consortium, W.W.W.: Web accessibility for older users. http://www.w3.org/TR/wai-age-literature/
13. Di Nuovo, A., Broz, F., Belpaeme, T., Cangelosi, A., Cavallo, F., Dario, P., Esposito, R.: A web based multi-modal interface for elderly users of the robot-era multi-robot services. In: Proceedings of IEEE Conference on System, Man and Cybernetics, pp. 1–6. IEEE, October 2014
14. Dumas, B., Lalanne, D., Oviatt, S.: Multimodal interfaces: a survey of principles, models and frameworks. In: Lalanne, D., Kohlas, J. (eds.) Human Machine Interaction. LNCS, vol. 5440, pp. 3–26. Springer, Heidelberg (2009)
15. Jaimes, A., Sebe, N.: Multimodal human-computer interaction: a survey. Comput. Vis. Image Underst. 108(1), 116–134 (2007)
16. Jian, C., Shi, H., Schafmeister, F., Rachuy, C., Sasse, N., Schmidt, H., Hoemberg, V., von Steinbüchel, N.: Touch and speech: multimodal interaction for elderly persons. In: Gabriel, J., Schier, J., Van Huffel, S., Conchon, E., Correia, C., Fred, A., Gamboa, H. (eds.) BIOSTEC 2012. CCIS, vol. 357, pp. 385–400. Springer, Heidelberg (2013)

17. Kline, P.: The Handbook of Psychological Testing, 2nd edn. Routledge, London (2000)

18. Kristoffersson, A., Coradeschi, S., Loutfi, A.: A review of mobile robotic telepresence. Adv. Hum. Comput. Interact. **2013**, 3 (2013)

19. Mayer, P., Beck, C., Panek, P.: Examples of multimodal user interfaces for socially assistive robots in ambient assisted living environments. In: 2012 IEEE 3rd International Conference on Cognitive Infocommunications (CogInfoCom), pp. 401–406. IEEE (2012)

20. Panek, P., Edelmayer, G., Mayer, P., Beck, C., Rauhala, M.: User acceptance of a mobile LED projector on a socially assistive robot. In: Wichert, R., Eberhardt, B. (eds.) Ambient Assisted Living. ATSC, vol. 2, pp. 77–92. Springer, Heidelberg (2012)

21. Petzold, M., Barbabella, F., Bobeth, J., Kern, D., Mayer, C., Morandell, M.: Towards an ambient assisted living user interaction taxonomy. In: CHI 2013 Extended Abstracts on Human Factors in Computing Systems, pp. 49–54. ACM (2013)

22. Tapus, A., Maja, M., Scassellatti, B.: The grand challenges in socially assistive robotics. IEEE Robot. Autom. Mag. **14**(1), 1–7 (2007)

23. Wang, N., Broz, F., Di Nuovo, A., Belpaeme, T., Cangelosi, A.: A user-centric design of service robots speech interface for the elderly. In: Proceedings of the International Conference on Non-Linear Speech Processing, NOLISP 2015 (2015)

24. Werner, F., Werner, K., Oberzaucher, J.: Tablets for seniors – an evaluation of a current model (iPad). In: Wichert, R., Eberhardt, B. (eds.) Ambient Assisted Living. ATSC, vol. 2, pp. 177–184. Springer, Heidelberg (2012)

25. Zaphiris, P., Kurniawan, S., Ghiawadwala, M.: A systematic approach to the development of research-based web design guidelines for older people. Univ. Access Inf. Soc. **6**(1), 135–136 (2007)

Enhancing Autonomy in VTOL Aircraft Based on Symbolic Computation Algorithms

James A. Douthwaite$^{(\boxtimes)}$, Lyudmila S. Mihaylova, and Sandor M. Veres

Department of Automatic Control Systems Engineering,
University of Sheffield, Sheffield, United Kingdom
{jadouthwaite1,l.s.mihaylova,s.veres}@sheffield.ac.uk

Abstract. Research into the autonomy of small Unmanned Aerial Vehicles (UAVs), and especially on Vertical Take Off and Landing (VTOL) systems has intensified significantly in recent years. This paper develops a generic model of a VTOL UAV in symbolic form. The novelty of this work stems from the designed Model Predictive Control (MPC) algorithm based on this symbolic model. The MPC algorithm is compared with a state-of-the-art Linear Quadratic Regulator algorithm in attitude rate acquisition and its more accurate performance and robustness to noise is demonstrated. Results for the controllers designed for each of the aircraft's angular rates are presented in response to input disturbances.

1 Introduction

Vertical Take-off and Landing (VTOL) aircraft are unique in having a propulsion system that allows the generation of lift independently of the aircraft's velocity. This affords the vehicle to conduct controlled manoeuvres in scenarios where other vehicles may be unable to operate. More recently, the growth of micro-Unmanned Aerial Vehicle (UAV) technology has seen a rise in interest in developing compact VTOL systems as a platform for a range of applications, such as search and rescue, ordinance surveying and aerial cinematography [1]. The current applications of these systems are however, limited by the level of autonomy that has been achieved to allow the system to handle events that could otherwise compromise the vehicle. Robust control regimes able to handle events such as gusts, or rotor loss are highly desirable in enhancing the future of autonomous vehicles.

One of the most powerful methods for control is the Model Predictive Control (MPC) [14]. Recent advances for MPC algorithms are presented in the recent survey [10]. Although widely applied to industrial systems, partially to fixed-wing aircraft and UAV formations [3], the MPC application to VTOL is still limited. One of the main advantages of the MPC is that it can provide both the desired level of performance and safety. This is especially important for small VTOL aircraft, who's applications may be limited by their resilience to disturbances. Efficient numerical methods for non-linear MPC and moving horizon estimation are presented in [4]. Other efficient algorithms for linear small scale control are

© Springer International Publishing Switzerland 2016
L. Alboul et al. (Eds.): TAROS 2016, LNAI 9716, pp. 99–110, 2016.
DOI: 10.1007/978-3-319-40379-3_10

presented in [8]. Although a significant efforts have been devoted to both linear and non-linear MPC, including in [12] its application to VTOL UAVs is still limited.

Hence, in this paper we explore the advantages on the MPC approach in the light of VTOL craft. The MPC performance is compared with a Linear Quadratic Gaussian (LQG) regulator and its accuracy is demonstrated.

The main contributions of this paper stem from: (i) the developed symbolic model of the VTOL UAV. The model is general and comprises all possible motions and changes in 3D manoeuvres. (ii) the designed MPC algorithm to conduct attitude changes taking into account the design constraints of the VTOL aircraft. As a VTOL we consider the quadrotor UAV shown on Fig. 1.

The rest of the paper is as follows. The second section proposes a symbolic approach to modelling a generic VTOL aircraft is proposed. In the third section, the flight dynamics of a quadrotor micro-UAV are introduced; which uses two rotor pairs to generate a body thrust and torque vector. The fourth section then outlines the design of the two control regimes; a Model Predictive Controller (MPC) and Linear Quadratic Regulator (LQR) for angular rate acquisition. Section five evaluates the performance of the two controllers through discussion and comparison. Finally, conclusive remarks on the effectiveness of the controllers are provided, and the implications each controller's use are discuss in conclusion.

Fig. 1. F450 quadrotor layout and 3D CAD assembly.

2   Rigid Body Analysis and Modelling

The aircraft is free to move in all six Degrees of Freedom (DOF) with linear and angular velocities; $\mathbf{v}_1 = [u, v, w]^T$ and $\omega_{CG} = \mathbf{v}_2 = [p, q, r]^T$, respectively. The aircraft's body can be approximated as a point mass (m) with a symbolic inertia matrix ($\mathbf{I}_{CG}$). Forces due to the atmosphere and gravitation act on the airframe in the global axes, these require that they be translated into the body axes before they can be introduced into the model:

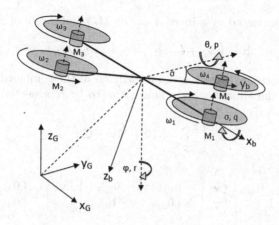

Fig. 2. The Earth and body reference axes.

Using standard aerospace convention, the body frame rotation can be described in Earth coordinates in Euler angles (as shown in Fig. 2). The rotational arguments can be combined to form the directional co-sine matrix DCM [2,5,6]:

$$\mathbf{DCM}_{GB} = \begin{bmatrix} c(\theta)c(\psi) & c(\theta)s(\psi) & -s(\theta) \\ s(\phi)s(\theta)c(\psi) - c(\phi)s(\psi) & c(\phi)c(\psi) + s(\phi)s(\theta)s(\psi) & s(\phi)c(\theta) \\ s(\phi)s(\psi) + c(\phi)s(\theta)c(\psi) & c(\phi)s(\theta)s(\psi) - s(\phi)c(\psi) & c(\phi)c(\theta) \end{bmatrix},$$
(1)

where $c(\theta)$ and $s(\theta)$ denotes the *cos* and *sin* functions, respectively. This allows gravity and other external forces to be mapped onto the body axis reference frame, as shown in Eqs. (1) and (2) [7,13]

$$\mathbf{f}_g = m\mathbf{g} = \mathbf{DCM}_{GB} * \begin{bmatrix} 0 \\ 0 \\ mg \end{bmatrix}.$$
(2)

The resultant force acting on the aircraft, currently neglecting the VTOL propulsion system, is in the form:

$$m\dot{\mathbf{v}}_1 = \mathbf{f} - (\mathbf{v}_2 \times \mathbf{v}_1).$$
(3)

Here the resultant linear acceleration can be seen as the difference between the body momentum and the external force vector, $\mathbf{f}$. Similarly, the sources of torque are modelled by the vector, τ, which acts to induce an instantaneous change in angular momentum $\dot{\mathbf{h}}$ [6,7]:

$$\dot{\mathbf{h}} = \mathbf{I}_b\dot{\mathbf{v}}_2 = \tau - (\dot{\mathbf{v}}_2 \times \mathbf{h}).$$
(4)

The resultant angular momentum ($\mathbf{h}$) of the body is that of all the rotary elements of the system. VTOL aircraft typically operate using fans or rotors.

These can be represented by a inertial matrix ($\mathbf{I}_r$) and a rate of rotation (Ω_i):

$$\mathbf{h} = \mathbf{I}_b\mathbf{v}_2 + \mathbf{I}_r(\Omega_1 + \Omega_2 + ... + \Omega_{i,N}). \tag{5}$$

The above matrix expressions can be written in the combined form (6) and (7). This allows the aircraft's body kinematics to be represented as a series of matrix coefficients [7]:

$$\begin{bmatrix} \mathbf{f} \\ \tau \end{bmatrix} = \mathbf{M}\dot{\mathbf{v}} + \mathbf{N}\mathbf{v} + \mathbf{C} \tag{6}$$

or

$$\begin{bmatrix} \mathbf{f} \\ \tau \end{bmatrix} = \begin{bmatrix} m\mathbf{I}_{3x3}, \ \mathbf{0}_{3x3} \\ \mathbf{0}_{3x3}, \ \mathbf{I}_{CG} \end{bmatrix} \begin{bmatrix} \dot{\mathbf{v}}_1 \\ \dot{\mathbf{v}}_2 \end{bmatrix} + \begin{bmatrix} mS(\mathbf{v}_2), \ \mathbf{0}_{3x3} \\ \mathbf{0}_{3x3}, \ -S(\mathbf{h}_{CG}) \end{bmatrix} \begin{bmatrix} \mathbf{v}_1 \\ \mathbf{v}_2 \end{bmatrix} + \begin{bmatrix} \mathbf{0}_{3x3}, \ \mathbf{R} \\ \mathbf{0}_{3x3}, \ \mathbf{0}_{3x3} \end{bmatrix}. \tag{7}$$

Here the function S is used to map the enclosed vector to the matrix space, where $\mathbf{R}$ represents the constant gravitational force acting on the vehicle. Other forces and torques acting on the body, such as control inputs can be seen as an additional term ($\mathbf{C}$) which must also be converted into the body acceleration vector. The above expression can also be divided through by the inertial matrix to define the systems 6DOF acceleration vector. The final non-linear system is then seen in Eq. (8):

$$\begin{bmatrix} \dot{\mathbf{v}}_1 \\ \dot{\mathbf{v}}_2 \end{bmatrix} = \left(\begin{bmatrix} mS(\mathbf{v}_2), \ \mathbf{0}_{3x3} \\ \mathbf{0}_{3x3}, \ -S(\mathbf{h}_{CG}) \end{bmatrix} \begin{bmatrix} \mathbf{v}_1 \\ \mathbf{v}_2 \end{bmatrix} + \begin{bmatrix} \mathbf{0}_{3x3}, \ \mathbf{R} \\ \mathbf{0}_{3x3}, \ \mathbf{0}_{3x3} \end{bmatrix} + \mathbf{C} \right) \begin{bmatrix} m\mathbf{I}_{3x3} \ \mathbf{0}_{3x3} \\ \mathbf{0}_{3x3} \ \mathbf{I}_{CG} \end{bmatrix}^{-1}. \tag{8}$$

The VTOL aircraft's full non-linear equations of motion have derived symbolically. In this paper, the dynamics associated with a small quadrotor micro-UAV are now introduced to allow a control mechanism to be designed. In addition to the dynamics, several assumptions were introduced to simplify the resulting model [1]. We assume that the aircrafts body axis are parallel with the Euler axis of rotation, therefore the following relationships are true:

$$\dot{p} = \ddot{\phi}, \ p = \dot{\phi}, \ \dot{q} = \ddot{\theta}, \ q = \dot{\theta}, \ \dot{r} = \ddot{\psi}, \ r = \dot{\psi}. \tag{9}$$

Similarly, for the linear states:

$$\dot{u} = \ddot{x}, \ u = \dot{x}, \ \dot{v} = \ddot{y}, \ v = \dot{y}, \ \dot{w} = \ddot{z}, \ w = \dot{z}. \tag{10}$$

The aircraft was assumed to be symmetrical in two dimensions, reducing the inertia of the system to a diagonal matrix. The rotors are completely symmetrical and identical, with negligible blade flapping. The resulting thrust vector is also parallel to the quadrotor body z-axis.

3   Quadrotor Dynamics

So far, a generalised rigid body problem has been proposed. Using the Matlab toolbox for symbolic computation, the full 6 Degree of Freedom (DOF) system definition was derived. A quadrotor was chosen to example the characteristics of a VTOL micro-UAV. The quadrotor's propulsion mechanism consists of four groups of propellers, Brushless Motors and Electronic Speed Controllers (ESC). In this instance, each group generates a thrust that acts parallel to the vertical body axis vector at distance l from the center of mass (11).

Using blade element theory, the total body axis thrust generated by the rotor can be approximated symbolically [1,15]:

$$f_T = C_T \rho A (\Omega R)^2 = k_T . \Omega^2 \tag{11}$$

where C_T is the thrust coefficient, k_T is a constant described in detail below. Here ρ is the air density, A is cross-sectional area of the rotor of radius R. Similarly, the net hub force (H), can be written proportional to the rotor angular velocity (Eq. (12)) [1]. The hub force constant (k_H) is then expressed as:

$$f_H = C_H \rho A (\Omega R)^2 = k_H . \Omega^2. \tag{12}$$

The quadrotor's control matrix is then defined in equation

$$\mathbf{u} = \begin{bmatrix} U_1 \\ U_2 \\ U_3 \\ U_4 \end{bmatrix} = \begin{bmatrix} L \\ M \\ N \\ T_z \end{bmatrix} = \begin{bmatrix} 0 & -k_T \mid \bar{l}_2 \mid & 0 & k_T \mid \bar{l}_4 \mid \\ k_T \mid \bar{l}_1 \mid & 0 & -k_T \mid \bar{l}_3 \mid & 0 \\ -k_H \mid \bar{l}_1 \mid & k_H \mid \bar{l}_2 \mid & -k_H \mid \bar{l}_3 \mid & k_H \mid \bar{l}_4 \mid \\ k_T & k_T & k_T & k_T \end{bmatrix} \begin{bmatrix} \Omega_1^2 \\ \Omega_2^2 \\ \Omega_3^2 \\ \Omega_4^2 \end{bmatrix}. \tag{13}$$

This describes how the propulsion system acts to induce corrective body axis forces and torques. Inverting this matrix also provides the necessary mapping for the control mechanism to command a given body axis torque. In addition to the static thrust properties, the propulsion groups each have an associated rise-time (T_S) to achieve a set-point angular velocity (Ω_i). This relationship was found experimentally to be of first order (see Fig. 3), with area of proportionality sufficient to allow the approximation of k_T and k_H:

$$R(s) = \frac{k_T}{1 + T_s s} = \frac{0.0529}{1 + 0.108s}. \tag{14}$$

This expression then describes the relationship between set point and instantaneous angular velocity of the rotor. The dynamics of other VTOL aircraft can be approximated similarly, however these expressions were sufficient to represent the dynamics of the quadrotor.

4 Control Design

This approximation allows the representation of the aircraft's states in terms of the body axis coordinate system and Euler rotations. The complete symbolic non-linear model can then be represented by the following expressions:

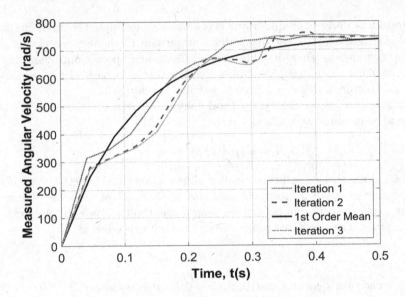

Fig. 3. Approximation of the rotor transient response to a set input at its maximum rotational velocity.

$$\frac{d\mathbf{x}}{dt} = \begin{bmatrix} \ddot{x} \\ \ddot{y} \\ \ddot{z} \\ \ddot{\psi} \\ \ddot{\theta} \\ \ddot{\phi} \end{bmatrix} = \begin{bmatrix} \dot{\theta}\dot{z} - \dot{\psi}\dot{y} - g\sin(\theta) \\ \dot{\psi}\dot{x} - \dot{\phi}\dot{z} + g\cos(\theta)\sin(\phi) \\ \dot{\phi}\dot{y} - \dot{\theta}\dot{x} + g\cos(\phi)\cos(\theta) - (\Omega_1 + \Omega_2 + \Omega_3 + \Omega_4)k_t/m \\ (I_{33}^b - I_{22}^b)\dot{\theta}\dot{\psi}/I_{11}^b + (I_{33}^r\Omega_T\dot{\theta})/I_{11}^b + (\Omega_4 - \Omega_2)k_t l/I_{11}^b \\ (I_{11}^b - I_{33}^b)\dot{\phi}\dot{\psi}/I_{22}^b - (I_{33}^r\Omega_T\dot{\phi})/I_{22}^b + (\Omega_1 - \Omega_3)k_t l/I_{22}^b \\ (I_{22}^b - I_{11}^b)\dot{\phi}\dot{\theta})/I_{33}^b + (-\Omega_1 + \Omega_2 - \Omega_3 + \Omega_4)k_h l/I_{33}^b \end{bmatrix}, \quad (15)$$

where I_{11}^b, I_{22}^b and I_{33}^b, respectively moments of inertia around the three axes of rotation.

To allow the application of linear control theory, a symbolic application of small perturbation theory is applied directly to Eq. (15). This technique assumes that the system is operating around a nominal state ($\mathbf{x}_0$) with small perturbations ($\mathbf{x}_d$). The completion of a symbolic linearisation allows parameters to be selected to represent different stages of operation. This paper focuses on the design of a linear stability controller operating around the hover state.

The resulting linear system is then of the form:

$$\frac{d\mathbf{x}}{dt} = \mathbf{A}\mathbf{x} + \mathbf{B}\mathbf{u} + \mathbf{w} \qquad (16)$$

$$\mathbf{y} = \mathbf{C}\mathbf{x} + \mathbf{D}\mathbf{u} \qquad (17)$$

where $\mathbf{x} = [x, y, z, \psi, \theta, \phi, \dot{x}, \dot{y}, \dot{z}, \dot{\psi}, \dot{\theta}, \dot{\phi}]^T$ is the state vector, $\mathbf{A}$ is the state transition matrix, $\mathbf{B}$ is the control matrix, the control vector containing the rotor angular velocities is defined $\mathbf{u} = [\Omega_1^2, \Omega_2^2, \Omega_3^2, \Omega_4^2]^T$. A source of system noise is also added, $\mathbf{w}$, with a predefined standard deviation of 0.2. The system output vector $\mathbf{y}$ can then be given by Eq. (17), as a function of state and the output matrices $\mathbf{C}$ and $\mathbf{D}$.

Direct substitution of the F450 quadrotor design parameters and considering an attitude hold scenario allowed a numeric state-space representation to be obtained. This linear approximation then provides the basis necessary to design linear controller mechanisms to introduce stability around the straight and level condition.

4.1 Model Predictive Control

A Model Predictive Controller (MPC) was devised as a comparative controller for tracking the error in the aircraft's attitudes. The linearised system model is used directly in the prediction of the future system state across a defined discrete horizon H_N [14]:

The predicted output of the system and control deflections are defined by $\mathbf{G}$ and $\mathbf{F}$, respectively:

$$\mathbf{F} = \begin{bmatrix} \mathbf{A} \\ \mathbf{A}^2 \\ \mathbf{A}^3 \\ \vdots \\ \mathbf{A}^N \end{bmatrix}, \mathbf{G} = \begin{bmatrix} \mathbf{B} & 0 & \cdots & 0 \\ \mathbf{AB} & \mathbf{B} & \cdots & \vdots \\ \vdots & \vdots & \ddots & \vdots \\ \mathbf{A}^{H_N-1}\mathbf{B} & \mathbf{A}^{H_N-2}\mathbf{B} & \cdots & \mathbf{B} \end{bmatrix}. \tag{18}$$

Using this formulation, the horizon output predictions take the form:

$$\mathbf{z}(k) = \mathbf{C}[\mathbf{Fx}(k) + \mathbf{Gu}_{ss}]. \tag{19}$$

We assume that at a given reference point $\mathbf{r}(k)$, exists a steady state $\mathbf{x}_{ss}$ with a steady-state input $\mathbf{u}_{ss}$. It is then possible to use the system dynamics to define the input required to maintain the reference (20)

$$\begin{bmatrix} \mathbf{I}_{12x12} - \mathbf{A} & -\mathbf{B} \\ \mathbf{C} & \mathbf{D} \end{bmatrix} \begin{bmatrix} \mathbf{x}(k)_{ss} \\ \mathbf{u}(k)_{ss} \end{bmatrix} = \begin{bmatrix} \mathbf{0}_{12x1} \\ \mathbf{r}(k) \end{bmatrix}. \tag{20}$$

Solving Eq. (20) for a given reference output $\mathbf{r}(k)$ allows the steady state $\mathbf{x}(k)_{ss}$ and input $\mathbf{u}(k)_{ss}$ to be determined. Calculation of this value allows the absolute input to be defined as a difference between the reference control and the control signal generated by the controller.

The predicted error over the horizon is defined by:

$$\mathbf{e}(k) = \mathbf{r}(k) - \mathbf{z}(k). \tag{21}$$

Here, the error is that between the reference signal and the system output over the prediction horizon H_N. The tracking error and control weightings over the horizon are also defined as the following:

$$\mathbf{Q}_{H_p} = \begin{bmatrix} \mathbf{Q} & 0 & \cdots & 0 \\ 0 & \mathbf{Q} & \cdots & \vdots \\ \vdots & \vdots & \ddots & \vdots \\ 0 & \cdots & \cdots & \mathbf{N} \end{bmatrix}, \mathbf{R}_{H_p} = \begin{bmatrix} \mathbf{R} & 0 & \cdots & 0 \\ 0 & \mathbf{R} & \cdots & \vdots \\ \vdots & \vdots & \ddots & \vdots \\ 0 & \cdots & \cdots & \mathbf{R} \end{bmatrix}, \tag{22}$$

where $\mathbf{Q}$, $\mathbf{R}$ and $\mathbf{N}$ are weighting matrices.

The cost function can then be derived to evaluate the optimal control inputs as a function of the predicted error, control effort $\mathbf{u}(k)$ and a terminal error at the horizon (23) [11,14]:

$$V(k) = \sum \left[||(\mathbf{r}(k) - \mathbf{z}(k))||^2\mathbf{Q} + ||\mathbf{u}(k)||^2\mathbf{R}\right] + ||(\mathbf{r}(k) - \mathbf{z}(k))||^2\mathbf{N}. \qquad (23)$$

Here $||u(k)||$ is used to describe the Euclidean norm of the input. Substituting the horizon prediction matrix expressions allows Eq. (23) to be redefined as quadratic coefficients of the control input $\mathbf{u}(k)$. The terminal cost matrix $\mathbf{N}$ can be seen neglected, with no value assigned to a terminal angular rate:

$$V(k) = \mathbf{u}(k)^T[\Theta^T\mathbf{Q}\Theta + \mathbf{R}]\mathbf{u}(k) - 2\mathbf{u}(k)^T\Theta^T\mathbf{Q}\mathbf{e}(k) + \mathbf{e}(k)^T\mathbf{Q}\mathbf{e}(k), \qquad (24)$$

where:

$$\Theta = \mathbf{CG}. \qquad (25)$$

If $\mathbf{H} = \Theta^T\mathbf{Q}\Theta + \mathbf{R}$ and $\mathbf{G} = -2\Theta^T\mathbf{Q}\mathbf{e}(k)$ then the cost function takes the reduced form [14]:

$$V(k) = \mathbf{u}(k)^T\mathbf{H}\mathbf{u}(k) + \mathbf{u}(k)^T\mathbf{G} + \mathbf{e}(k)^T\mathbf{Q}\mathbf{e}(k). \qquad (26)$$

The optimal control sequence then occurs where Eq. (26) is minimal, subject to the dynamics of the system and design constraints. This optimisation operation was computed directly using the Matlab function *quadprog*. The first of the optimal control inputs is then used to instigate a change in the system state (16).

A series of design constraints are also introduced to represent the performance limits of the quadrotor aircraft:

$$-35^0 \leq \phi, \theta \leq 35^0, \qquad (27a)$$
$$0rad/s \leq \Omega_{1-4} \leq 580rad/s. \qquad (27b)$$

A regime was then applied to define a set of maximum deflection angles to prevent inversion and to aid in maintaining stability (27a). The above input conditions (27b) were then selected to represent the physical limits of the actuators on board the F450 quadrotor.

Initially, the state and input weighting matrices $\mathbf{Q}$ and $\mathbf{R}$ of the LQR controller were used as a benchmark value in tuning the MPC characteristics. A step response was then used to observe the performance over different prediction horizons.

4.2   Linear Quadratic Regulation (LQR)

The Linear Quadratic Regulator (LQR) controller acts to resolve the optimum feedback gain $\mathbf{K}$ which symbolises the control effort and the transient response of the system. The optimal solution is found by evaluating the minimum value of a cost function J_{QR}; containing weightings of the plant ($\mathbf{Q}$) and the control inputs ($\mathbf{R}$). The cost function is then of the form:

$$J_{QR} = min \int \left[\mathbf{x}(t)^T \mathbf{Q} \mathbf{x}(t) + \mathbf{u}(t)^T \mathbf{R} \mathbf{u}(t) \right] dt. \tag{28}$$

Here the input vector $\mathbf{u}(t)$ represents the rotor speeds applied to the system as a result of the control mapping, derived from inverting the expression in Eq. (13) or directly. The result of optimal control problem is then a state feedback gain ($\mathbf{u} = -\mathbf{Kx}$) where J_{QR} is minimal, subject to the dynamics $\dot{\mathbf{x}} = \mathbf{Ax} + \mathbf{Bu}$. The system input is then the sum of the state feedback and a reference set point $\mathbf{r}(t)$:

$$\mathbf{u}(t) = \mathbf{r}(t) - \mathbf{Kx}(t). \tag{29}$$

Here the matrix $\mathbf{K}$ is the solution to the algebraic Riccati equation

$$\mathbf{A}^T \mathbf{P} + \mathbf{P} \mathbf{A} - \mathbf{P} \mathbf{B} \mathbf{R}^{-1} \mathbf{B}^T \mathbf{P} + \mathbf{Q} = 0, \tag{30}$$

where the feedback is defined as:

$$\mathbf{K} = \mathbf{R}^{-1} \mathbf{B}^T \mathbf{P}. \tag{31}$$

The LQR feedback gain is then used to instigate state feedback in the aircraft's dynamic expressions (16).

5   Performance Evaluation

The designed MPC algorithm is evaluated by observing the systems transient response to a step input. The results are then directly compared to the response of the LQR, subject to the same inputs. Both controllers are designed to obtain a reference roll, pitch and yaw rate $\dot{\phi}, \dot{\theta}, \dot{\psi}$, respectively, as preliminary attitude controllers for the VTOL aircraft.

The gain matrices $\mathbf{Q}$ and $\mathbf{R}$ are to be selected to instigate the desired transient responses in each axis. The LQR gain matrices $\mathbf{Q}$ and $\mathbf{R}$ are initially defined by identity matrices of the form $\mathbf{Q} = \mathbf{C}^T \mathbf{C}$ and $\mathbf{R} = \mathbf{I}_{4x4}$. Each LQR gain was then tuned heuristically until a critically damped step response, with an amplitude of 0.5 rad/s was observed. The relative settling times, overshoot and sensitivity to noise could then be compared, on Fig. 4.

The MPC gain matrices are initially set to be the resultant LQR gains. The prediction horizon is then varied to optimise the response to a step input.

The responses of the two control regimes (LQR and MPC) controllers shown in Fig. 4. Both controllers can be seen to stabilise the system about a reference axis rate (Ref) of 0.5 rad/s. Some steady-state error can be seen in the LQR output as a result of the Gaussian noise added to the system. This is likely due to the deviation away from the linear system and some integral feedback is required. The response of the MPC controller can then also be seen to achieve the desired rates in equivalent conditions. The mean settling times of the two controllers were then calculated based on one hundred Monte-Carlo experiments (see Table 1).

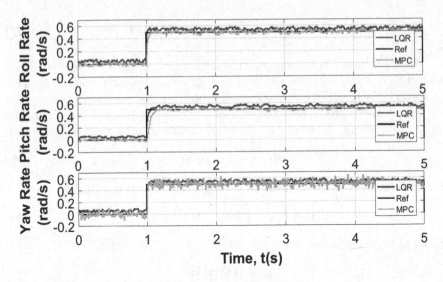

Fig. 4. The aircraft's response to a step input of 0.5 rad/s under the MPC and LQR control regimes.

As seen in Table 1, the LQR and MPC algorithms demonstrate similar performance in the presence of the Gaussian noise signal. The MPC controller is however shown to be able to settle the system faster, with no overshoot or steady state error. The constraints applied with in the MPC demonstrated further tolerance to noise through minimal steady-state error. The optimal inputs of the MPC are also found considering the physical limits of system which the LQR cannot do directly. This allows the algorithm to plan the inputs around the possibility of the VTOL system reaching actuator saturation or a limit on the physical output.

Table 1. A comparison of the LQR/MPC mean settling times, in seconds, following one hundred Monte-Carlo iterations.

Controller	Roll Rate $t_{set}(s)$	Pitch Rate $t_{set}(s)$	Yaw Rate $t_{set}(s)$
LQR	0.10	0.22	0.11
MPC	0.09	0.20	0.05

6   Conclusions

In this paper, a symbolic approach to modelling Vertical Takeoff and Landing (VTOL) vehicles was presented. The dynamics and numeric definition of a small quadrotor UAV are also introduced as a basis for the design of both a Linear

Quadratic Regulator and Model Predictive attitude rate controllers. Both controllers were successful in stabilising the aircraft about a given reference rate of 0.5 rad/s in the presence of white noise signal. A comparison of the controllers demonstrated that the MPC was able to achieve the desired state quicker than the LQR on average, but also demonstrated minimal steady-state error and overshoot.

The proposed approach has the potential in a number of applications, especially in adaptive control algorithms for complex UAV topologies. Derivation of both the linear and non-linear systems in such cases facilitate their stability analysis. Future work in this area will focus be on the derivation of non-linear control strategies on non-linear representations of other UAV systems. Scope also exists for the symbolic calculation of the Lyapunov functions and their application to further VTOL systems.

Acknowledgements. The authors gratefully acknowledge the support from the UK EPSRC under grant number EP/M506618/1.

References

1. Bouabdallah, S., Siegwart, R.: Full control of a quadrotor. In: IEEE/RSJ International Conference on Intelligent Robots and Systems, IROS 2007, pp. 153–158. IEEE (2007)
2. Bouffard, P.: On-board model predictive control of a quadrotor helicopter: Design, implementation, and experiments. Technical report, DTIC Document (2012)
3. Chao, Z., Zhou, S.-L., Ming, L., Zhang, W.-G.: UAV formation flight based on nonlinear model predictive control. Math. Probl. Eng. **2012**(261367), 15 (2012)
4. Diehl, M., Ferreau, H.J., Haverbeke, N.: Efficient numerical methods for nonlinear MPC and moving horizon estimation. In: Magni, L., Raimondo, D.M., Allgöwer, F. (eds.) Nonlinear Model Predictive Control. LNCIS, vol. 384, pp. 391–417. Springer, Heidelberg (2009)
5. ESDU. Quaternion representation of aeroplane attitude. Technical report, The Royal Aeronautrical Society (2002)
6. ESDU. Introduction to aerodynamic derivatives, equations of motion and stability. Technical report 86021c, The Royal Aeronautrical Society (2003)
7. Fossen, T.I.: Mathematical models for control of aircraft and satellites. Department of Engineering Cybernetics Norwegian University of Science and Technology (2011)
8. Frison, G., Srensen, H.H.B., Dammann, B., Jrgensen, J.B.: High-performance small-scale solvers for linear model predictive control. In: 2014 European Control Conference (ECC), pp. 128–133, June 2014
9. Grancharova, A., Grøtli, E.I., Ho, D., Johansen, T.A.: UAVs trajectory planning by distributed MPC under radio communication path loss constraints. J. Intell. Robot. Syst. **79**(1), 115–134 (2014)
10. Johansen, T.I.: Toward dependable embedded model predictive control. IEEE Syst. J. **PP**(99), 1–12 (2014)
11. Kunz, K., Huck, S.M., Summers, T.H., Lygeros, J.: Fast model predictive control of miniature helicopters. In: Proceedings of the IEEE European Control Conference, pp. 1377–1382 (2013)

12. Landry, B., Deits, R., Florence, P., Tedrake, R.: Aggressive quadrotor flight through cluttered environments using mixed integer programming (2015)
13. Liu, C., Chen, W.-H., Andrews, J.: Explicit non-linear model predictive control for autonomous helicopters. Proc. Inst. Mech. Eng. Part G J. Aerosp. Eng. **226**, 1171–1182 (2011)
14. Maciejowski, J.M.: Predictive Control with Constraints, 1st edn. Pearson Education, Harlow, England (2002)
15. Moness, M., Bakr, M.: Development and analysis of linear model representations of the quad-rotor system. In: Proceedings from the 16th International Conference on Aerospace Siences & Aviation Technology (2015)

Development of an Intelligent Robotic Rein for Haptic Control and Interaction with Mobile Machines

Musstafa Elyounnss[1,2(✉)], Alan Holloway[1,2], Jacques Penders[1,2], and Lyuba Alboul[1,2]

[1] Materials and Engineering Research Institute,
Sheffield Hallam University, Sheffield, UK
[2] Sheffield Robotics, Sheffield, UK
b2051861@my.shu.ac.uk

Abstract. The rescue services face numerous challenges when entering and exploring dangerous environments in low or no visibility conditions and often without meaningful auditory and visual feedback. In such situations, rescue-personnel may have to rely solely on their own immediate haptic feedback in order to make their way in and out of a burning building by running their hands along the wall as a means of navigation. Consequently, the development of technology and machinery (robot) to support exploration and aid navigation would provide a significant benefit to the search and rescue operation; enhancing the capabilities of the fire and rescue personal and increasing their ability to exit safely [1]. A brief review and analysis of the previous published literature on exploring environments in low or no visibility conditions where haptic feedback has been utilized is provided and the design of a new intelligent haptic rein is proposed.

Keywords: Haptic feedback · Haptic rein · Navigation

1 Introduction

Search and rescue operations are often undertaken in complex and hazardous high risk situations where factors such as limited or no visibility, noise and time constraints impede progress into an unknown environment. Robots with sensing and navigation capabilities have been successfully used in such dangerous or hazardous situations [2]. Inevitably, however, human intervention is often still required in order to complete the required search and rescue operation.

In these situations a human may have some of their senses severely impaired; for example, the vision of the rescue personnel may be limited to detect obstacles and distinguish hazards. Consequently humans could be aided by a machine that can sense the local environment and provide tactile and haptic information in a similar manner to the interaction between the visually impaired person and a guide dog. The visually impaired person follows the dog through the signals being transmitted to their hand and interprets them into information about the environment and how to navigate the route [2]; the proposed design takes inspiration from this approach.

© Springer International Publishing Switzerland 2016
L. Alboul et al. (Eds.): TAROS 2016, LNAI 9716, pp. 111–115, 2016.
DOI: 10.1007/978-3-319-40379-3_11

2 Investigation and Evaluation of the Previous Designs

Only limited examples targeted to specific application areas such as fire and rescue are available; however several research works in related topics have been undertaken; in most instances the primary end users are the visually impaired and or/mobility impaired. This draws many parallels with the proposed application; however a significant difference in the degree of risk is observed [3].

A number of the proposed solutions share much in common with the traditional white cane and appear to be an evolution of the fundamental concept. Examples include the electronic guides cane such as proposed by Michigan University, which provides environmental data using active scanning and the servomotor system that adds motorized guidance to the cane [4]. The robot walker which has been designed as an assistive device for the elderly/physically weak and/or persons with cognitive impairment is described in [5]. In most walkers the focus is on safety but in the walker described in this paper in addition to safety and navigation the stability of the user is considered. The additional capabilities are achieved by a suite of software for localization and navigation combined with a shared-control haptic interface. A solution to this has been also proposed in [5]. The mobile robot is equipped with two force-sensing handle bars that resemble the grippers of conventional walkers. Forces asserted through this haptic interface are mediated with control from the navigation system in such a way as to maximize a person's perceived freedom while still achieving point to point navigation.

Mixed modes of user assistance in the form of controlled robot motion and visual cues are examined to assist the user navigation without becoming intrusive to the users desires. This mobile robotic platform is implemented as a shared control system. Navigation and guidance must be provided by the robot while maintaining a natural motion response. Shared control can be defined as a combination of two or more independent control systems to achieve common goals [6].

The REINS project[1] utilized a physical stiff rein attached directly to the mobile robot [7]; the project focused on navigation where limited or no visual sensory information can be used and the predominant feedback method is using haptic interfaces. The stiff rein was utilized to avoid complex issues of location and orientation with respect to the robot and the human user and to provide direct feedback to the human follower. The stiff rein prototype was assembled using a crutch-like handle with a joint at the base-connecting the pole to the robot. This consisted of a ball joint mechanism to allow movement in all directions [7] as shown in Fig. 1.

In the experiments conducted a problem appeared when the robot made sharp movements, the human user had significant difficulty following the trajectory of the robot. In such cases the system could be improved through pre-emptive indications to the user and a mutual adaption of both the robot and user's response (speed and turning rate) should be taken into consideration in order to maintain consistent fluid locomotion.

From the previous review it is proposed that an intelligent stiff rein system with feedback of the environment and perceptual capabilities can enable and enhance

---

[1] Research was supported by the UK Engineering and Physical Sciences Research Council (EPSRC) grant no. EP/I028757/1.

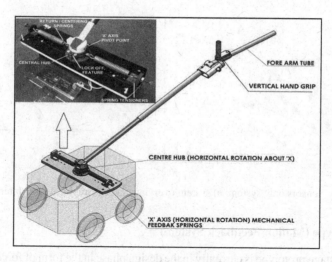

Fig. 1. Stiff rein prototype

navigation in complex environments. Additionally the use of haptic communication through force feedback guiding the user can be considered as a suitable approach of providing navigation information and is the least affected mode of communication [7] in noisy environments.

3 Proposed Rein Prototype Description

In order to develop a prototype intelligent rein, detailed information about the relative positioning and compliance/resistance of the user to the robots responses must be known. This information can then be processed by the shared control system to provide adaptive control and force feedback. Due to complexity the system design has been separated into distinct prototype systems sensors and monitoring, motion/feedback and a combined system with adaptive shared control.

3.1 Prototype (Sensors Only System)

In this prototype a range of sensors has been embedded on the rein to collect data about robot and user position. Digital Encoders are placed at all horizontal and vertical rotating joints and on-board robot sensors such as speed sensor and odometer have been utilized. The sensors are connected to a data acquisition system which logs and displays all the relevant parameters. Initial tests have demonstrated accurate tracking of the user/robot position data to within $\pm 1°$. Figure 2a shows a diagram of the sensor system and Fig. 2b shows the Graphical User Interface (GUI) which is used to visualize the data during real time experiments. The GUI shows the relative angles of the rein joints.

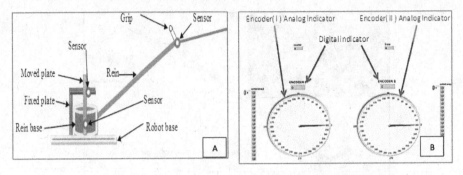

Fig. 2. Sensors only system: a) system overview and b) Graphical User Interface

3.2 Prototype (Motion/Feedback System)

The second stage prototype is currently in the design phase in the form of force feedback between the user and rein. Actuators placed on the rotational joints allow the rein to provide variable force on the users forearm guiding/steering the user in the desired trajectory, the amount of force applied is proportional to the rotation needed. The level of compliance/resistance force exerted on the user from the actuator must be continuously monitored by the control system in order to maintain safe operation.

3.3 Prototype (Shared Control Intelligent Rein System)

A final prototype system is proposed combining the previous systems with a shared intelligent control system. The shared control relationship is complex as the requirements of two independent systems (the user and the robot) with different behaviours and control algorithms must be processed and executed in real time. Work continues to establish suitable control strategies which aim to provide a level intelligence and user experience emulating the natural and adaptable control relationship observed between guide dog and human user.

4 Conclusion

A first stage prototype system has been developed, which focuses on the deployment of suitable sensors to allow accurate and reliable measurement of the robot, rein and users relative positions. The data are required to enable for further stages of the intelligent robotic rein design. The stiff rein solves the issues of robot localization and orientation with respect to the user and provides a direct method of haptic feedback. The first prototype was tested and completed as the first part of the research. We now proceed to develop and construct the motion/feedback system. The overall system aims to mimic the complex shared control relationship observed between guide dog and human user. A similar testing strategy as defined in [7] is proposed to evaluate the design.

Acknowledgement. This work was supported by Libyan Embassy.

References

1. Jones, P., Ghosh, A., Penders, J. Reed, H.: Towards human technology symbiosis in the haptic mode. In: International Conference on Communication, Media, Technology and Design, Famagusta, North Cyprus, 2–4 May 2013, pp. 307–312 (2013)
2. Penders, J., Holloway, A., Reed, H.: REINS (2011). http://gow.epsrc.ac.uk/ NGBOViewGrant.aspx?GrantRef=EP/I028757/1
3. Brown, J., Stickford, J.: Physiological Stress Associated with Structural Firefighting Observed in Professional Firefighters. Indiana University, Bloomington (2007)
4. Ulrich, I., Borenstein, J.: The GuideCane-applying mobile robot technologies to assist the visually impaired. Trans. Sys. Man Cyber. Part A **31**(2), 131–136 (2001)
5. Morris, A., et al: A robotic walker that provides guidance. In: International Conference on Robotic & Automation, Taipei, Taiwan. School of Computer Science Carnegie Mellon University, Pittsburgh (2003)
6. Wasson, G., Gunderson, J.: Variable autonomy in a shared control pedestrian mobility aid for the elderly. In: Proceedings of the IJCAI 2001 Workshop on Autonomy, Delegation, and Control (2001)
7. Ghosh, A., Alboul, L., Penders, J., Jones, P., Reed, H.: Following a robot using a haptic interface without visual feedback. In: Seventh International Conference on Advances in Computer-Human Interactions, Barcelona, Spain (2014)

Infrastructure Mapping in Well-Structured Environments Using MAV

Yuantao Fan[✉], Maytheewat Aramrattana, Saeed Gholami Shahbandi, Hassan Mashad Nemati, and Björn Åstrand

School of Information Technology, Halmstad University, Halmsatd, Sweden
{yuantao.fan,maytheewat.aramrattana,saeed.gholami_shahbandi,
hassan.nemati,bjorn.astrand}@hh.se

Abstract. In this paper, we present a design of a surveying system for warehouse environment using low cost quadcopter. The system focus on mapping the infrastructure of surveyed environment. As a unique and essential parts of the warehouse, pillars from storing shelves are chosen as landmark objects for representing the environment. The map are generated based on fusing the outputs of two different methods, point cloud of corner features from *Parallel Tracking and Mapping* (PTAM) algorithm with estimated pillar position from a multi-stage image analysis method. Localization of the drone relies on PTAM algorithm. The system is implemented in Robot Operating System(ROS) and MATLAB, and has been successfully tested in real-world experiments. The result map after scaling has a metric error less than 20 cm.

1 Introduction

To achieve an industrial automatic warehouse environment, where all the autonomous vehicles can localize themselves, perform task planning and interact with each other, a map that represents the infrastructure and goods in the environment is required. One of the approaches to achieve such warehouse environment is to create the map automatically using unmanned vehicle. To explore an unknown environment, simultaneous localization and mapping (SLAM) technique is often used. It is the joint estimation of the unmanned vehicle's position in the environment and a model of its surroundings i.e., the map. It is a key requirement for fully autonomous mobile vehicles operating in an unknown workspace. A vast number of SLAM implementations have been proposed for aerial or ground vehicles. In this paper, a quadrotor helicopter was used to create the map of a warehouse.

1.1 Related Works

Compared to ground vehicles, the micro aerial vehicles (MAVs) have advantages in form of mobility. For instance, they can operate in three-dimensional space and do not require the ground to run on. While MAVs enable opportunities for performing new tasks, they contains more challenges compared to the ground

L. Alboul et al. (Eds.): TAROS 2016, LNAI 9716, pp. 116–126, 2016.
DOI: 10.1007/978-3-319-40379-3_12

robots. Especially, weight and power constraints, as discussed in [15]. Many studies have been performed on SLAM using MAVs in indoor [7,16,20,22] and outdoor [3,17] environments. In these works different type of sensors such as camera [9,13], and lightweight laser scanner are commonly used for mapping and tracking purposes. In contrast to laser rangefinders, cameras are affordable, small, and light. Furthermore, they can be used in both indoor and outdoor environments with less limitations. These characteristics make cameras more suitable choice for aerial vehicles.

Parallel Tracking and Mapping (PTAM) [14] is one of the popular keyframe-based algorithms. By using image streams from a camera, one can construct point cloud of corner features in the environment. PTAM can be used for tracking-while-mapping purposes [3,10]. By using PTAM, no prior model of the scene is required, and the algorithm provide a 3D map of corner features in the observed image frames.

Majority of the performed researches were using MAV platform from Ascending Technologies GmbH (AscTec). This platform was used mostly to perform autonomous exploration and navigation tasks such as the work from *Shen et al.* [21,22], *Weiss et al.* [23], *Bachrach et al.* [4–6] and *Pravitra et al.* [20]. The platform offers high performance on-board processors with around 600–650 g payload. Another alternative is the platform from MikroKopters [2], which as an example is used in *Piratla et al.*'s work [18]. Finally, *Parrot AR.Drone 2.0*, is a quadrocopter platform with cheaper price. It does not offer any payload. Only a few research have been done using *AR.Drone*. For example, autonomous indoor flight from *Bills et al.* [7], vision based navigation from *Blosch et al.* [8] and *Engel et al.* [11]. The interesting feature of the AR.Drone is the cost; it doesn't have a pay load, which makes it more suitable for exploration mission and human robot interaction (HRI) [19].

In this work we employ a cheap and easy to use platform, AR.Drone, to explore and map the infrastructure in a real warehouse environment. The system provides a map, representing the structure of the surveyed environment. Our experiments are done in a warehouse where pillars from storing shelves are desired as landmark objects. In order to detect pillars, we combine the information from PTAM with a multi-stage image analysis algorithm which uses the prior knowledge about the unique and uniform color of the pillars. This information fusion provides us a robust and accurate position estimation of pillars location in the warehouse. The system is implemented in Robot Operating System (ROS) and MATLAB, and has been successfully tested in a real-world experiments. The map generated after scaling has a metric error less than 20 cm.

The remaining of this paper is structured as follows. In Sect. 3.1, our system architecture and modifications are elaborated. Section 2 explain the proposed method, and we describe our experiments and results in Sect. 3. We summarize our contribution and discuss future work in Sect. 4.

2   Method

System overview is elaborated in Fig. 1. Based on image stream from *AR.Drone*, Parallel Tracking and Mapping (PTAM) [14] provides position of the drone and point cloud of corner features in the environment. Localization of drone by means of the *PTAM* and required modifications are described in Sect. 2.1. With the same image stream, image analysis techniques are applied to obtain pillars' position (projection) in two-dimensional world coordinates. In Sect. 2.2, the process of pillars detection and corresponding assumptions are presented. Finally, the map of pillars in world coordinates is created using the correspondence between PTAM's point cloud and the two-dimensional position of the pillars (described in Sect. 2.3).

2.1   Localization and Point Cloud Map

Localization of the drone is carried out by employing an implementation of PTAM, which is a vision based tracking system designed for augmented reality.

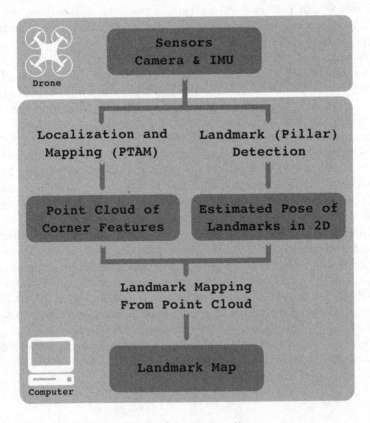

Fig. 1. System overview

It provides position of the drone in the environment as well as point cloud of corner features. The package employs Extended Kalman Filter, of which the control gains are used for EKF prediction, location of extracted corner features and navigation data are used for EKF update. After the initialization of PTAM, a point cloud map $m = \{m_1, m_2, ..., m_k\}$ that contains the location of corner features in the image is created.

2.2 Landmark Detection

Pillars of the shelves capture structural layout of a warehouse. They are common in warehouse environment and usually painted with unique and uniform color, as shown in Fig. 2. Therefore, they are chosen as representative objects in this work. In this work, a multi-stage image analysis and a grouping algorithm is employed to obtain pillars' position $\Lambda = \{\lambda_1, \lambda_2, ..., \lambda_l\}$ from image sequences and drone's odometry [12].

Initially the image acquired from *AR.Drone* is rotated to align the vertical axis of the image with the corridor. This rotation will facilitate correlating the image coordinate and the point cloud data (provided by PTAM). The first stage of the image analysis algorithm is color segmentation. This task is performed in HSV color space. Pixels belonging to pillars are extracted by thresholding the HSV component. Afterwards, the edges of pillars are detected using *Canny edge detector*. At last, Hough transform is employed to extract lines that represent the edges of the pillars. Since the correlation between image coordinate and the point cloud is provided in an open loop, only horizontal edges are accepted to improve the accuracy of the result. The pillar positions Λ in the global frame are simultaneously published into ROS topic and stored. After all the image sequences are processed, the estimated pillar positions will be grouped into sev-

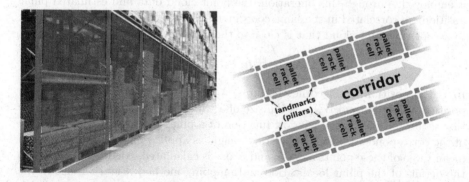

Fig. 2. (left) A scene in the warehouse, where pillars are common infrastructure with uniformed and unique colour. (right) Structural layout of the warehouse: pillar can be considered as landmark of the environment.

Fig. 3. (left) shows the view of bottom camera. Red lines in (right) are detected edges of pillars. (Color figure online)

eral clusters, each cluster representing a pillar. Then, weighted average mean is applied on all pillar projections and the mean position will be considered as the estimated pillar position of each cluster (Fig. 3).

2.3   Mapping Pillars Using Data Fusion

Creation of pillar map relies on two sources, point cloud $\{m_1, m_2, ..., m_k\}$ from PTAM and estimated pillar position $\{\lambda_1, \lambda_2, ..., \lambda_l\}$ from pillar extraction method. The mapping algorithm [12] fuses the two sources and generates a two-dimensional map of pillar's position $\hat{\boldsymbol{\lambda}} = \{\hat{\lambda}_1, \hat{\lambda}_2, ..., \hat{\lambda}_l\}$. Point cloud map from PTAM includes 3D position of corner features in image frames, some of the corners extracted from the pillars (since there are structural rectangle hole patterns on every pillar) and therefore indicate pillar's two-dimensional position $(x_{\lambda_t}, y_{\lambda_t})$. Estimated pillar position from pillar detection could be utilized to locate these points and filter out the irrelevant points (extracted from other object). To find the dominant orientation of the point cloud. Radon transform is employed. Through this operation, the point cloud data and estimated pillar position are correlated in the same coordinates. At the end, a filter (1) is applied to accept the point cloud that is close to the estimated pillar position.

$$\hat{\boldsymbol{\lambda}}_{\boldsymbol{n}} = f(\boldsymbol{m}, \lambda_n, W_{size}) \qquad (1)$$

$\boldsymbol{m}$ is the position of all corner features, λ_n is the estimated two-dimensional position of a pillar and W_{size} is the size of a window function. Any corner point that is located within the window function of a pillar is selected as a candidate of the correspondent pillar. An average weight is applied on each candidate and mean value of these points along x and y axis is calculated. Algorithm 1 describe the details of the pillar localization and mapping method. Figure 4a shows the point cloud that aligned to the corridor and Fig. 4b shows the point cloud that is belong to the pillars filtered by the Algorithm 1. The red points are considered as the estimated two-dimensional position of the pillar $\hat{\boldsymbol{\lambda}}$.

Data: *pointcloud, pillar$_{pos}$*
Result: pillar_filtered_pos
angle = *pointcloud*.radontransform();
pointcloud = *pointcloud*.rotate(angle);
for *All point in pointcloud* **do**
  | *pillar$_{cloud}$* = Filter(*point, pillar_pos*);
end
for *All pillar$_{cloud\_cluster}$ in pillar$_{cloud}$* **do**
  | *pillar$_{filtered\_pos}$[i]* = *pillar$_{cloud\_cluster[i]}$*.average();
end

Function Filter(*point, pillar_pos*)
for *All pos in pillar_pos* **do**
  | **if** *point.inrage(pos, filter_range)* **then**
  | | *pillar$_{cloud}$*.add(*point*);
  | | *pillar$_{cloud\_cluster}$[i]*.add(*point*);
  | | **break;**
  | **end**
end
return *pillar$_{cloud}$*;

Algorithm 1. Procedure of mapping algorithm

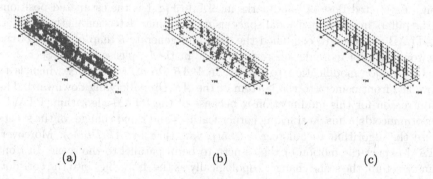

(a)   (b)   (c)

Fig. 4. (a) shows point cloud from PTAM, (b) shows filtered point cloud and the mapping result is shown in (c)

3   Experiments and Results

3.1   System Architecture and Modifications

Overview of our system is shown in Fig. 1. It contains two major parts. First, *Parrot AR.Drone 2.0*, a quadrotor helicopter platform that carries two cameras, one looking forward (front camera) and the other facing downward (bottom camera). Second, a computer, which gather data from the *AR.Drone* into robot operating system (ROS). The computer and the *AR.Drone* is connected with a wireless connection. The data consists of (*a*) image stream from a selected camera and

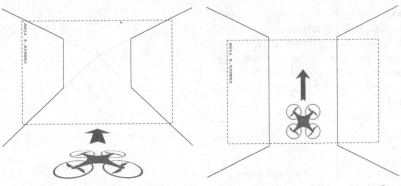

(a) Scene changes morphologically as drone drive forward.

(b) Scene only translate as drone flies forward.

Fig. 5. Illustration of scenes from original and modified camera

(b) position of the *AR.Drone* estimated by inertial measurement unit (IMU). ROS uses the data to run PTAM algorithm, which provide three-dimensional coordinates of corner features detected in the environment, referred to as point cloud in Fig. 1. Moreover, pillars are detected from images using color segmentation. "Estimated Pose of Landmarks in 2D" in Fig. 1 is the estimated positions of the pillars in two-dimensional space using this pillar detection method. Later, MATLAB was used to combined the data and generate a map of pillars of the environment. For more details regarding the methods, please refer to Sect. 2.

One major modification to the original *AR.Drone* was made, which is to move the front camera to the bottom of the *AR.Drone* looking downward. The main reason for this modification is because of the PTAM algorithm. PTAM's performance depends on tracking corner features and rapid changes of the scenes cause the algorithm to fail (e.g. a sharp yaw turn by *AR.Drone*). Moreover, PTAM expect the motion of the camera to be in parallel to the scene. In front camera set up, the scene changes topologically as the drone flies into the corridor, hence the features' motion is not in parallel to the camera. Figure 5 illustrates the view comparison between front and down camera configurations. However, the bottom camera has a considerably small field of view and low resolution which doesn't provide enough feature for the algorithm to track. Therefore, to satisfied the requirements, the front camera set up on the *AR.Drone* was modified. This camera setup was inspired by *ethzasl_ptam* package [1], where with PTAM is employed with a similar camera setup mounted on a high altitude aerial vehicle.

3.2 Results

The experiment was conducted in a warehouse. The AR.Drone flew straight through a corridor in a warehouse, start from $(0,0)$ going in $+y$ direction. It was manually controlled using joystick. Following data are collected from the *AR.Drone* through ROS: (a) image stream; (b) IMUs data; and (c) control

Table 1. Mean and variance of distance between pillars in meters

Pillar	D_x (m)		D_y (m)	
	Mean	Variance	Mean	Variance
Left side	0.0044	0.0071	2.7554	0.8973
Right side	−0.0091	0.0307	2.6766	0.1767

Table 2. Estimation error after scaling

Scaling base	Error (%)		
	Left	Right	Width
Left pillar distance	0	−2.8592	−6.9931
Right pillar distance	2.9397	0	−4.2550
Corridor width	7.5189	4.4469	0

command that sent to the *AR.Drone*. Then, the recorded data was used in localization, pillar detection method and generate point cloud map. Finally, outputs from the methods were used to generate an object map in MATLAB. The map is illustrated in Fig. 4c. Pillars are represented with (x, y) value in world coordinates. Corridor width is calculated based on taking the distance between two peaks of point cloud projection at the dominant orientation of Radon transform, which is 2.8 m for this dataset. Distances between adjacent pillars is calculated in x and y direction. After that, mean and variance of distances, shown in Table 1, are calculated for further evaluation.

In order to evaluate the result, the map is scaled up to the actual distance in the environment. In this case, we scale the map based on three criteria: (a) distance between pillars on left side of the corridor; (b) distance between pillars on right side of the corridor; and (c) width of the corridor. Table 2 presents errors compared to real distances after scaling.

4   Conclusion

Towards the goal of industrial automation in warehouses, a surveying system based on a low cost platform, *Parrot AR.Drone 2.0*, is proposed in this paper. Based on image stream and IMUs, the drone employs *Parallel Tracking and Mapping* (PTAM) to localize itself in the environment. Moreover, PTAM generates point cloud of corner features in the environment, which is used in pillar localization and mapping. Concurrently, our pillar detection method detects and provide 2D pillar projections. Finally, a pillar map is created by finding the correspondence between the point cloud and pillar projections.

As a result, the system provides a pillar map with the biggest error of 20.3 cm (7.5189 %). Cost of the system is comparatively low and doesn't need any prior setup on the environment, e.g. pre-mounted tags.

4.1 Discussion

One problematic issue of the map generated by PTAM is the scale. It is greatly depends on the initialization of the first two image frame (key frames), which is currently done manually. Therefore the map generated from two different trails can be inconsistent.

In addition, image sequences from *AR.Drone* are not always smooth and sometimes do not have sufficient quality to detect enough corner features. As a result, the algorithm eventually lose tracking and fail the localization. Later, image sequences were improved to be more smooth by reducing the control speed of *AR.Drone*. Pillar detection method is based on the color of pillars, which is known and pre-defined before the experiment. Our test environment has uniform color of pillars, which might be different in other environment. Thus, a more general approach is desired.

4.2 Future Work

Long-term autonomy of intelligent vehicles within warehouse environment requires autonomous exploration, obstacle avoidance and self-charging functionality. The proposed work can be integrated into autonomous warehouse for surveillance purpose, e.g. MAVs can be deployed to modelling traffics in the warehouse, provide useful information for planning the path of ground vehicles as well as detect anomalies in the warehouse.

References

1. Ethzasl ptam, 25 September 2013. http://ros.org/wiki/ethzasl_ptam
2. Mikrokopters, 25 September 2013. http://www.mikrokopter.de/en/home
3. Achtelik, M., Achtelik, M., Weiss, S., Siegwart, R.: Onboard imu and monocular vision based control for mavs in unknown in-and outdoor environments. In: 2011 IEEE International Conference on Robotics and Automation (ICRA), pp. 3056–3063. IEEE (2011)
4. Bachrach, A., He, R., Roy, N.: Autonomous flight in unknown indoor environments. Int. J. Micro Air Veh. **1**(4), 217–228 (2009). http://multi-science.metapress.com/index/80586KML376K2711.pdf
5. Bachrach, A., Prentice, S., He, R., Henry, P., Huang, A.S., Krainin, M., Maturana, D., Fox, D., Roy, N.: Estimation, planning, and mapping for autonomous flight using an rgb-d camera in gps-denied environments. Int. J. Robot. Res. **31**(11), 1320–1343 (2012). http://ijr.sagepub.com/content/31/11/1320.short
6. Bachrach, A., Prentice, S., He, R., Roy, N.: Range-robust autonomous navigation in gps-denied environments. J. Field Robot. **28**(5), 644–666 (2011). http://onlinelibrary.wiley.com/doi/10.1002/rob.20400/full
7. Bills, C., Chen, J., Saxena, A.: Autonomous mav flight in indoor environments using single image perspective cues. In: 2011 IEEE International Conference on Robotics and Automation (ICRA), pp. 5776–5783. IEEE (2011). http://ieeexplore.ieee.org/xpls/abs_all.jsp?arnumber=5980136

8. Blosch, M., Weiss, S., Scaramuzza, D., Siegwart, R.: Vision based mav navigation in unknown and unstructured environments. In: 2010 IEEE International Conference on Robotics and Automation (ICRA), pp. 21–28. IEEE (2010). http://ieeexplore. ieee.org/xpls/abs_all.jsp?arnumber=5509920

9. Davison, A.J., Reid, I.D., Molton, N.D., Stasse, O.: Monoslam: real-time single camera slam. IEEE Trans. Pattern Anal. Mach. Intell. **29**(6), 1052–1067 (2007)

10. Engel, J., Sturm, J., Cremers, D.: Camera-based navigation of a low-cost quadro-copter. In: 2012 IEEE/RSJ International Conference on Intelligent Robots and Systems (IROS), pp. 2815–2821. IEEE (2012)

11. Engel, J., Sturm, J., Cremers, D.: Camera-based navigation of a low-costquadrocopter. In: 2012 IEEE/RSJInternational Conference on Intelligent Robots and Systems (IROS), pp. 2815–2821. IEEE (2012). http://ieeexplore.ieee. org/xpls/abs_all.jsp?arnumber=6385458

12. Fan, Y., Aramrattana, M.: Exploration and mapping of warehouses using quadro-torhelicopters (2013)

13. Kerl, C., Sturm, J., Cremers, D.: Dense visual slam for rgb-d cameras. In: 2013 IEEE/RSJ International Conference on Intelligent Robots and Systems (IROS), pp. 2100–2106. IEEE (2013)

14. Klein, G., Murray, D.: Parallel tracking and mapping for small ar workspaces. In: 6th IEEE and ACM International Symposium on Mixed and Augmented Reality, ISMAR 2007, pp. 225–234. IEEE (2007). http://ieeexplore.ieee.org/xpls/abs_all. jsp?arnumber=4538852

15. Kumar, V., Michael, N.: Opportunities and challenges with autonomous micro-aerial vehicles. Int. J. Robot. Res. **31**(11), 1279–1291 (2012). http://ijr.sagepub.com/content/31/11/1279.short

16. Morris, W., Dryanovski, I., Xiao, J.: 3d indoor mapping for micro-uavs using hybrid range finders and multi-volume occupancy grids. In: RSS 2010 Workshop on RGB-D: Advanced Reasoning with Depth Cameras, Zaragoza, Spain (2010)

17. Newman, P., Cole, D., Ho, K.: Outdoor slam using visual appearance and laser ranging. In: Proceedings 2006 IEEE International Conference on Robotics and Automation, ICRA 2006, pp. 1180–1187. IEEE (2006). http://ieeexplore.ieee.org/ xpls/abs_all.jsp?arnumber=1641869

18. Piratla, V., Malode, S.B., Saini, S.K., Jakhotia, A., Sao, A.K., Rajpurohit, B.S., Haobijam, G.: Autonomous navigation in gps denied indoor environment using rgbd sensor, kinect. https://www.researchgate.net/publication/265111081_ Autonomous_Navigation_in_GPS_Denied_Indoor_Environment_Using_RGBD_ Sensor_Kinect

19. Pourmehr, S., Monajjemi, V.M., Vaughan, R., Mori, G.: you two! take off!: Cre-ating, modifying and commanding groups of robots using face engagement and indirect speech in voice commands. In: 2013 IEEE/RSJ International Conference on Intelligent Robots and Systems (IROS), pp. 137–142. IEEE (2013)

20. Pravitra, C., Chowdhary, G., Johnson, E.: A compact exploration strategy for indoor flight vehicles. In: 2011 50th IEEE Conference on Decision and Control and European Control Conference (CDC-ECC), pp. 3572–3577. IEEE (2011). http:// ieeexplore.ieee.org/xpls/abs_all.jsp?arnumber=6161200

21. Shen, S., Michael, N., Kumar, V.: Autonomous multi-floor indoor navigation with a computationally constrained mav. In: 2011 IEEE International Conference on Robotics and Automation (ICRA), pp. 20–25. IEEE (2011). http://ieeexplore.ieee. org/xpls/abs_all.jsp?arnumber=5980357

22. Shen, S., Michael, N., Kumar, V.: Autonomous indoor 3d exploration with a micro-aerial vehicle. In: 2012 IEEE International Conference on Robotics and Automation (ICRA), pp. 9–15. IEEE (2012). http://ieeexplore.ieee.org/xpls/abs_all.jsp?arnumber=6225146
23. Weiss, S., Achtelik, M., Kneip, L., Scaramuzza, D., Siegwart, R.: Intuitive 3d maps for mav terrain exploration and obstacle avoidance. J. Intell. Robot. Syst. **61**(1–4), 473–493 (2011). http://link.springer.com/article/10.1007/s10846-010-9491-y

Probabilistic Model Checking of Ant-Based Positionless Swarming

Paul Gainer[✉], Clare Dixon, and Ullrich Hustadt

Department of Computer Science,
University of Liverpool, Liverpool L69 3BX, UK
{P.Gainer,CLDixon,U.Hustadt}@liverpool.ac.uk

Abstract. Robot swarms are collections of simple robots cooperating without centralized control. Control algorithms for swarms are often inspired by decentralised problem-solving systems found in nature. In this paper we conduct a formal analysis of an algorithm inspired by the foraging behaviour of ants, where a swarm of flying vehicles searches for a target at some unknown location. We show how both exhaustive model checking and statistical model checking can be used to check properties that complement the results obtained through simulation, resulting in information that would facilitate the logistics of swarm deployment.

1 Introduction

A robot swarm is a multi-robot system comprised of some number of simple, autonomous, homogeneous robots, working together to achieve objectives in some environment without centralised control [17]. Coordination between members of the swarm is achieved through self-organisation and local interactions [2]. The aim of the design of decentralised control algorithms is to produce robot swarms that are *scalable* and *fault tolerant*.

Swarm behaviours are generally analysed through simulation and observations of real implementations. The formal analysis of swarm behaviours can complement the design of swarm algorithms by revealing potential problems that may go unnoticed by empirical analysis [7]. Formal verification is the process by which a property expressed in a suitable formalism (usually some form of temporal logic) is exhaustively checked against every possible run of a system.

Temporal verification has been applied to robot swarms. In [1] deductive verification was applied to prove properties of the foraging behaviour of a swarm of robots; algorithmic verification techniques helped to analyse and refine swarm aggregation in [7]; statistical runtime verification combined with agent-based simulation was used to determine the likelihood of emergent swarm behaviours in [11]; an agent-based temporal-epistemic approach is used in [14] to specify and verify emergence in swarms, and in [13] a probabilistic analysis of population-based swarm models was conducted. Whilst this work has clearly demonstrated that formal verification can be used to exhaustively analyse swarm behaviours, there are still many problems that need to be addressed. Notable issues include

© Springer International Publishing Switzerland 2016
L. Alboul et al. (Eds.): TAROS 2016, LNAI 9716, pp. 127–138, 2016.
DOI: 10.1007/978-3-319-40379-3_13

the state space explosion that occurs with naive modelling of swarms, and the need for a general framework applicable to a range of swarm algorithms.

Designing control mechanisms for swarms is a challenging problem. Individual robot behaviours must be formulated at the *microscopic* level and should result in the emergence of complex desired group behaviours at the *macroscopic* level. There are many examples found in nature of decentralised systems that solve complex problems [3]. A common approach in swarm robotics has been to develop control algorithms based on abstractions of these natural systems. In particular, much work has been conducted to develop control algorithms based on the behaviours of social insects, such as foraging for food [5,15], cooperative nest building [19], and efficient distribution of labour [4].

In [10] a swarm of micro air vehicles (MAVs) attempts to form a communication pathway between multiple ground users in a disaster area. The control algorithm used by the MAVs is inspired by the stigmergic foraging behaviour of army ants which lay and maintain pheromone paths from their nest to sources of food. A model of this behaviour was originally developed in [6], where the results of running Monte Carlo simulations of ants moving through a discrete network of points were analysed. These findings were later discussed in detail in [3].

In this paper we apply probabilistic temporal verification to the scenario presented in [10]. For its verification we generate parameterised formal models for the probabilistic model checker PRISM, which we use to either exhaustively or statistically test probabilistic reachability and reward-based properties. While exhaustive model checking checks a property against all possible runs of the system, statistical model checking performs statistical analysis over a subset of the possible runs of the system. We validate our models by comparing the results of checking temporal properties in our models to results obtained from simulations in [10]. We demonstrate how values pertaining to the logistics of deployments of swarms of MAVs, that would be unobtainable through simulation alone, can be calculated a priori by exhaustively checking reward-based properties against every possible execution of our models. This work is an initial step towards the development of a generic probabilistic verification approach that can be applied to other swarm algorithms inspired by the pheromone-based foraging behaviour of social insects.

Section 2 introduces the ant-based swarming scenario described in [10]. In Sect. 3 we detail the generation of our parameterised input models for the probabilistic model checker PRISM, and discuss the abstractions used and assumptions made when designing the discrete formal model. The results of checking probabilistic temporal logic properties in our models are given in Sect. 4. Concluding remarks and suggestions for further work are given in Sect. 5.

2   The Ant-Based Swarming Scenario

The scenario to which we apply probabilistic model checking techniques is presented in [10]. Here, a simulated swarm of *positionless* fixed-wing Micro

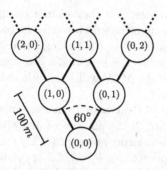

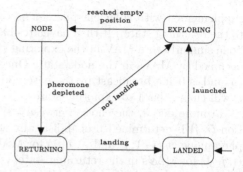

Fig. 1. The Y-junction grid illustrating the ideal positions for MAVs. Each node is 100 m distant from each of its neighbours.

Fig. 2. A finite state machine describing the behaviour of a MAV.

Aerial Vehicles (MAVs) is deployed by a human operator in order to establish a robust emergency communication network between a *target user*, situated at some unknown location, and the base station wherefrom the swarm is launched. Each MAV is positionless in that it relies solely upon proprioceptive sensors and local neighbourhood communication to position itself [18]. The establishment and maintenance of this communication network is studied in detail in [10], however in this paper we focus our analysis on the exploration behaviour of the MAVs.

Figure 1 shows a Y-junction grid consisting of possible positions that MAVs will ideally adopt in their search for the target user, and the paths that connect them, while Fig. 2 shows the finite state machine describing the behaviour of an individual MAV. A MAV begins in landed state at the *base node*, denoted as $(0,0)$ on the grid. MAVs are launched at regular intervals and are then in the exploring state. In the exploring state a MAV navigates through the grid, travelling at a velocity of $10\,m/s$. When a MAV reaches a position in the grid where there is no other MAV it will change to node state and remain at that position, acting as a platform upon which other MAVs can "deposit" virtual pheromone. Upon changing to node state a MAV will initialise its pheromone levels to some given amount. A MAV in the node state at some position (i, j) is considered to be an *internal* node if there is a MAV in the node state at either of $(i+1, j)$ or $(i, j+1)$. Internal nodes supplement their levels of deposited pheromone at each time step by some amount. While in the node state each MAV broadcasts its pheromone level to other MAVs within its communication range of $100\,m$. When a MAV in the exploring state reaches a position in the grid where there is already a MAV in the node state, it continues moving outward and makes a probabilistic choice which branch to take, determined by the levels of pheromone deposited at the next positions on the left and right branches, as transmitted by the MAV in the node state.

Pheromone levels dissipate gradually over time and when they are depleted a MAV in the node state changes to returning state (internal nodes are supplemented with sufficient pheromone to guarantee that they cannot change to the returning state.) It then navigates back through the grid towards the base node similarly to a MAV in the exploring state but only moving along positions occupied by MAVs in the node state. Once it reaches the base node, if a signal to land is being broadcast by the base node then the MAV will land, otherwise it will change back to exploring state.

In more detail, the choice between the left and right path from some position (i, j) is determined probabilistically according to the amount of deposited pheromone at $(i+1, j)$ and $(i, j+1)$ for MAVs in exploring state, or $(i-1, j)$ and $(i, j-1)$ for MAVs in the returning state. Given pheromone levels of $\phi_{(i+1, j)}$ and $\phi_{(i, j+1)}$, the probability of a MAV choosing the left or right path is calculated using (1) to (4), where μ is a constant which determines the attractiveness of unexplored paths and is set to 0.75 for the simulations in [10]. If there is no MAV in the node state at (i, j) then $\phi_{i,j} = 0$. Equations (3) and (4) are the calculations of the probabilities of taking the left or right path at (i, j) where the correction factor $c_L(i, j)$ defined in [10] is applied to the original probability calculation $p_L(i, j)$ given in [6]. This correction ensures that positions equidistant from the base node have an equal chance of being eventually reached, given equal amounts of pheromone on every path.

$$p_L(i, j) = \frac{\left[\mu + \phi_{i+1, j}\right]^2}{\left[\mu + \phi_{i+1, j}\right]^2 + \left[\mu + \phi_{i, j+1}\right]^2} \tag{1}$$

$$c_L(i, j) = \frac{i + 1}{i + j + 2} \tag{2}$$

$$\pi_L(i, j) = \frac{p_L(i, j) \cdot c_L(i, j)}{p_L(i, j) \cdot c_L(i, j) + (1 - p_L(i, j)) \cdot (1 - c_L(i, j))} \tag{3}$$

$$\pi_R(i, j) = 1 - \pi_L(i, j) \tag{4}$$

3   Modelling the Scenario

Next we discuss the design and automatic generation of discrete, parameterised models of the scenario to which we can apply probabilistic analysis.

3.1   The PRISM Model Checker

Models of the scenario were constructed using the *probabilistic model checker* PRISM [12]. Given a probabilistic model of a system, PRISM can be used to analyse both temporal and probabilistic properties of the input model by exhaustively checking some logical requirement against all possible behaviours. Properties to be checked can be specified using *probabilistic temporal logics* such as Probabilistic Computation Tree Logic (PCTL) [9]. PCTL consists of classical logical operators ($\wedge, \vee, \neg$), temporal operators $\square\phi$ (at all points in the future ϕ holds), $\Diamond\phi$ (at some point in the future ϕ holds), $\phi\,\mathcal{U}\,\psi$ (ϕ holds until ψ holds), and the probabilistic operator $P_{\bowtie\gamma}(\phi)$ where $\bowtie \in \{<, \leq, >, \geq\}$ is a relational operator and γ is a probability threshold. PCTL can therefore be used to specify properties such as $P_{\geq 0.5}(\Diamond \phi)$, meaning "$\phi$ holds at some future point with a probability of at least 0.5". PRISM allows properties to be expressed which evaluate to a numerical value, for instance $P_{=?}(\Diamond \phi)$, "the probability of ϕ being true at some point in the future", and also supports metric temporal operators where the property is bound by time, for example $\Diamond^{\leq T} \phi$, "ϕ holds at some point in the future within T units of time".

3.2   Discretisation

Simulations were conducted in [10] using a time-step of $50\,ms$. Since a MAV travels at $10\,m/s$, and ideal positions for nodes are $100\,m$ distant from their neighbours, a MAV in the exploring or returning state takes $10\,s$ to move from one position to the next. In the probabilistic models we construct using PRISM we consider one transition in the model to be equivalent to a time-step of $10\,s$. In the original scenario MAVs are launched from the base node by a human operator every $15 \pm 7.5\,s$, giving an interval in seconds of possible durations between launches of $[7.5, 22.5]$. By rounding the endpoints of the interval to the nearest 10 (since one transition in our model is equivalent to $10\,s$), we determined that a MAV is launched from the base in our model once every 1 or 2 transitions. This non-determinism in the model resulted in state spaces too large to verify properties for, so the model was further simplified so that exactly one MAV was launched per transition. A comparison of the results obtained from applying probabilistic model checking to two models differing only by this additional simplification showed that there were only very minor differences between the results.

When a MAV switches to node state at time t and position (i, j) it initialises the pheromone level $\phi_{i,j}(t)$ to ϕ_{init}. The evolution of the pheromone levels at some position (i, j) is defined given by the equation

$$\phi_{i,j}(t+1) = \min[\phi_{i,j}(t) - \Delta\phi_{\text{dec}} + n \cdot \Delta\phi_{\text{ant}} + \Delta\phi_a, \phi_{\text{max}}] \qquad (5)$$

where ϕ_{max} is the maximum amount of pheromone that can be deposited at any node, $\Delta\phi_{\text{dec}}$ is the rate at which deposited pheromone dissipates, $\Delta\phi_{\text{ant}}$ is the rate at which pheromone is deposited on a MAV in the node state by a MAV in the exploring state or the returning state, n is the number of MAVs in the exploring state or the returning state at (i, j), $\Delta\phi_a = \Delta\phi_{\text{int}}$ if there is some MAV in the node state at $(i+1, j)$ or $(i, j+1)$, or 0 otherwise, and $\Delta\phi_{\text{int}}$ is the rate at which extra pheromone is deposited on internal nodes.

The simulations conducted in [10] used values $\phi_{\text{init}} = 0.7$ and $\phi_{\text{max}} = 1$. The rates at which pheromone was deposited, or dissipated, given in terms of units per time-step with one time-step corresponding to $50\,ms$, were given as $\Delta\phi_{\text{ant}} = 0.002$, $\Delta\phi_{\text{int}} = 0.001$ and $\Delta\phi_{\text{dec}} = 0.001$; in our model we consider one transition to be equivalent to $10\,s$ and therefore multiply these values by 200 to get the pheromone deposition/dissipation rates per transition in the model. To decrease the size of our models we use a range of discrete integer values to model pheromone levels. Given some $Ph \in \mathbb{N}$, the number of discrete values to be used to model pheromone levels, we define a mapping $\tau : [0, \phi_{max}] \to \mathbb{N}$ that maps a pheromone value in $[0, \phi_{max}]$ to some integer value such that $\tau(\phi) = [\phi \cdot Ph]$. In our models we use a value of $Ph = 5$; pheromone is deposited by MAVs in the exploring state or the returning state at a rate of $\tau(200 \cdot \Delta\phi_{ant}) = 2$ per transition, internal nodes supplement their own pheromone levels at a rate of $\tau(200 \cdot \Delta\phi_{int}) = 1$, and pheromone dissipates at a rate of $\tau(200 \cdot \Delta\phi_{dec}) = 1$.

3.3 Abstractions and Assumptions

Modelling each MAV individually would result in intractable models. We therefore take advantage of the following abstractions. First, since one transition in the models is equivalent to the duration of a flight between two adjacent nodes, and since all MAVs begin at the base node (position $(0,0)$, see Fig. 1), after each transition we can assume that the location of each MAV is always at some position (i,j), instead of in-between positions. Second, as demonstrated in [13] a *counting abstraction* can also be used when modelling the behaviours of multiple identical processes. Since all MAVs are behaviourally identical, and their action decisions depend solely on their immediate state and percepts, when appropriate we can associate a counter with every position (i,j) that records the number of MAVs at that location.

In the simulations in [10] if the signal to land has not been given, then MAVs that have returned to the base node instead resume exploration. Here we constrain the number of MAVs that may leave the base node at any moment in time to a single MAV. This simplification allows us to greatly reduce the size of the model. Since each MAV is considered to land upon returning, and at most one MAV is launched each round, we can conclude that at most one exploring MAV is at any given position at any time. This can be modelled using Boolean variables to record if an exploring MAV has moved from some position (i,j) to either of $(i+1,j)$ or $(i,j+1)$.

A strategy is given in [10] to automatically assign altitudes to individual MAVs, ensuring that MAVs in the exploring or the returning state maintain an altitude higher than MAVs in the node state. While this strategy did not prove to be successful in all cases the chance of a collision occurring was sufficiently low (2.6 % of 7500 MAVs collided over 500 trials) for us to assume that altitude differentiation always avoids collisions.

3.4 Prism Models

Each generated PRISM model can be defined as a set of m *modules* $M_1, \ldots, M_m$ [16]. Each module M_i is a tuple $(\mathcal{V}_i, \mathcal{I}_i, \mathcal{C}_i)$, where $\mathcal{V}_i = \{v_1, \ldots, v_{k_i}\}$ is a set of local variables over the domain consisting of finitely bound integers and booleans, $\mathcal{I}_i$ is a mapping of variables to initial values, and $\mathcal{C}_i = \{c_1, \ldots, c_{n_i}\}$ is a set of commands that define the behaviour of the module. With each local variable $v \in \mathcal{V}_i$ we associate a variable v' denoting the state of v in the next moment of time. The set of all local variables in the model is denoted as $\mathcal{V} = \bigcup_{i=1}^{m} \mathcal{V}_i$.

For a module M_i every command $c_j \in C_i$ is a pair $(g, \mathcal{U})$ where g is a predicate over $\mathcal{V}$ and $\mathcal{U} = \{(p_1, u_1), \dots, (p_t, u_t)\}$ is a set of possible transitions for M_i. For a pair $(p_j, u_j) \in \mathcal{U}$, u_j is an assignment of values to each of the local variables $v_1, \dots, v_{k_i}$ and is of the form $\bigwedge_{a=1}^{k_i} (v'_a = ex_a)$ where ex_a is an expression in terms of V and the domain of variables, and $p_i \in \mathbb{R}^+$ is a constant defining the probability of that update occurring. Since our model is a DTMC it is required that for every command we have $p_i \in (0, 1]$ for $1 \le i \le t$, and $\sum_{i=1}^{t} p_i = 1$.

The semantics of a PRISM model can be defined in terms of a DTMC. A DTMC is a tuple $(\mathcal{S}, \bar{s}, \mathbf{P})$ where $\mathcal{S}$ is a set of states, $\bar{s} \in \mathcal{S}$ is the initial state, and $\mathbf{P} : \mathcal{S} \times \mathcal{S} \to [0, 1]$ is the probability transition matrix. The state space S_i of a module M_i is the set of all valuations of $\mathcal{V}_i$, and the global state space of a model is the product of the local state spaces of all modules. The local state of a module M_i is denoted as s_i, and a global state $s \in \mathcal{S}$ is a tuple $(s_1, \dots, s_m)$ of local states. The initial state $\bar{s}$ is determined by $\mathcal{I}$. For brevity, we omit the details of the calculation of $\mathbf{P}$ and refer the reader to [16], noting that while many different parallel compositions of modules can be defined in PRISM, in our model all modules synchronise over all transitions.

3.5 Model Generation

We automate the generation of our models by using the PRISM Preprocessor to construct a parameterised model of the system. Then, given values for the parameters $N \in \mathbb{N}$, the number of MAVs in the swarm, $D \in \mathbb{N}$, the maximum distance between the target user and the base node (in hundreds of metres), and Ph, the number of discrete values used to record pheromone levels, we can automatically generate a PRISM model $\mathcal{M}$. The model is given by

$$\mathcal{M} = \{B\} \cup \{E_{i,j}, R_{i,j} \mid i, j \in 0 \dots D \text{ and } 0 < i + j < D\}$$
$$\cup \{F_{i,j} \mid i, j \in 0 \dots D \text{ and } i + j = D\},$$

where B is a module that models the movement of MAVs in exploring state from the base node, each $E_{i,j}$ is a module that models the movement of MAVs in the exploring or node states at (i, j), each $R_{i,j}$ is a module that models the movement of MAVs in the returning state at (i, j), and each $F_{i,j}$ is a module that models the movement of MAVs in the exploring, node or returning states, at or beyond (i, j). We now define each module in the model, however due to space limitations we simply provide an informal description of the behaviour of each module. A full formal description of the model is given in [8].

The base module is a tuple $B = (\mathcal{V}_B, \mathcal{I}_B, \mathcal{C}_B)$, where $\mathcal{V}_B = \{bc, \diagdown_{0,0}, \diagup_{0,0}\}$ and $\mathcal{I}_B = \{bc \mapsto N, \diagdown_{0,0} \mapsto false, \diagup_{0,0} \mapsto false\}$. The finitely bound integer variable bc records the number of MAVs at the base node. The boolean variables $\diagdown_{0,0}$ and $\diagup_{0,0}$ record the movement of MAVs in the exploring state, and are $true$ iff in the last moment in time a MAV moved from the base node to $(1, 0)$ and $(0, 1)$, respectively. Should there be one or more MAVs at the base node then one will be launched and will move to $(1, 0)$ with probability $\pi_L(0, 0)$, or to $(0, 1)$ with probability $\pi_R(0, 0)$.

For every $E_{i,j} = \{\mathcal{V}_E^{i,j}, \mathcal{I}_E^{i,j}, \mathcal{C}_E^{i,j}\}$ we have $\mathcal{V}_E^{i,j} = \{p_{i,j}, n_{i,j}, \diagdown_{i,j}, \diagup_{i,j}\}$ with $\mathcal{I}_E^{i,j} = \{p_{i,j} \mapsto 0, n_{i,j} \mapsto 0, \diagdown_{i,j} \mapsto false, \diagup_{i,j} \mapsto false\}$, where $p_{i,j}$ is a finitely bound integer variable recording the levels of pheromone deposited at (i,j), $n_{i,j}$ is a boolean variable that is *true* if there is a MAV in the node state at (i,j), and $\diagdown_{i,j}$ and $\diagup_{i,j}$ are boolean variables which are *true* iff a MAV moved from (i,j) respectively to $(i+1,j)$ or $(i,j+1)$ in the last moment in time. If no MAVs in the exploring state have moved to (i,j) then in the next moment in time no MAVs in the exploring state will be moving from (i,j); if $n_{i,j} = true$ then pheromone updates will be applied, and if $p_{i,j} \leq 0$ then in the next moment in time the MAV will be in the returning state, otherwise the MAV remains in the node state. If a MAV in the exploring state has moved to (i,j) when $n_{i,j} = false$ then in the next moment in time $n_{i,j}$ will be *true* and $p_{i,j}$ will be initialised to $\tau(\phi_{init})$; no MAV in the exploring state will be moving from (i,j) in the next moment in time. If a MAV in the exploring state has moved to (i,j) when $n_{i,j} = true$, then in the next moment in time the exploring MAV will have moved to $(i+1,j)$ with probability $\pi_L(i,j)$, or to $(i,j+1)$ with probability $\pi_R(i,j)$, and remains in the exploring state; pheromone updates are applied and if $p_{i,j} \leq 0$ then in the next moment in time the MAV in the node state will be in the returning state, otherwise this MAV remains in the node state.

For every $R_{i,j} = \{\mathcal{V}_R^{i,j}, \mathcal{I}_R^{i,j}, \mathcal{C}_R^{i,j}\}$ if $i \geq 0, j > 0$ we have a finitely bound integer variable $\diagup_{i,j} \in \mathcal{V}_R^{i,j}$ with $\mathcal{I}_R^{i,j}(\diagdown_{i,j}) = 0$, and if $i > 0, j \geq 0$ we have a finitely bound integer variable $\diagdown_{i,j} \in \mathcal{V}_R^{i,j}$ with $\mathcal{I}_R^{i,j}(\diagdown_{i,j}) = 0$, which record the number of MAVs in the returning state that moved from (i,j) respectively to $(i,j-1)$ and $(i-1,j)$ in the last moment in time. Unlike exploring MAVs, it is often the case that two or more returning MAVs will simultaneously move to the same position. Any MAV in the returning state at some location (i,j) will always move respectively to $(i-1,j)$ or $(i,j-1)$ if $i = 0$ or $j = 0$, otherwise it will move from (i,j) to $(i-1,j)$ with probability $\pi_L(i-1,j-1)$, or to $(i,j-1)$ with probability $\pi_R(i-1,j-1)$. We define $E_{k,l}^{i,j}$ to be the event where given k MAVs in the returning state at (i,j), l MAVs move from (i,j) to $(i-1,j)$ in the next moment in time, and $k-l$ MAVs move from (i,j) to $(i,j-1)$ in the next moment in time. Given values for i, j, k and l we can calculate the probability of $E_{k,l}^{i,j}$ as

$$P(E_{k,l}^{i,j}) = \left(\pi_L(i-1,j-1)^l \cdot \pi_R(i-1,j-1)^{k-l}\right)\binom{k}{l},$$

since if we have k MAVs then we must consider all distinct subsets of size l. If no MAVs in the returning state have moved to (i,j), and there is no MAV in the node state whose pheromone levels have depleted, then in the next moment in time no MAVs in the returning state will be moving to either of $(i-1,j)$ or $(i,j-1)$. We then consider each case where there are k MAVs at (i,j) where $0 < k \leq N$. In each case calculating $P(E_{k,l}^{i,j})$ for $l = 0, \ldots, k$ gives a discrete probability distribution since for any $k \in \mathbb{N}$ we have that $\sum_{l=0}^{k} P(E_{k,l}^{i,j}) = 1$.

For every $F_{i,j} = \{\mathcal{V}_F^{i,j}, \mathcal{I}_F^{i,j}, \mathcal{C}_F^{i,j}\}$ we have a finitely bound integer variable $m_{i,j} \in \mathcal{V}_F^{i,j}$ with $\mathcal{I}_F^{i,j}(m_{i,j}) = 0$ that records the number of MAVs that are at or beyond this position and a finitely bound integer variable $r_{i,j} \in \mathcal{V}_F^{i,j}$ with $\mathcal{I}_F^{i,j}(r_{i,j}) = 0$ that is used to determine when MAVs should return from (i,j) or beyond. If $i \geq 0, j > 0$ we have a finitely bound integer variable $\diagup_{i,j} \in \mathcal{V}_F^{i,j}$ with $\mathcal{I}_F^{i,j}(\diagup_{i,j}) = 0$ and if $i > 0, j \geq 0$ we have a finitely bound integer variable $\diagdown_{i,j} \in \mathcal{V}_F^{i,j}$ with $\mathcal{I}_F^{i,j}(\diagdown_{i,j}) = 0$ which record the number of MAVs in the returning state that moved from (i,j) respectively to $(i,j-1)$ and $(i-1,j)$ in the last moment in time. The variable $r_{i,j}$ is an approximation of the average number of transitions we would expect between a MAV in the exploring state moving to (i,j) or beyond, and then returning from a position at or beyond (i,j). If no MAV moves to (i,j) then after each transition the variable $r_{i,j}$ is decremented by 1. If there are MAVs at (i,j) and $r_{i,j}$ has decreased to 0, then in the next moment in time $m_{i,j}$ is decremented by 1 and a single MAV returns to $(i-1,j)$ or $(i,j-1)$ with probability respectively. For final positions where $i=0, j=0$, all MAVs returning from (i,j) will move to $(i,j-1)$ or $(i-1,j)$ respectively.

4   Experiments

To validate our model we applied statistical model checking using the PRISM discrete-event simulator and compared our results to those obtained from the simulations conducted in [10]. The mean probability of establishing contact with a user within 30 min was calculated over a series of 500 simulations for varying swarm sizes. Users were located at some randomly determined location within a 60 degree arc in a known cardinal direction from the base node at a distance of ≈ 200–$500\,\mathrm{m}$. In our model we assumed that a MAV has established communication with a user if it has moved to a position at most $100\,m$ distant from the user. Since a user can be located up to $\approx 500\,m$ from the base node, models were generated with $D = 5$ and $Ph = 5$ for all models. We define $\mathcal{P}_{user} = \{(i,j) \mid i, j \in 0 \dots 5 \text{ and } 1 < i+j \leq 5\}$ to be the set of all possible locations at which a user may be located. For each $(i,j) \in \mathcal{P}_{user}$ we used PRISM to calculate the probability of a MAV moving to (i,j) within 30 min (equivalent to 180 transitions in our model) by formally specifying this as a probabilistic reachability property in PCTL. Since the grid of positions is symmetrical we only check PCTL properties for a subset of $\mathcal{P}_{user}$, as shown in Fig. 3; the probability of a MAV moving to some (i,j) is equivalent to the probability of a MAV moving to (j,i). For each (i,j) we define $moved_{i,j}$ as

$$moved_{i,j} \equiv \begin{cases} (\diagdown_{i-1,j} \vee \diagup_{i,j-1}) & \text{if } i, j > 0 \\ \diagup_{i,j-1} & \text{if } i = 0, j > 0 \\ \diagdown_{i-1,j} & \text{if } i > 0, j = 0 \end{cases} \qquad (6)$$

We then calculate λ, the mean probability over all locations, as

$$\lambda = \left(\sum_{(i,j) \in \mathcal{P}_{user}} P_{=?}(\lozenge^{\leq 180} moved_{i,j}) \right) / |\mathcal{P}_{user}| \qquad (7)$$

Figure 4 compares the results of calculating λ for values of $N \in 5, \dots, 20$ to the results presented in [10]. Statistical model checking results were obtained

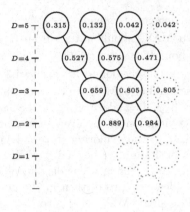

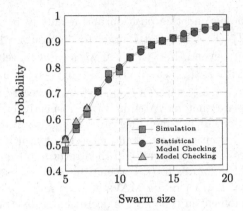

Fig. 3. The subset of $\mathcal{P}_{user}$ for which the PCTL property is checked. Here the results correspond to checking the property for $N = 5$.

Fig. 4. The mean probability of finding the user within 30 min over 500 trials for simulation and 500 samples for statistical model checking.

using 500 discrete-event simulation samples with an average confidence interval of $\pm 2\%$ based on a 99.0% confidence level. Experiments were conducted on a PC with a 2.20 GHz Intel Xeon E5-2420 CPU, 196 GB RAM, running Scientific Linux 6.6. There is clearly a strong correlation between both sets of results. For some swarm sizes, namely for $N = 5,9,10,16$, there was a more pronounced difference between the two values. However, some minor discrepancies were expected due to the relatively low number of simulations/samples used to obtain the results. Exhaustive model checking results are shown only for $N = 5, 6, 7$ since the reachable state space of models for $N \geq 8$ was too large to be calculated.

PRISM can be used to reason about other measurable aspects of model behaviours. Rewards can be associated with individual states, or groups of states, and properties relating to expected values for these rewards can be checked in models. PRISM also provides the R operator which allows properties to be expressed such as the reachability reward property $R_{=?}(\Diamond \phi)$, "what is the expected reward for reaching a state where ϕ is true". By associating a reward of one with each state in our models we can test the property $R_{=?}(\Diamond m_{i,j})$ for every $(i,j) \in \mathcal{P}_{user}$, which calculates the total time expected for the swarm to establish contact with a user at (i,j) with probability 1. These calculated values could facilitate the logistics of MAV swarm deployments where guaranteed contact with a user is required, given a limited number of MAVs. In Fig. 5 the four graphs on the left show the total expected time in hours for a deployment of N MAVs, depth D, and lateral distance of the target user from the base node, to establish communication with the target user with a probability 1. This is done by checking the property $R_{=?}(\Diamond moved_{i,j})$ for each (i,j) The four graphs on the right show the probability of establishing communication with the target user within 30 min by a deployment of N MAVs, depth D, and lateral distance of the target user from the base node. This is done by checking the property $P_{=?}(\Diamond^{\leq 180} moved_{i,j})$ for each (i,j). Results where $N > 7$ were obtained using statistical methods over 4000 samples.

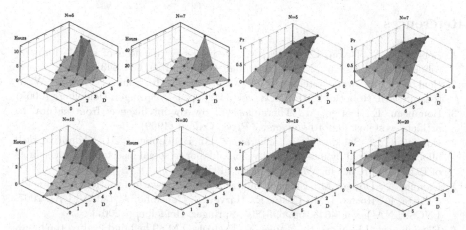

Fig. 5. Results for checking the properties $R_{=?}(\Diamond\, moved_{i,j})$ and $P_{=?}(\Diamond^{\leq 180}\, moved_{i,j})$ for each (i,j) in the generated models.

5   Conclusions and Further Work

We have constructed formal probabilistic models making some simplifying assumptions given in Sect. 3.3, and clearly shown a close correspondence between these models and the simulations conducted in the original scenario. We then used these models to verify both probabilistic and reward-based properties, where the resultant calculated values could be used to plan the deployment of a swarm of MAVs where establishing contact with a user must be guaranteed, or achieved with a probability that exceeds some given threshold. Since battery life greatly impacts the flight duration of MAVs the a priori calculation of the total expected flight time, or total expected distance travelled, for the swarm would ensure that sufficient resources could be made available to ensure that it achieves its objectives.

A natural extension of this work would be to use the parametric model checking functionality of PRISM to investigate more thoroughly how each parameter affects the results of checking properties in the model and to calculate optimum parameter values that maximise or minimise probabilistic or reward-based properties. We also aim to further abstract our approach so that the techniques that we have developed here can be applied to a broader range of swarm algorithms where stigmergic communication is used to coordinate the behaviour of the swarm.

Acknowledgments. The authors would like to thank the Networks Sciences and Technology Initiative (NeST) of the University of Liverpool for the use of their computing facilities. The first author would like to acknowledge the funding received from the Sir Joseph Rotblat Alumni Scholarship.

References

1. Behdenna, A., Dixon, C., Fisher, M.: Deductive verification of simple foraging robotic behaviours. Int. J. Intell. Comput. Cybern. **2**(4), 604–643 (2009)
2. Beni, G.: From swarm intelligence to swarm robotics. In: Şahin, E., Spears, W.M. (eds.) Swarm Robotics 2004. LNCS, vol. 3342, pp. 1–9. Springer, Heidelberg (2005)
3. Bonabeau, E., Dorigo, M., Theraulaz, G.: Swarm intelligence: from natural to artificial systems. Oxford University Press, Oxford (1999)
4. Bonabeau, E., Sobkowski, A., Theraulaz, G., Deneubourg, J.L.: Adaptive task allocation inspired by a model of division of labor in social insects. In: Proceedings of BCEC97, pp. 36–45. World Scientific (1997)
5. Campo, A., Dorigo, M.: Efficient multi-foraging in swarm robotics. In: Almeida e Costa, F., Rocha, L.M., Costa, E., Harvey, I., Coutinho, A. (eds.) ECAL 2007. LNCS (LNAI), vol. 4648, pp. 696–705. Springer, Heidelberg (2007)
6. Deneubourg, J.L., Goss, S., Franks, N., Pasteels, J.M.: The blind leading the blind: modeling chemically mediated army ant raid patterns. J. Insect Behav. **2**(5), 719–725 (1989)
7. Dixon, C., Winfield, A.F., Fisher, M., Zeng, C.: Towards temporal verification of swarm robotic systems. Robot. Auton. Syst. **60**(11), 1429–1441 (2012)
8. Gainer, P., Dixon, C., Hustadt, U.: Probabilistic Model Checking of Ant-Based Positionless Swarming (Extended Version). Techical report ULCS-16-001, University of Liverpool, Liverpool, UK (2016)
9. Hansson, H., Jonsson, B.: A logic for reasoning about time and reliability. Formal Aspects Comput. **6**(5), 512–535 (1994)
10. Hauert, S., Winkler, L., Zufferey, J.C., Floreano, D.: Ant-based swarming with positionless micro air vehicles for communication relay. Swarm Intell. **2**(2–4), 167–188 (2008)
11. Herd, B., Miles, S., McBurney, P., Luck, M.: Approximate verification of swarm-based systems: a vision and preliminary results. In: Proceedings of 23rd Safety-Critical Systems Symposium, pp. 361–378. SCSC (2015)
12. Hinton, A., Kwiatkowska, M., Norman, G., Parker, D.: PRISM: a tool for automatic verification of probabilistic systems. In: Hermanns, H., Palsberg, J. (eds.) TACAS 2006. LNCS, vol. 3920, pp. 441–444. Springer, Heidelberg (2006)
13. Konur, S., Dixon, C., Fisher, M.: Analysing robot swarm behaviour via probabilistic model checking. Robot. Auton. Syst. **60**(2), 199–213 (2012)
14. Kouvaros, P., Lomuscio, A.: Verifying emergent properties of swarms. In: Proceedings of AAAI 2015, pp. 1083–1089. AAAI Press (2015)
15. Liu, W., Winfield, A.F.T., Sa, J., Chen, J., Dou, L.: Strategies for energy optimisation in a swarm of foraging robots. In: Şahin, E., Spears, W.M., Winfield, A.F.T. (eds.) SAB 2006 Ws 2007. LNCS, vol. 4433, pp. 14–26. Springer, Heidelberg (2007)
16. Parker, D.A.: Implementation of symbolic model checking for probabilistic systems. Ph.D. thesis, University of Birmingham (2002)
17. Şahin, E., Winfield, A.: Special issue on swarm robotics. Swarm Intell. **2**(2), 69–72 (2008)
18. Støy, K.: Using situated communication in distributed autonomous mobile robotics. In: Proceedings of SCAI 2001, pp. 44–52. IOS Press (2001)
19. Theraulaz, G., Bonabeau, E.: Coordination in distributed building. Science **269**, 686–686 (1995)

A Control Structure for Bilateral Telemanipulation

William Harwin[✉]

School of Systems Engineering, University of Reading, Reading, UK
w.s.harwin@reading.ac.uk

Abstract. A framework for considering the stability of bilateral tele-manipulator systems is considered. The approach adapts the work of Lawrence [3] to use a state-space formulation thus simplifying the iden-tification of the stability conditions from the eigenvalues of the feedback system. Both numerical and symbolic stability conditions are considered.

Keywords: Bilateral telemanipulator · Control · Stability

1 Background

Telemanipulator theory forms the basis for applications that include remote handling, surgical robots, exoskeletons and haptic systems [5]. In many cases, to avoid the complexity of the control system, the system operates as a strict master-slave system where position coordinates are generated by the master and the slave is servoed to follow these positions. In such cases the operator must rely on visual feedback when contact is made between the slave and the remote environment. Better control can be achieved when position and force information available at the slave is directed back to the master. One example is the OOEC Magpie [1] which is a foot operated assisted dining device that give direct mechanical feedback of forces encountered by the eating utensil to the operator's foot. A similar principal applies to orthotic devices such as the Wilmington Robotics Exoskeleton (WREX) [4] where there is a close coupling between the orthosis and the arm.

In more complex telemanipulator systems some form of closed-loop control around the actuator mechanisms is needed. In most cases this is based on position feedback control of small permanent magnet motors. Where there is backdrivable transmission between the motor and the linkage joint it is possible to estimate the external forces applied to the linkage [6], and these forces can then be reflected into the associated master or slave.

Lawrence [3] identified four connections that can be made between the master and slave of a backdrivable linkage telemanipulator system that can then allow permutations of position, force, impedance or admittance to be reflected into the master and slave systems (Fig. 1).

© Springer International Publishing Switzerland 2016
L. Alboul et al. (Eds.): TAROS 2016, LNAI 9716, pp. 139–145, 2016.
DOI: 10.1007/978-3-319-40379-3_14

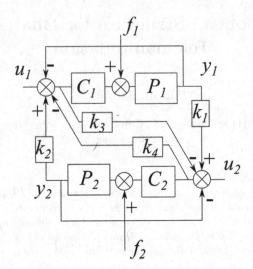

Fig. 1. Bilateral telemanipulator as a coupled control system. The upper system (C_1 P_1) will be considered as the master and the lower as the slave although in general the master and slave can be exchanged without loss of generality

2   Simplified structure of a master-slave telemanipulator

A variation of the Lawrence [3] control structure is shown in Fig. 2. It consists of two control systems that are assumed to be backdrivable. Thus the upper part of the figure (the master) consists of a plant P_1 that is assumed to contain the robot linkage and transmission, and the controller C_1 is assumed to also include the characteristics of the actuator. Evidently with this structure in a unity closed-loop feedback system as shown it is possible to show that the position gain y_1/u_1 is given as

$$y_1 = (1 + P_1 C_1)^{-1} P_1 C_1 u_1$$

Thus under a restricted set of conditions $P_1 C_1$ can be considered to dominate with respect to unity and the output will try to match the input. These conditions require a high controller gain without local instability so are likely to be realised when the system is tuned with an algorithm such as Zeigler-Nichols. In a similar argument the gain due to the 'disturbance' f_1 can be expressed in the form

$$y_1 = (1 + P_1 C_1)^{-1} P_1 f_1$$

and in this case with the same restricted set of conditions the admittance of the mechanism, the position response to the disturbance force f_1, can be considered to be C_1. The controller gain can be seen to relate directly to the impedance (stiffness) of the closed-loop system and inversely to the admittance.

The four gains k_1, k_2, k_3, k_4 in the controller structure shown in Fig. 2 can be considered in two parts. Gains k_1 and k_2 simply set the position demand, master to slave and slave to master. To assess the effect of the remaining two gains k_3 and k_4 it should be noted that the gains $k_1 k_4$ supplement the controller C_1 and the gains $k_2 k_3$ supplement the controller C_2. With this adaption the impedance of the master can be considered to be set by $C_1(1 + k_4 k_1)$ and the stiffness of the slave is set by $C_2(1 + k_2 k_3)$. Alternatively the two controller gains k_1, k_3 can be chosen to be an estimate of f_1 and scaled to be a force demand of the slave with a similar argument for the gains k_2, k_4 [6].

Thus it can be seen that forces or positions from the master can be reflected into the slave, and vice versa.

2.1   A Simplified State-Space Representation of a Bilateral Telemanipulator

If we assume each 'plant' is a mass and damper so $\dot{x} = Ax + B\epsilon$ and $y = Cx$. This assumption provides a minimal system that uses Newton's second law along with the damper to provide a channel for the energy dissipation.

If each system has two states and decoupled we can generate a combined state matrix

where $x = [x_1 \ \dot{x}_1 \ x_2 \ \dot{x}_2]^T$ and $\epsilon = [\epsilon_1 \ \epsilon_2]^T$

For example the figure without the gain terms k_1 to k_4 could be considered as

$$A = \begin{pmatrix} 0 & 1 & 0 & 0 \\ 0 & -b_1 & 0 & 0 \\ 0 & 0 & 0 & 1 \\ 0 & 0 & 0 & -b_2 \end{pmatrix} \qquad B = \begin{pmatrix} 0 & 0 \\ c_1 & 0 \\ 0 & 0 \\ 0 & c_2 \end{pmatrix} \tag{1}$$

The output matrix C then selects outputs y_1 and y_2

$$C = \begin{pmatrix} 1 & 0 & 0 & 0 \\ 0 & 0 & 1 & 0 \end{pmatrix}$$

A gain matrix can be identified from

$$\begin{bmatrix} 1 & -k_4 \\ -k_3 & 1 \end{bmatrix} \begin{bmatrix} \epsilon_1 \\ \epsilon_2 \end{bmatrix} = \begin{bmatrix} -1 & k_2 \\ k_1 & -1 \end{bmatrix} \begin{bmatrix} y_1 \\ y_2 \end{bmatrix}$$

That is $\epsilon = Ky$ where

$$K = \begin{pmatrix} -\frac{k_1 k_4 - 1}{k_3 k_4 - 1} & -\frac{k_2 - k_4}{k_3 k_4 - 1} \\ -\frac{k_1 - k_3}{k_3 k_4 - 1} & -\frac{k_2 k_3 - 1}{k_3 k_4 - 1} \end{pmatrix}$$

The feedback gain and original system can be combined to for a new state-space system of the form.

$$\dot{x} = (A - BKC)x + BKr$$

So the revised A matrix is

$$A' = \begin{pmatrix} 0 & 1 & 0 & 0 \\ \frac{c_1\,(k_1\,k_4-1)}{k_3\,k_4-1} & -b_1 & \frac{c_1\,(k_2-k_4)}{k_3\,k_4-1} & 0 \\ 0 & 0 & 0 & 1 \\ \frac{c_2\,(k_1-k_3)}{k_3\,k_4-1} & 0 & \frac{c_2\,(k_2\,k_3-1)}{k_3\,k_4-1} & -b_2 \end{pmatrix} \tag{2}$$

We can crosscheck see it is still stable by setting the k gains to 0 and testing the eigenvalues, which require negative real part of the expression $-\frac{b_1}{2} \pm \frac{\sqrt{b_1{}^2 - 4\,c_1}}{2}$ Setting any one of the gains k_1, k_2, k_3, k_3 to be non zero makes no change to the eigenvalues and hence to the system stability.

The eigenvalues of the full system 2 where all the gains k_1, k_2, k_3, k_3 are set, can be computed in Matlab but are too large to be of any value.

2.2 A Strict Symmetrical Bilateral Telemanipulator

By setting the master and slave to have identical components as well as ensuring $k_2 = k_1$ and $k_4 = k_3$ we get a symmetrical bilateral telemanipulator. In this case the eigenvalues relatively simple, that is

$$\lambda = -\frac{b_1}{2} \pm \frac{1}{2}\sqrt{4\,C_1\frac{1+k_1}{k_3-1} + b_1{}^2}$$

$$\lambda = -\frac{b_1}{2} \pm \frac{1}{2}\sqrt{4\,C_1\frac{k_1-1}{k_3+1} + b_1{}^2}$$

The four values come about because k_1 and k_3 range from 0 to ∞ Evidently setting $k_3 = \pm 1$ would result in an unstable response.

Essentially need the C_1 term to be negative. C_1 itself must be positive so k_1 must be less than 1 and k_3 greater than 1. This strict arrangement is of limited value.

2.3 A Simplified Identical Bilateral Telemanipulator

A further simplification is to make the master and slave identical so that the controllers c_1 and c_2 are the same proportional gain, and both plants have a the same damping term. Under these conditions the eigenvalues are considerably simpler and the four values can be computed as.

The revised A matrix becomes

$$A' = \begin{pmatrix} 0 & 1 & 0 & 0 \\ -\frac{c_1\,(k_1\,k_4-1)}{k_3\,k_4-1} & -b_1 & -\frac{c_1\,(k_2-k_4)}{k_3\,k_4-1} & 0 \\ 0 & 0 & 0 & 1 \\ -\frac{c_1\,(k_1-k_3)}{k_3\,k_4-1} & 0 & -\frac{c_1\,(k_2\,k_3-1)}{k_3\,k_4-1} & -b_1 \end{pmatrix}$$

so the eigenvalues are

$$\lambda = -\frac{b_1}{2} \pm \frac{1}{2}\sqrt{\frac{4\,c_1 + 2\,c_1\,\beta - b_1{}^2 - 2\,c_1\,k_1\,k_4 - 2\,c_1\,k_2\,k_3 + b_1{}^2\,k_3\,k_4}{k_3\,k_4 - 1}}$$

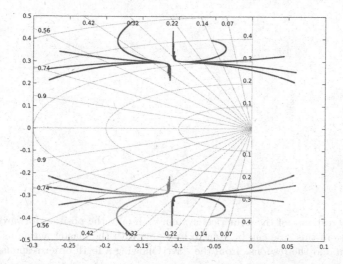

Fig. 2. Movement of eigenvalues towards the positive real axis. Test conditions are $m_1 = m_2 = 1$, $b_1 = .2$ $b_2 = .24$, $c_1 = c_2 = .1$, $k_1 = 2$, k_3 is set to the values -1.5 0 1 and 1.5. k_4 is varied between -0.1 and 0.5. Note that for values of k_3 below 1.5 the eigenvalues all move across into the positive half plane. Thus unusually a higher gain results in a more stable system, however this is at the expense of the forward position gain of the master-slave system.

where

$$\beta = \pm\sqrt{k_1{}^2\,k_4{}^2 - 2\,k_1\,k_2\,k_3\,k_4 + 4\,k_1\,k_2 - 4\,k_1\,k_4 + k_2{}^2\,k_3{}^2 - 4\,k_2\,k_3 + 4\,k_3\,k_4}$$

3   Numerical simulation

Further insight is possible by numerical simulation. It is first assumed that the master and slave are independently stable via the linear controllers c_1 and c_2. Although the numerical simulation could enforce the master and slave to have identical plant and controller gains, a slightly less restrictive condition is investigated where the master and slave are simply close (in the sense of the damping values b_1 and b_2 having similar values. The simulations are all done with k_1 set to 2 so there is a position magnification of 2.

4   Discussion

In reality exoskeletons and telerobotics are non-linear thus further complicating the stability analysis. Local stability can be considered by linearising around a set of operating points however it is unlikely that a completely general stability condition can be set, in particular once other nonlinear effects start to manifest, in particular the discontinuous forces that result from the slave making contact

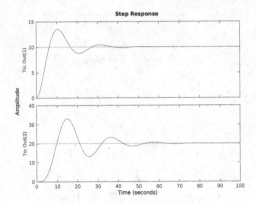

Fig. 3. Step response of the numerical system with only the position forward gain k_1. The response in this case can be seen to be stable with the master acting as a second order system and the response from the slave following a forth order response.

with the environment. It is possible that gain scheduling via Linear parameter-varying control, or passivity estimators [2] may allow changing the gains to ensure stability of complete system across all operation modes (Fig. 3).

5   Conclusions

This paper outlines the control considerations for a bilateral force reflecting feedback mechanisms, in particular the eigenvalues of a simple telemanipulator with no attempt to convey slave forces back to the operator, and a force feedback approach that uses the controller error term as an indication of impedance. The paper outlines a convenient state-space representation of a simple master-slave telemanipulator that facilitates analysis.

Acknowledgements. The authors is pleased to acknowledge the help of Gareth Barnaby and Rory Mangles who helped to highlight the structure of bilateral telemanipulation during their final year project.

References

1. Bajcsy, R., Kumar, V., Harwin, W., Harker, P.: Rapid design and prototyping of customised rehabilitation aids. Commun. ACM **39**(2), 55–61 (1996). Special Section on Computers in Manufacturing
2. Hannaford, B., Ryu, J.: Time domain passivity control of haptic interfaces. IEEE Trans. Rob. Autom. **18**(1), 1–10 (2002)
3. Lawrence, D.: Stability and transparency in bilateral teleoperation. IEEE Trans. Rob. Autom. **9**(5), 624–637 (1993)
4. Rahman, T., Sample, W., Jayakumar, S., King, M.M., et al.: Passive exoskeletons for assisting limb movement. J. Rehabil. Res. Dev. **43**(5), 583 (2006). http://www.rehab.research.va.gov/jour/06/43/5/pdf/Rahman.pdf

5. Salisbury, K., Conti, F., Barbagli, F.: Haptic rendering: introductory concepts. IEEE Comput. Graph. Appl. **24**(2), 24–32 (2004)
6. Thomas, R., Harwin, W.: Estimation of contact forces in a backdrivable linkage for cognitive robot research. In: Natraj, A., Cameron, S., Melhuish, C., Witkowski, M. (eds.) TAROS 2013. LNCS, vol. 8069, pp. 235–246. Springer, Heidelberg (2014)

Testing, Verification and Improvements of Timeliness in ROS Processes

Mohammed Y. Hazim, Hongyang Qu[✉], and Sandor M. Veres

Department of Automatic Control and Systems Engineering,
University of Sheffield, Sheffield, UK
{myhazim1,h.qu,s.veres}@shef.ac.uk

Abstract. This paper addresses the problem improving response times of robots implemented in the Robotic Operating System (ROS) using formal verification of computational-time feasibility. In order to verify the real time behaviour of a robot under uncertain signal processing times, methods of formal verification of timeliness properties are proposed for data flows in a ROS-based control system using Probabilistic Timed Programs (PTPs). To calculate the probability of success under certain time limits, and to demonstrate the strength of our approach, a case study is implemented for a robotic agent in terms of operational times verification using the PRISM model checker, which points to possible enhancements to the operation of the robotic agent.

Keywords: ROS · Verification · PTP · LISA

1 Introduction

The Robot Operating System (ROS [10]) is an open-source operating system used to develop control software for robots. It has become popular due to its capabilities in perception, object detection, navigation, etc., and the increasing demand for a uniform platform for programmable robots. The correctness of a ROS program then attracts serious attention as the deployment of ROS grows rapidly. An important way to guarantee correctness in software is formal verification and several attempts have been conducted to apply it to ROS programs, such as [4,9,11]. ROSRV in [4] is a runtime verification framework on top of ROS in order to address safety and security issues of robots. The work in [9] considered the problem of generating a platform-specific glue code for platform-independent controller code in ROS, and the code generation process is amenable to formal verification. In [11], formal verification was applied to a high-level planner/scheduler for autonomous personal robotic assistants (Care-O-bot). However, none of the attempts addresses the performance alongside the correctness of a ROS program via formal verification to ensure stringent constraints on timeliness and other properties in ROS programs. This assurance is crucial to correct system behaviour and uncertainty in their environment.

This work was supported by the EPSRC project EP/J011894/2.

© Springer International Publishing Switzerland 2016
L. Alboul et al. (Eds.): TAROS 2016, LNAI 9716, pp. 146–157, 2016.
DOI: 10.1007/978-3-319-40379-3_15

This paper is concerned with methods which can improve the performance of ROS based robot control systems. One of the difficulties in robot programming is to ensure that the robot responds to environmental challenges in a timely manner, let it be a threat approaching, to avoid something or the execution of a command which should not be delayed. Physical actions make the robot primarily depend on suitable speed of sensor signal processing, e.g., recognition and interpretation of relationships of static and moving objects in the environment, making sense of a command issued by a trusted human based on the context the robot and the human share, or planning of an action sequence to achieve a goal in a timely manner which does not render the goal outdated by the time the plan is ready, etc.

The above computational challenges are addressed in the computational processes of ROS while a number of nodes are running, each in possibly several threads that communicate with each other between nodes. Broadcast of topics often interrupts subscriber nodes, and services requested from other nodes need to be waited for in order to be able to make use the data returned. For instance, sensing and recognition by computer vision may require some fixed or variable time, depending on the significant number of objects of the environment. Discovering relationships in the environment may however take even more variable time to compute. Clearly action taking can suffer delays as planning cannot start before relationships are modelled. We propose that improvements to ROS-based computational performance can be analysed and carried out in three phases:

1. Statistical modelling of computational times in various categories and complexities of perception (including sensing and analysis), planning and execution of planned actions.
2. Formal analysis of the statistically modelled given ROS system using probabilistic timed programs (PTPs) [2] by answering PCTL queries on unacceptable delays in computation in operations by model checker PRISM [7].
3. Revision of procedures used in the ROS system to reduce the chance of computational delays.

In this paper, we first design a ROS system in a rational agent framework LISA (Limited Instruction Set Architecture) [5], which is based on AgentSpeak expansions such as Jason and Jade, with more focusing on external planning process, abstraction from planning and optimisation from decision making. The LISA model is then compiled into a PTP model for the formal analysis.

The structure of the paper is as follows. Section 2 gives a brief introduction to ROS and Sect. 3 provides a solid definition for modelling a ROS system. We present the details of collecting timing statistics of a ROS system in Sect. 4 and the design of LISA in Sect. 5. Translation from a LISA model into a PTP model is defined in Sect. 6. We conduct a case study to demonstrate the usage of our method in Sect. 7 and conclude the paper in Sect. 8.

2   The Robot Operating System

ROS is not a traditional Operating System. Rather it provides a structured communications layer in which individual processes can interact [10]. It simplifies the task of programming robots by providing a robust framework where the designer is provided a declarative programming environment for parallel computational processes of a robot. A ROS implementation of a robotic software has three typical components:

- *Nodes* - Nodes are basic processes that perform the sensing, computation and control tasks. Typically each node can contain several computational threads, although it may have additional sub-threads which the programmer is responsible for designing. Typical systems are formed from many nodes, each of which does a portion of the overall task.
- *Services* - Services provide a strict communication model where there is an established request and response message between two nodes. In a process similar to web services, a node may subscribe and subsequently request information via a service and then be supplied back with the information on demand.
- *Topics* - In order to publish messages any node can establish a topic and publish messages to it, as and when necessary. Any other node within the network may also publish to this topic. In order to receive messages, the other nodes may subscribe, wherein they can receive any message sent via a callback. A topic is a broadcast messaging stream and so does not provide any synchronous message transfer.

A fundamental difference between services and topics is that services are requester/receiver initiatied while topics are sender/provider initiated and the receivers are immediately notified, asynchronously. Both are however many-to-many communications as there can be several providers and receivers of any service or topic. Topics are inefficient when a node only needs some data from another node occasionally, when it needs it; while services are inefficient when a node needs some data supplied on a continuous, "as soon as possible" basis, though asynchronously. In their own way both are efficient ways to comunicate for different purposes. Care needs to be taken however that a subscriber to a topic does not receive more data than it needs as otherwise it is wasting its computational resources on handling redundant messages from the topic. For instance sensor messages are to be published to a topic only with a frequency which is needed by other nodes, thereby resulting in less latency than if a service were doing the same job.

3   Mathematical Model of a ROS Package

One way to describe a ROS based system is a tri-partite graph with vertices for nodes, topics and services. These vertex types are not interchangeable in graph matching algorithms. New topics and services can be easily introduced that can allow reconfiguration of the system to provide agents with the information they required, albeit sourced from different locations. All node communication must occur through topics or services.

Definition 3.1. A ROS-graph is $G = (N, T, S, E, D, C, X, \lambda)$, where N are the set of vertices representing ROS nodes, T are a set of topics and S are a set of services, C is a partially order set of object classes and X is a set of labels to name all vertices. $E \subset (N \times T) \cup (T \times N) \cup (N \times S) \cup (S \times N)$ is a set of directed edges to represent publishing of, and subscription to, topics and provision of, and subscription to, services, respectively. $D : E^- \to C^*$, $E^- = T \cup (N \times S) \cup (S \times N)$, is a data descriptor function where C^* is a notation for finite sequences of entries from the set of data object classes C, which are used in services and topics to send information between nodes. Each of N, T, S are labelled by a surjective labelling function $\lambda : N \cup T \cup S \to X$.

A ROS system enables the nodes to advertise or use services, and to publish or subscribe to topics. G represents the maximum ability of the robot when the system has all nodes, topics and services nominally functioning. If some nodes are not available due to sensor, actuator or computational hardware breakdown, then G needs sufficient redundancy to enable continued functioning of the robot or at least some of its functionality. The ROS graph G defines all the possible data flows for sensor readings, signal processing and control action in the environment. A detailed description is not within the scope of this paper and we refer the reader to [1].

4   Statistics of ROS Nodes

When ROS based robot control system's programming is completed, the robot is ready to be tested in a series of scenario tests. Performance may not acceptable due to a few factors:

1. When a plan of an agent is triggered due to environmental change the computational times of perception modelling and planning are excessive and delay action taking in some environmental scenarios.
2. In some environmental scenarios scene interpretation and planning is several times faster than typical response time requires. The question arises whether more complex model of the scene could have been built to more fully grasp an environmental situation.

Overall the performance problem of the robot is to discover scenarios which are not favourable for the robots computational system. These are searched and synthesised based on sensor and perception statistics derived in practical use of the ROS system. This section provides a formal model of statistical estimation of computation and communication times in a given ROS system already operating on a hardware platform. Consequent application of probabilistic model checking can guide us to introduce improvements in the choice of computational processes involved in reasoning.

4.1   Performance Evaluator Node

To estimate the processing and communication time across the ROS system and additional runtime statistics node Σ can be introduced, which collects runtime data from all the robots functional nodes. Each of the functional nodes i has a data array D_i recording timed-performance of services and topics in the node. Let denote $s_k \in S$ a service in a ROS-graph $G = (N, T, S, E, D, C, X, \lambda)$. The following timed data are recorded about a service call.

1. When a request is to be made from node j for service s_k, then a data entry $(n_j \xrightarrow{\text{req}} s_k, t^j)$ is added to D_j just before the service command is issued from node j to node i with time stamp t^j in node j.
2. Upon request, and before any execution of service actions, a data entry $(n_j \xrightarrow{\text{req}} s_k, t^i)$ is added to D_i with time stamp t^i in node i.
3. Upon completion of the computational processes or physical controls performed, a data entry $(n_j \xrightarrow{\text{ans}} s_k, t^i)$ is added to D_i with time stamp in node i.
4. Upon answer data received in node j for service s_k, then a data entry $(n_j \xrightarrow{\text{ans}} s_k, t^j)$ is added to D_j with time stamp t^j in node j.

For topics recording of runtime data is slightly different:

1. When a topic is to be published by node j for topic p_k, then a data entry $(n_j \xrightarrow{\text{pub}} p_k, t^j)$ is added to D_j just before the topic broadcast is issued from node j with time stamp t^j in node j.
2. Upon receiving the broadcast, and before any execution of actions due to the topic broadcast, a data entry $(n_j \xrightarrow{\text{rec}} p_k, t^i)$ is added to D_i with time stamp t^i in node i.
3. Upon completion of the computational processes or physical controls performed, a data entry $(n_j \xrightarrow{\text{top}} s_k, t^i)$ is added to D_i with time stamp in node i.

Note that there are other ways to collect statistics on execution time and latency, such as in [3], but our method suits our need better because it does not depend on the header of messages, which is not always available.

4.2   Estimation of Operations

From each node i the data containers D_i are sent to the runtime statistics node Σ, which can compute the following amongst others:

- Probability distribution of the request communication times $t^i - t^j$ from $(n_j \xrightarrow{\text{req}} s_k, t^j)$ and $(n_j \xrightarrow{\text{req}} s_k, t^i)$.
- Probability distribution of the service execution times $t^{si} - t^{ei}$ f from $(n_j \xrightarrow{\text{req}} s_k, t^{si})$ and $(n_j \xrightarrow{\text{ans}} s_k, t^{ei})$.
- Probability distribution answer communications times $t^j - t^i$ from $(n_j \xrightarrow{\text{ans}} s_k, t^j)$ and $(n_j \xrightarrow{\text{ans}} s_k, t^i)$.

- Probability distribution of communication broadcast times $t^i - t^j$ from of $(n_j \xrightarrow{\text{pub}} p_k, t^j)$ and $(n_j \xrightarrow{\text{rec}} p_k, t^i)$.
- Probability distribution of topic interruption times , $t^{si} - t^{ei}$ from $(n_j \xrightarrow{\text{rec}} p_k, t^{si})$ and $(n_j \xrightarrow{\text{top}} s_k, t^{ei})$.

Performance tuning of a ROS based computational system is carried out iteratively through a series of trial runs, during which the average runtime probabilities (or conditional runtime probabilities of the duration events are evaluated), followed by a ROS system. This is followed by algorithmic adjustments made to the ROS system and the iteration continues by another trial run. The series of iterations consisting of (1) trial-run (2) compilation to PRISM model (3) running of PCTL queries (4) algorithmic amendments are cyclically repeated on the ROS system until satisfactory computational performance is achieved.

5   A Rational Agent Framework LISA

Comparing with Jason in terms of plan selection function, LISA [5] proved to enhance the architecture with a runtime probabilistic model checking by predicting the outcomes of applicable plan and selections. The LISA structure is simpler than its predecessors and can easily lend itself to design time and runtime verification. Now we give the detail about LISA. By analogy to previous definitions [8,12] of AgentSpeak-like architectures, we define our agents as a tuple: $\mathcal{R} = (\mathcal{F}, B, L, \Pi, A)$, where:

- $\mathcal{F} = \{p_1, p_2, \ldots, p_{n_p}\}$ is the set of all predicates.
- $B \subset \mathcal{F}$ is the total set of belief predicates. The current belief base at time t is defined as $B_t \subset B$. Beliefs that are added, deleted or modified can be either called *internal* or *external* depending on whether they are generated from an internal action, in which case are referred to as "mental notes", or from an external input, in which case they are called "percepts".
- $L = \{l_1, l_2, \ldots l_{n_l}\}$ is a set of logic-based implication rules.
- $\Pi = \{\pi_1, \pi_2, \ldots, \pi_{n_\pi}\}$ is the set of executable plans or *plans library*. Current applicable plans at time t are part of the subset applicable plan $\Pi_t \subset \Pi$ or "desire set".
- $A = \{a_1, a_2, \ldots, a_{n_a}\} \subset \mathcal{F} \setminus B$ is a set of all available actions. Actions can be either *internal*, when they modify the belief base or data in memory objects, or *external*, when they are linked to external functions that operate in the environment.

AgentSpeak like languages, including LISA, can be fully defined and implemented by specifying initial beliefs and actions, and reasoning cycles:

- *Initial Beliefs.* The initial beliefs and goals $B_0 \subset F$ are a set of literals that are automatically copied into the *belief base* B_t (that is the set of current beliefs) when the agent mind is first run.

– *Initial Actions.* The initial actions $A_0 \subset A$ are a set of actions that are executed when the agent mind is first run. The actions are generally goals that activate specific plans.

The following operations are repeated for each reasoning cycle in AgentSpeak.

– *Maintenance of Percepts.* This means generation of perception predicates for B_t and data objects such as the world model used here W.
– *Logic rules.* A set of logic based implication rules L describes *theoretical* reasoning to improve the agent current knowledge about the world.
– *Executable plans.* A set of *executable plans* or *plan library* Π. Each plan π_j is described in the form: $p_j : c_j \leftarrow a_1, a_2, \ldots, a_{n_j}$, where $p_j \in B$ is a *triggering predicate*, which allows the plan to be retrieved from the plan library whenever it comes true, $c_j \in B$ is a logic formula of a *context*, which allows the agent to check the state of the world, described by the current belief set B_t, before applying a particular plan sequence $a_1, a_2, \ldots, a_{n_j} \in A$ with a list of actions. Each a_j can be one of (1) predicate of an external action with arguments of names of data objects, (2) internal (mental note) with a preceding $+$ or $-$ sign to indicate whether the predicate needs to be added or taken away from the belief set B_t (3) conditional set of items from (1)–(2). The set of all triggers p_j in a program is denoted by E_{tr}

LISA enhanced the above reasoning cycle to allow multiple actions to be executed in parallel. The enhanced reasoning cycle consists of the following steps:

1. *Belief base update.* The agent updates the belief base by retrieving information about the world through perception and communication. Adding and removing beliefs from the belief base is carried out by the function *Belief Update Function (BUF)*.
2. *Application of logic rules.* The logic rules in L are applied in a round-robin fashion (restarting at the beginning of the list) until there are no new predicates generated for B_t. This means that rules need to be verified not to lead to infinite loops.
3. *Trigger Event Selection.* For every reasoning cycle a function called *Belief Review Function* $S_t : \wp(B_t) \rightarrow \wp(E_t)$ selects the current event set E_t, where $\wp(\cdot)$ is the so called *power operator* and represents the set of all possible subset of a particular set. We call the current selected trigger event $S_t(B) = T_t$ and the associated plans the *Intention Set*.
4. *Plan Selection.* All the plans in T_t are checked for their context to form the *Applicable Plans* set Π_t by function $S_O : E_t \rightarrow \wp(S_t)$. We will call the current selected plan $S_O(\Pi_t) = \pi_t$.
5. *Plan Executions.* All plans in $S_O : E_t$ are started to be executed concurrently by going through the plan items $a_1, a_2, \ldots, a_{n_j}$ one-by-one sequentially.

6   Modelling of Agent Operational Times in PRISM

In this section we assume that the response of the physical environment of the agent is modelled as a probabilistic timed program (PTP) E in terms of the predicates feed back to the belief base of the agent under various environmental states. E is composed of environmental states, and transitions which under each state through the conditional probabilities of the environment corresponds to triggering of predicates through the sensor system of the robotic agent. Given that the agent has well defined decision structures as described in the previous subsection, the environment-agent model will also be a PTP. This section describes how the combination of probability distributions, which were estimated in the previous section, when combined with the environmental PTP and the logic based decision making of the agent, can be modelled in PRISM.

6.1 Probabilistic Timed Programs (PTP)

Probabilistic timed programs [6] are an extension of Markov Decision Processes (MDPs) with state variables and realtime clocks. We only give a brief introduction to PTP in this paper. The complete definition can be found in [2,6].

Given a set $\mathcal{V}$ of variables, let $Asrt(\mathcal{V})$, $Val(\mathcal{V})$ and $Assn(\mathcal{V})$ be a set of *assertions*, *valuations* and *assignments* over $\mathcal{V}$ respectively. Given a set S, let $\mathcal{P}(S)$ be the set of subsets of S and $\mathcal{D}(S)$ the set of discrete probability distributions over S. A set $\mathcal{X}$ of *clock* variables represents the time elapsed since the occurrence of various events. The set of *clock valuations* is $\mathbb{R}^{\mathcal{X}}_{\geq 0} = \{t : \mathcal{X} \rightarrow \mathbb{R}_{\geq 0}\}$. For any clock valuation t and any $\delta \geq 0$, the *delayed* valuation $t + \delta$ is defined by $(t + \delta)(x) = t(x) + \delta$ for all $x \in \mathcal{X}$. For a subset $Y \subseteq \mathcal{X}$, the valuation $t[Y := 0]$ is obtained by setting all clocks in Y to 0: $t[Y := 0](x)$ is 0 if $x \in Y$ and $t(x)$ otherwise. A (convex) *zone* is the set of clock valuations satisfying a number of clock difference constraints, i.e. a set of the form: $\rho = \{t \in \mathbb{R}^{\mathcal{X}}_{\geq 0} \mid t_i - t_j \lesssim b_{ij}\}$. The set of all zones is $Zones(\mathcal{X})$.

Definition 1 (PTP). *A PTP is a tuple* $P = (L, l_0, \mathcal{X}, \mathcal{V}, v_i, \mathcal{I}, \mathcal{T})$ *where:*

- L *is a finite set of* locations *and* $l_0 \in L$ *is the* initial location;
- $\mathcal{X}$ *is a finite set of* clocks *and* $\mathcal{I} : S \rightarrow Zones(\mathcal{X})$ *is the* invariant condition;
- $\mathcal{V}$ *is a finite set of* state variables *and* $v_i \in Val(\mathcal{V})$ *is the* initial valuation;
- $\mathcal{T} : S \rightarrow \mathcal{P}(Trans(L, \mathcal{V}, \mathcal{X}))$ *is the* probabilistic transition function, *where* $Trans(L, \mathcal{V}, \mathcal{X}) = Asrt(\mathcal{V}) \times Zones(\mathcal{X}) \times \mathcal{D}(Assn(\mathcal{V}) \times \mathcal{P}(\mathcal{X}) \times L)$.

A step from a state (l, v, t) consists of the elapse of a certain amount of time $\delta \in \mathbb{R}_{\geq 0}$ followed by a *transition* $\tau = (\mathcal{G}, \mathcal{E}, \Delta) \in \mathcal{T}(l)$. The transition comprises a *guard* $\mathcal{G} \in Asrt(\mathcal{V})$, *enabling condition* $\mathcal{E} \in Zones(\mathcal{X})$ and probability distribution $\Delta = \lambda_1(f_1, r_1, l_1) + \cdots + \lambda_k(f_k, r_k, l_k))$ over triples containing an *update* $f_j \in Assn(\mathcal{V})$, clock resets $r_j \subseteq \mathcal{X}$ and *target location* $l_j \in L$.

The delay δ must be chosen such that the invariant $\mathcal{I}(l)$ remains continuously satisfied; since $\mathcal{I}(l)$ is a (convex) zone, this is equivalent to requiring that both t and $t + \delta$ satisfy $\mathcal{I}(l)$. The chosen transition τ must be *enabled*, i.e., the guard $\mathcal{G}$ and the enabling condition $\mathcal{E}$ in τ must be satisfied by v and $t + \delta$, respectively. Once τ is chosen, an assignment, set of clocks to reset, and successor location are selected at random, according to the distribution Δ in τ.

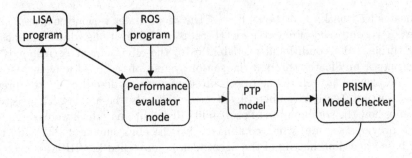

Fig. 1. The verification process

6.2   Performance Queries

Given a PTP, we can use the following PCTL queries to check its properties:

- $P_{\bowtie=?}[F\ a]$,
- $P_{\bowtie=?}[F_{\leq T}\ a]$,

where $\bowtie\in\{max,\ min\}$, a is Boolean expression that does not refer to any clocks and T is an integer expression. The first query asks what is the maximum/minimum probability that a is satisfied, and the second one inquires the probability that a can be satisfied within time bound T. Based on these queries, we can compute the maximum/minimum probability of all target states that satisfy a without time limit or within a bound T. For example, we can ask what is the minimum probability for a robot moving to a specific location within certain time. A concrete example will be shown in the next section.

6.3   Verification Process

Figure 1 illustrates the whole process in our method. A system is first written in LISA and then translated into ROS. A performance evaluator node is generated for this system. After the evaluator node collects sufficient statistics on the time delay, it computes the probability distribution. A PTP model is then constructed using this information and the LISA program, although it is feasible to build the PTP model from the ROS program directly. The reason that we build the PTP model from the LISA program is that it provides a high level abstraction of the system, which can make the PTP model compact. The PTP model is fed to PRISM for verification. The result is then used as a reference when improving the design of the ROS program.

7   Case Study

In this section we demonstrate the strength of our approach using the following scenario. An autonomous ground vehicle (AGV) is exploring a remote area with a vision system consisting two cameras (primary and secondary camera). The system merges two images, one from each camera, to look for an object in the area. Here we are mainly interested in two ROS nodes: one for receiving images from the cameras and the other for processing these images. The statistics shows that time for receiving one image respect the following probability distribution:

- With probability 0.3, it take less than 4 units of time, but more than 3 units to receive one image;
- With probability 0.6, the receiving time locates in the interval $(4, 6)$;
- With probability 0.1, the receiving time locates in $(6, 8)$.

It takes less than 16 units of time but more than 12 units to process two images, and the probability of successfully finding the object in the images is 0.91. When the system fails to find the object, it will take two new images from the cameras and repeat the process.

Figure 2 illustrates the PTP model for the system, where x is a clock, which is used to count the time elapse for each step. The timing constraints in a node (which represents a state), such as $x < 4$, is the upper bound and the constraints on an edge (which represents a transition), such as $x > 3$, is the lower bound. This figure shows that the system receives the image from the primary camera first (states s_1, s_2 and s_3), and then receives the one from the secondary camera (states s_4, s_5 and s_6). In state s_7, the system processes the images. We can ask a query that at what probability the system successfully find the object within 35 units of time, which can formulated in PCTL as follows:

$$P_{max=?}[F_{\leq 35} \text{ "Success"}]. \tag{1}$$

The result returned by PRISM is 0.91. One problem in this system is that it has to wait for two images before it can start to look for the object. If the image processing and receiving can be performed in parallel by different hardware, we may be able to increase the performance of the system, which is possible if the object can be found from one image, even if at a lower probability, e.g., 0.7. One

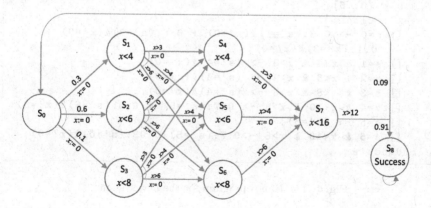

Fig. 2. The PTP for the system

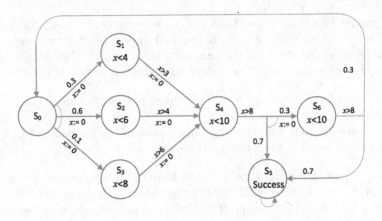

Fig. 3. The PTP for the new system

way to achieve it as follows. The system starts to process the first image immediately after it arrives. Here we assume that processing one image is between 8 and 10 units. As the processing time exceeds the time required for receiving an image, the system does not need to wait once it finishes processing the first image. Instead, it can immediately process the second images. Although it is slightly slower to process the images separately than processing them altogether, eliminating the waiting time for the second image makes the system able to receive more images within the time limit and thus, find the object at higher probability. Figure 3 illustrates the improved system design. The result for the query in Equation (1) is 0.9724, which shows a big improvement from the previous design. Figure 4 shows the PRISM program for this improved system.

```
1   pta
2   module M
3     s : [0..6];
4     x : clock;
5     [] s=0 -> 0.3:(s'=1)&(x'=0) + 0.6:(s'=2)&(x'=0) +
          0.1:(s'=3)&(x'=0);
6     [] s=1 & x<4 & x>3 -> (s'=4)&(x'=0);
7     [] s=2 & x<6 & x>4 -> (s'=4)&(x'=0);
8     [] s=3 & x<8 & x>6 -> (s'=4)&(x'=0);
9     [] s=4 & x<10 & x>8 -> 0.7:(s'=5) + 0.3:(s'=6)&(x'=0);
10    [] s=5 -> (s'=5);
11    [] s=6 & x<10 & x>8 ->0.7:(s'=5)+0.03:(s'=0)&(x'=0);
12  endmodule
```

Fig. 4. The PRISM program for the new system.

8 Conclusions

This paper presented a method for formal verification of timeliness properties of robots implemented in ROS. The LISA framework was used to design a robotic agent as LISA provides a solution for the verification of robotic agents through the PRISM model checker. Statistical estimation was applied to robot operations under the ROS system to detect and collect information about the latency in the system. The LISA model was then associated with runtime probabilities and translated into a PTP model and verified in PRISM. It has been illustrated how to apply the methods to improve the design of a ROS system in a case study.

In the future we intend to bring the methods nearer to industrial applicability by improving their timing performance analysis, which might require the development of more efficient model checking algorithms for PTPs in the case of very large models. Another direction of future work is to search for other modelling formalisms, which can handle continuous probability distributions on timing variances, as PTP can only deal with discrete probability distributions.

References

1. Aitken, J.M., Veres, S.M., Judge, M.: Adaptation of system configuration under the robot operating system. In: Proceedings of the 19th World Congress of the International Federation of Automatic Control (2014)
2. Dräger, K., Kwiatkowska, M.Z., Parker, D., Qu, H.: Local abstraction refinement for probabilistic timed programs. Theor. Comput. Sci. **538**, 37–53 (2014)
3. Forouher, D., Hartmann, J., Maehle, E.: Data flow analysis in ROS. In: Proceedings of the 41st International Symposium on Robotics, pp. 1–6. VDE (2014)
4. Huang, J., Erdogan, C., Zhang, Y., Moore, B., Luo, Q., Sundaresan, A., Rosu, G.: ROSRV: runtime verification for robots. In: Bonakdarpour, B., Smolka, S.A. (eds.) RV 2014. LNCS, vol. 8734, pp. 247–254. Springer, Heidelberg (2014)
5. Izzo, P., Qu, H., Veres, S.M.: Reducing complexity of autonomous control agents for verifiability. arXiv:1603.01202 [cs.SY], March 2016
6. Kwiatkowska, M., Norman, G., Parker, D.: A framework for verification of software with time and probabilities. In: Chatterjee, K., Henzinger, T.A. (eds.) FORMATS 2010. LNCS, vol. 6246, pp. 25–45. Springer, Heidelberg (2010)
7. Kwiatkowska, M., Norman, G., Parker, D.: PRISM 4.0: verification of probabilistic real-time systems. In: Gopalakrishnan, G., Qadeer, S. (eds.) CAV 2011. LNCS, vol. 6806, pp. 585–591. Springer, Heidelberg (2011)
8. Lincoln, N.K., Veres, S.M.: Natural language programming of complex robotic BDI agents. Intell. Robotic Syst. **71**(2), 211–230 (2013)
9. Meng, W., Park, J., Sokolsky, O., Weirich, S., Lee, I.: Verified ROS-based deployment of platform-independent control systems. In: Havelund, K., Holzmann, G., Joshi, R. (eds.) NFM 2015. LNCS, vol. 9058, pp. 248–262. Springer, Heidelberg (2015)
10. Quigley, M., Conley, K., Gerkey, B.P., Faust, J., Foote, T., Leibs, J., Wheeler, R., Ng, A.Y.: ROS: an open-source robot operating system. In ICRA Workshop on Open Source Software, vol. 3 (2009)
11. Webster, M., Dixon, C., Fisher, M., Salem, M., Saunders, J., Koay, K., Dautenhahn, K.: Formal verification of an autonomous personal robotic assistant, pp. 74–79. AAAI (2014)
12. Wooldridge, M.: An Introduction to MultiAgent Systems. Wiley, Chichester (2002)

The Psi Swarm: A Low-Cost Robotics Platform and Its Use in an Education Setting

James Hilder[✉], Alexander Horsfield, Alan G. Millard, and Jon Timmis

York Robotics Laboratory, University of York, Heslington YO10 5DD, UK
james.hilder@york.ac.uk
http://www.york.ac.uk/robot-lab

Abstract. The paper introduces the Psi Swarm robot, a platform developed to allow both affordable research in swarm robotics and versatility for teaching programming and robotics concepts. Motivated by the goals of reducing cost and construction complexity of existing swarm platforms, we have developed a trackable, sensor-rich and expandable platform which needs only a computer with internet browser to program. This paper outlines the design of the platform and the development of a tablet-computer based programming environment for the robot, intended to teach primary school aged children programming concepts.

Keywords: Swarm robotics · Educational robotics · Teaching · Android

1 Introduction

Simple robotic platforms are ideal tools for teaching a wide range of science and engineering disciplines, encompassing electronics, embedded programming, control theory, signal processing and energy management. When the scope of the individual robot is expanded to allow communication and interaction with other similar robots, their potential value as a tool for research expands to allow investigation of swarm intelligence: the study of complex swarm behaviours and emergent behaviours.

The motivation for this research was to create a new platform to replace a similar robot previously developed by the authors (the Pi Swarm robot [2]) with the goal of reducing cost, improving functionality and specifically enhancing its strengths as a versatile robot for use in a classroom environment. It was desired to have a robotic platform flexible enough to allow the tuition of core robotic disciplines to a wide range of age groups, whilst being scalable enough to permit large swarm research to also take place. Such platforms often come at a cost which is prohibitive to education and research, both in pure monetary terms, and also in the context of the time needed to learn how to use and program a system, and regarding general maintenance such as recharging and repairing.

© Springer International Publishing Switzerland 2016
L. Alboul et al. (Eds.): TAROS 2016, LNAI 9716, pp. 158–164, 2016.
DOI: 10.1007/978-3-319-40379-3_16

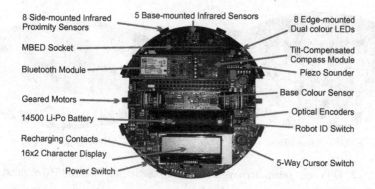

8 Side-mounted Infrared Proximity Sensors

5 Base-mounted Infrared Sensors

8 Edge-mounted Dual colour LEDs

MBED Socket

Bluetooth Module

Geared Motors

14500 Li-Po Battery

Recharging Contacts

16x2 Character Display

Power Switch

Tilt-Compensated Compass Module

Piezo Sounder

Base Colour Sensor

Optical Encoders

Robot ID Switch

5-Way Cursor Switch

Fig. 1. Photo of the main Psi Swarm PCB with additional modules

2   Hardware Design

The Psi Swarm robot is based around a dual-layer 108 mm diameter PCB, shown annotated in Fig. 1. A pair of 40-pin sockets allow for a MBED LPC1768 rapid prototyping board to be attached, which functions as the main controller for the robot. The MBED is a small module based around an ARM Cortex-M3 microcontroller. One of the core features of the MBED board is it is designed for use with a special set of online compiler tools, which allow for the development of programs from any machine with an internet connection and compatible browser; this eliminates the need for purchasing and installing a specific tool chain[1].

2.1   Sensors

A set of 8 infrared (IR) sensors equally spaced around the edge of the robot can be used to detect obstacles and other robots. Based around a Vishay TCRT1000 emitter:phototransistor pair connected to a 10-bit ADC, they can detect white surfaces up to approximately 15 cm and estimate the distance to that surface. On the base of the robot there are 5 additional IR sensors that are primarily designed to allow the robot to discriminate between light and dark surfaces allowing for tasks such as line-following.

Provisions are made on the PCB to allow a number of other sensors to be attached. These include the AMS TCS3472x series colour sensors, which can be attached to the base of the robot in a backlit configuration, and also facing up on the robot unlit, allowing the robot to discriminate between different zones of colour either on the arena surface or projected onto the robot from above - the latter configuration allowing dynamically changing arenas to be easily created for specific experiments. An SRF02 ultrasonic module can attach to the front of the robot, which allows obstacles to be detected between 15 cm and 200 cm from the robot, as seen in Fig. 2a. A socket for a CMPS11 tilt-compensated

---

[1] Projects created in the online tools can also be exported to a number of different tool-chains, such as Keil μVision and GCC if desired.

(a) SRF02 Ultrasonic sensor   (b) Optitrack tracking hull

Fig. 2. Different setup arrangements and plastic hulls for the Psi Swarm

compass module, which combines a 3-axis gyroscope, 3-axis accelerometer and a dedicated microcontroller to perform the Kalman filtering of raw data to return a compensated bearing is provided. Optical wheel encoders can be attached using special daughterboards which can provide up to 2 degree precision for each wheel. The robot also includes a precision digital temperature sensor, current measurement and voltage measurement. The use of existing modules for many of the sensors helps to simplify construction, minimise processing overheads and reduce costs as only the required modules need to be attached.

2.2   Actuators

The drive for the robot is provided by a pair of micro-metal geared motors attached to 32 mm diameter wheels with rubber tyres. Different gearboxes are available, with 102:1 being the default ratio, which provides a top speed of approximately $0.25\,ms^{-1}$. The motors each have dedicated regulated power supplies and are driven by PWM control using H-Bridge drivers, ensuring that speed remains constant as the battery discharges.

The robot also has a 16×2 character display which is useful for providing user feedback when testing algorithms, a set of 8 dual-colour LEDs around the perimeter of the robot, a high-brightness dual-colour LED in the center of the robot facing upwards and a pair of white LEDs on the robot to provide backlighting for the colour sensor. A piezo sounder allows for simple tunes and other sounds to be generated. A 5-direction cursor switch can be used for user input, and a 4-way DIL switch is provided to allow a robot ID to be set in hardware[2].

2.3   Power

The robot is powered using a single 14500-sized Lithium Polymer cell. These cells, similar in size to AA cells, typically provide 800 mAH of charge at 3.7 V

---

[2] This allows swarms of up to 15 to be simply set in hardware; larger swarms can be created by using the on-board EEPROM to store additional bits of index information.

Fig. 3. A swarm of Psi Swarm robots on a charging table with IR beacon

nominal voltage. The battery supplies power to a number of voltage regulation circuits to separate supplies for the motors, sensors and the MBED. Unless low-power states are used, the MBED will typically draw 250 mA at 5 V, as such the battery life for the robot varies between approximate 30 min at full-load[3] to approximately 120 min at idle.

One of the design goals for the robot was to allow for long-term swarm experiments to be run. The robots contain a recharging circuit which allows the battery to be charged and the robot to remain powered when a 6 V DC supply is provided to two contacts on the base of the robot. This supply is rectified such that the polarity is not important. Utilising this design, arenas can be created that either use bus-strips (as seen in Fig. 3) or chequerboard arrangements of power pads on the surface that allow the robot to recharge.

2.4 Construction

A set of 3D-printed parts form a protective sandwich around the main PCB, providing a reflective surface to allow robots to sense other robots. On the base of the robot, a pair of ball-bearings minimise wobble whilst allowing the robot to negotiate uneven surfaces. Various shells have been designed to allow different sensor configurations and tracking systems to be accommodated, including AprilTags [4] and the Optitrack system [3] as seen in Fig. 2b.

3 Software Design

The API for the platform is written in C++ and is designed for use with the MBED online compiler. Functions to simplify interaction with all the available sensors and actuators are written. The API includes a built-in demonstration mode, in which various testing routines and simple code-demos can be activated through a menu-based system when the robot is turned on. A simulator model for the robot using the ARGoS simulation package is currently in development [5].

---

[3] Full load considered to be all LEDs on, IR sensors active at 10 Hz, Bluetooth active and motors at 100% on a smooth surface.

A set of API functions work in conjunction with a special IR beacon (seen in Fig. 3) that emits 25 mS pulses of light at 4 Hz. Swarms of robots can synchronise with this beacon, and based on relative sensor strength estimate the heading of the beacon. The robots then use an ID-based TDMA system to ensure that only one robot is emitting IR at any one time. Using this system it is possible for swarms of up to 15 robots to can approximate range and bearing measurements for every other robot within line of sight.

3.1 Communication

The robot has two different RF-based communication modules available. A BlueSMIRF Bluetooth to RS-232 module can be attached which allows simple interaction between the robots and a host computer or other device. Inter-swarm communication is possible using the Bluetooth interface, or a 433 MHz transceiver module, which allows for low bitrate communications between robots using code previously developed for the Pi Swarm robot. The API includes functions that allow instructions to be sent to the robots using either communication interface. The functions allow sensor data to be communicated between robots and a computer and allow remote control of the robot hardware. The API also includes functionality to allow file-handling of the flash memory on the MBED using the Bluetooth interface; it is possible to send updated controller binaries to multiple robots at once using this method, which allows large swarms to be reprogrammed wirelessly.

4 Case Study: The Psi Swarm as an Education Platform

As of September 2014, there has been a dramatic change in the English national curriculum, shifting the focus from Information, Communication and Technology (ICT) to Computing. Where previously teaching has been focussed on office based skills such as Microsoft PowerPoint and Access, they will now be educated in programming skills and fundamentals such as algorithm design and logical flow [1]. Teachers are being encouraged to experiment with different possible teaching methods and materials to aid in the choice of a more standardised approach. To take advantage of this renewed interest in computing, we have developed an application that can educate primary school students in programming and logical flow through the control of a Psi Swarm robot using a Android based tablet.

4.1 Background

We have focussed on Key Stage 2 (pupils aged 7 to 11) and the following 4 learning goals outlined in the curriculum:

1. Design, write and debug programs that accomplish specific goals
2. Solve problems by decomposing them into smaller parts

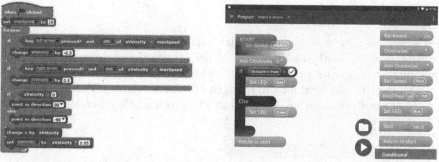

(a) Scratch Language   (b) Screenshot from Android Application

Fig. 4. Screenshots from Scratch and the Android Tablet Application

3. Use sequence, selection, and repetition in programs
4. Use logical reasoning to explain how some simple algorithms work

There are a number of existing projects for the education of programming skills, ranging from the older packages such as Logo to more recent examples such as the BBC Micro:Bit. Logo, a LISP based language, developed by MIT in 1967, was designed to allow children to learn to program by controlling the movement of a robot called a Turtle. Scratch, a graphical programming suite that allows the user to define the activity of a program using a set of pre-made building blocks, allows students to learn programming using the concepts of Logo while making interaction more media oriented to make it more appealing [6].

4.2 Android Application

Based on the basic design concepts of Scratch we have developed an Android-based tablet application utilising an intuitive drag-and-drop user interface; examples of both are shown in Fig. 4. These concepts include the use of colour codes and shaped blocks that have distinct areas to snap together, allowing the students to focus on the function and logic of the code without having to be concerned about syntax.

The user generated code can be saved and loaded on the tablet, and multiple methods to run the code on the robots are implemented. The tablet application can connect with the robot using Bluetooth and send program instructions directly, allowing for the quick testing and step-through debugging of programs. Test lessons for Year 3 and a Year 6 classes have been designed. The students are given a target course, printed onto an A1 acrylic-sheet, which contains a start location for the robot, a target location and a path they should try to follow. The younger students are instructed to move the robot from the start to the target using combinations of turtle-style commands. The older students are to use conditional statements and sensor data from the line-following IR sensors.

5 Conclusions

This paper has discussed the design and implementation of a low-cost robotics platform, designed with the dual-goals of being suitable for demonstrating programming and robotics principles to a wide-variety of age groups, whilst also being a powerful platform for the research of swarm-robotics. Further information about the Psi Swarm can be found at www.york.ac.uk/robot-lab/psiswarm.

References

1. National curriculum in England: computing programmes of study, September 2013
2. Hilder, J., Naylor, R., Rizihs, A., Franks, D., Timmis, J.: The pi swarm: a low-cost platform for swarm robotics research and education. In: Mistry, M., Leonardis, A., Witkowski, M., Melhuish, C. (eds.) TAROS 2014. LNCS, vol. 8717, pp. 151–162. Springer, Heidelberg (2014)
3. Millard, A.G., Hilder, J., et al.: A low-cost real-time tracking infrastructure for ground-based robot swarms. In: ANTS: 9th International Conference on Swarm Intelligence (2014)
4. Olson, E.: AprilTag: a robust and flexible visual fiducial system. In: Proceedings of the IEEE International Conference on Robotics and Automation (ICRA), pp. 3400–3407. May 2011
5. Pinciroli, C., Trianni, V., et al.: ARGoS: a modular, parallel, multi-engine simulator for multi-robot systems. Swarm Intell. 6(4), 271–295 (2012)
6. Resnick, M., Maloney, J., et al.: Scratch: programming for all. Commun. ACM 52(11), 60–67 (2009)

Multi Robot Cooperative Area Coverage, Case Study: Spraying

Alireza Janani[(✉)], Lyuba Alboul, and Jacques Penders

Sheffield Hallam University, Sheffield, UK
ajanani@my.shu.ac.uk, {l.alboul,j.penders}@shu.ac.uk

Abstract. Area coverage is a well-known problem in multi robotic systems, and it is a typical requirement in various real-world applications. A common and popular approach in the robotic community is to use explicit forms of communication for task allocation and coordination. These approaches are susceptible to the loss of communication signal, and costly with high computational complexity. There are very few approaches which are focused on implicit forms of communication. In these approaches, robots rely only on their local information for task allocation and coordination. In this paper, a cooperative strategy is proposed by which a team of robots perform spraying a large field. The focus of this paper is to achieve task allocation and coordination using only the robots' local information.

Keywords: Multi robotic system · Cooperative behaviour · Cooperative area coverage

1 Introduction

In area coverage, a team of robots is cooperatively trying sweep an entire area, possibly containing obstacles. The goal is to achieve coverage with efficient paths for each robot which jointly ensure that every single point in the environment is visited by at least one of the robots while performing the task [6]. Many real world applications require systematic area coverage including search in forested areas, demining, distribution of beacons and line searching. In this paper we focus on application of agricultural robotics.

Recently, there has been an increase of interest in performing agricultural tasks by a team of autonomous robots. One of the main reasons is shortage of labor force. Over the years, various approaches have been suggested to reduce the need of labour force. A conventional approach is to use larger machineries to process larger portions of the field at a time. However, deploying heavy and large machineries results in soil compaction [14]. Soil compaction has devastating outcomes on the crop, and it costs up to 90 % of the cultivation cost to recover [2].

Another trend in reducing input labour force is automation. However, single robotic approaches are expensive, and still require occasional human supervision [12]. Deploying a team of smaller and lighter agricultural machineries prevents

© Springer International Publishing Switzerland 2016
L. Alboul et al. (Eds.): TAROS 2016, LNAI 9716, pp. 165–176, 2016.
DOI: 10.1007/978-3-319-40379-3_17

soil compaction, reduces the cost of cultivation, and increases the fault-tolerance of the overall system [11]. Furthermore, multi robotic approach has the promise to reduce dependency on labour force [12].

If a team of robots is applied, the main question is how the robots should execute the task to cooperatively achieve the global goal? To answer this question, the task of spraying has to be studied in detail.

1.1 Spraying

Spraying is the process of dispensing Plant Protection Products (PPP) on the crop at different stages during cultivation. Conventionally, a tractor with spraying unit is driven throughout the field and the PPPs are gradually dispensed on the crop.

Spraying is distinguished from other agricultural tasks in that of redundancy of processing. In other agricultural tasks (e.g. ploughing, seeding, and harvesting), even though redundancy in processing (that is processing a point in the field more than once) increases the cost of execution, the final result is still acceptable. In spraying, any location in the field has to be processed only once, since excessive PPPs dispensing will destroy the crop. Moreover, in spraying, the direction of field processing is fixed. The sprayer unit is allowed to navigate through the field via gaps inbetween the crop rows. Any other motions or manoeuvres are prohibited.

In a single robotic approach, the sprayer unit simply starts the task from any location of the field and processes one track after another. However, if multiple robots are deployed to execute spraying, the field has to be divided among robots so that no two robots spray the same region in the field. According to the taxonomy presented in [4], this requirement expects the team to be strongly coordinated.

1.2 Problem Statement

As in any multi robotic system, there are various parameters that have to be set. For example: team architecture, communication structure (centralised vs decentralised), control structure (centralised vs distributed), task partitioning, task allocation (self-organising versus market-based), and coordination (not coordinated, weakly coordinated, strongly coordinated). In order to identify the appropriate choices, the scenario has to be described in more details.

Assuming that robots are capable of carrying out multiple tasks, it is possible that, by the time of spraying, not all robots are positioned near each other. Therefore, the initial position of the robots are unknown at the initiation of spraying. As a result, robots cannot rely on direct forms of communication (for example using Bluetooth, WiFi, etc.) to perform task allocation since robots could be out of range of the communication signal. Since no direct form of communication is possible, central-based approaches become inapplicable.

The aim of this paper is to develop a self-organising robotic system that would require little computational effort, no central-based control and communication, which at the same time is efficient and adaptable to various fields. The main focus of the proposed strategy is to perform task allocation using only local information of the robots.

In this paper, we are not considering the problem of localization and electro-mechanical aspect of the spraying. The problem of localization has been addressed in many researches in single robotic approaches in precision farming.

1.3 Assumptions and Initial Conditions

In summary, we assume the following:

Assumption 1. Robots are not capable of communicating with each other using explicit forms of communication.

Assumption 2. Robots are initially scattered around the field, hence field accessing is asynchronous.

Assumption 3. Robots are not allowed to perform any manoeuvring motion in the middle of the field, since they will run over the crop.

In addition, the following are considered as simplifying conditions:

Condition 1. All robots are equipped with an accurate localization system.

Condition 2. The robots have limited 180 degrees field of view.

Condition 3. The environment is a large known 2-dimension rectangle. Prior to execution, four coordinates of four corners of the field along with distance between two consecutive tracks are given to robots.

1.4 Overview of the Paper

We first look into related works in cooperative area coverage. Next, we describe and analyse the proposed strategy for task partitioning, task allocation and coordination. And finally, we look at the impact it has on the team performance.

2 Related Works

Cooperative area coverage has been studied for the past decades. This resulted in suggestion of numerous strategies for different applications. We classify the reviewed approaches into three categories: *Static approaches,* *Central-based approaches* (from control point of view), and *Explicit Communication-based approaches*. We provide examples for each category and describe why these approaches are inapplicable in cooperative spraying.

In static approaches, task allocation is carried out manually prior to execution [5]. In these approaches, the initial positions of the robots are known, and the field is divided among robots in a way to minimize the cost of execution using various known theories (e.g. graph theory). Static approaches are not appropriate for cooperative spraying as the initial positions of the robots are unknown by which robots attend their share of task at different instances of time. With this, robots, which are processing two adjacent regions, could fall into congestion in the middle of the field. A logical conclusion is that task allocation has to be performed in real-time which requires some form of interaction among robots.

One common method is to use a central-based unit. The central unit could be fixed in a place somewhere around the field or it could be one of the robots participating in the task. In this paper, we classified central-based approaches into two categories: decision-maker, and data-pool.

In decision-maker approaches, the central-base unit collects necessary information from individuals in the team, and based on the preprogrammed algorithms it organises the robots so that the global task is achieved. [3,9,10,15] are few examples of this type of approach in which the central unit based on the collected information performs task allocation and coordination if necessary.

In data-pool approaches, the central-unit is used as a shared computational space in which robots exchange information to make their decisions. In most cases, the central unit does not perform any coordination or task allocation. Anil et al. [1] demonstrated a low cost, multi-functional team of robots which are capable to perform various agricultural tasks. In this example, first few robots are sent to observe the environment. Once returned, the robots share their findings with others through a central unit via the attached ZigBee modules.

In communication based approaches, the success of the team is susceptible to the loss of the central unit. Besides, robots have to be within range of the central-unit communicating signal. Moreover, the cost and complexity of the system grows as the number of participating robots increases. In spraying, robots are initially scattered around a large field, therefore no central unit could guarantee task allocation and coordination throughout the execution. This requires a task allocation which is only based on local information of the robots. In [8], this form of task allocation is referred to as threshold-based methods. Robots could achieve the required information from two sources: (I) the sharing environment (stigmergy), (II) local interaction between robots.

Ranjbar et al. [13] present a cooperative area coverage model based on stigmergy. In this example, each robot dispenses pheromone-like material in the environment to mark its territory. Upon detection of another robot's trail, robots manoeuvre to a predefined direction. In [7], it is demonstrated how a team of robots could perform ploughing cooperatively on a large field using the state of the soil. The result of spraying is detectable only for a short period of time and stigmergic approach are not reliable for spraying. Therefore, the proposed solution has to be based on local interaction among robots.

3   System Description

In any team of robots, the main problem to be addressed is task allocation. In spraying, task allocation has to guarantee that each location in the field is visited only once. In addition, task allocation has to be carried out in real-time, and only the local information of the robots has to be taken into consideration. Therefore, a mechanism is required by which robots are informed about the state of the task. Note that this information cannot be conveyed with explicit forms of communication.

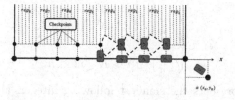

Fig. 1. A team of 8 robots are performing task allocation. r_1, r_2, and r_3 have found unoccupied checkpoints. In the meantime, other robots are examining every checkpoint.

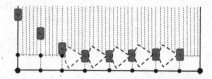

Fig. 2. Task initiation stage for a team of 8 robots; Spraying task starts whenever r_8 reaches checkpoint on the last region. r_7 starts spraying once it perceives r_8 decision.

The proposed strategy is that the field is divided into regions (each region consists of few tracks), with the aim that each region is processed by only one robot $(reg_i \rightarrow r_i)$. Each robot claims a region by occupying a particular location, which is referred to as checkpoint, outside the region. Checkpoints are set to be the last track of each region. Spraying a region starts from this location and tracks are processed consecutively to the first track in the region. Robots have to check each checkpoint to see if the region is occupied. If an unoccupied location is found, the robot proceeds and occupies the location (see Fig. 1).

Once the last robot, r_n, occupied the last region, other robots have to be informed otherwise they will never start spraying the field. To solve this, when a robot occupies a checkpoint, it poses itself in a way that it continuously monitors next checkpoint. The last robot does not need to comply with this, and instead it starts spraying the field right after it occupies the last checkpoint. Then, r_{n-1} starts spraying once r_n is no longer detected. With this, robots starts spraying one after another, and the first robot will be the last that starts the spraying (see Fig. 2).

3.1 Task Partitioning

In spraying, the field can be processed from multiple locations independently. To take advantage of this capability, the field is divided into regions and divided among robots. For this, the number of regions has to be equal to the number of robots in the team. Each region consists of few tracks. If there are n robots and K tracks in the field, the number of tracks in each region, k_i, can be determined as follows:

$$k_i = \begin{cases} \left\lceil \frac{K}{n} \right\rceil & \text{if } i \leq K \mod n \\ \\ \left\lfloor \frac{K}{n} \right\rfloor & \text{if } i > K \mod n \end{cases} \tag{1}$$

In here, $\left\lceil \frac{K}{n} \right\rceil$ represents the smallest following integer, and $\left\lfloor \frac{K}{n} \right\rfloor$ represents the largest previous integer.

Next, it is important to identify the track numbers that are within each region. Robots require this to locate the checkpoints. Let's assume that TR is a set of track numbers in the field, $TR = \{tr_l | l \in \{1, 2, ..., K\}\}$, and R is set of robots, $R = \{r_i | i \in \{1, 2, ..., n\}\}$, then track numbers allocated to r_i, can be obtained as follows:

$$G_i = \{tr_j | j \in \{m+1, m+2, ..., m+k_i\}\} \tag{2}$$

where m is the last track number assigned to previous robots, and it can be calculated as follows:

$$m = \sum_{j=1}^{i-1} k_j \tag{3}$$

3.2   Task Allocation

Since robots do not communicate with each other, robots cannot share their understanding of the environment with others; hence robots have to perform redundant checkpoint analysis. As the number of robots increases, the number of locations that the robot has to check also increases. For time analysis, let's first define duration of task allocation for a robot.

Definition 1. *Task allocation duration for a robot is the period from initial position until the time that the robot detects an unoccupied checkpoint.*

With this definition, time analysis becomes complex since robots are initially scattered around the field. To simplify calculations, let's assume that robots have formed a queue behind a location outside of the field. This location is referred to as *alpha* (see Fig. 3). Before robots access the first checkpoint, they first have to access *alpha*. However, distance that a robot has to travel to reach *alpha* depends on the robot's position in the queue. The length of the queue for each robot is $(\lambda + \epsilon)(i + 1)$ meters. In here, λ is the length of a robot, $\epsilon(epsilon)$ is the minimum distance between two consecutive robots.

Once a robot reaches *alpha*, it has to analyse the first checkpoint, and hence all robots have to travel a fixed distance between *alpha* and the first checkpoint, d_{α,l_1}. Also, except the first robot, other robots have to travel to other checkpoints, d_{l_{j-1},l_j}. For example, r_2 has to travel distance between the first and second checkpoint to reach its destination.

Once a robot reaches a checkpoint, it takes a period of time to draw conclusion on the status of the checkpoint. This period is denoted by τ and it will be

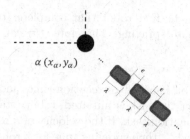

Fig. 3. Illustration of robots queueing for accessing *alpha*. Robots have the same length, λ, and the distances between robots, *epsilon*, are equal.

propagated in the queue since robots have to wait for the path to be cleared. The total delay for a robot is $(2i - 1)\tau$.

This is easy to see. For example in a team of three robots, r_1 will analyse only the first checkpoint, hence spends only a τ of time. Whereas r_2 will analyse checkpoint one and checkpoint two, but r_2 is already affected by the delay of the first robot. Therefore, the queue effect and the checkpoint analysis effect for r_2 sum to 3τ. Also, r_3 will analyse three checkpoints, plus the delay propagated by r_1. In addition another delay will be propagated as a result of analysis of r_2 on the first checkpoint. Any further checkpoint analysis of r_2 does not affect r_3 since both robots are performing their analysis simultaneously. Therefore, the overall queue effect on r_3 is 5τ.

If the velocities of all robots are equal and robots move with constant speed, task allocation duration for a robot (r_i) can be evaluated as follows:

$$t_{ta_i} = \frac{d_{\alpha l_1}}{v} + (2i - 1)\tau + \frac{(\lambda + \epsilon)(i - 1)}{v} + \sum_{j=2}^{i} \frac{d_{l_{j-1} l_j}}{v} \qquad (4)$$

3.3 Task Initiation

Task allocation for a robot is completed when it identifies the first unoccupied checkpoint, but the task is not yet initiated since the robot has to assure that all other robots are informed about the status of the occupied region.

Definition 2. *Task initiation duration for a robot is the period that a robot has to remain at a checkpoint until the initiation signal is detected.*

Since spraying will not start for all robots until the last robot has occupied its last region, the task initiation time for a robot will be affected by the last robots task allocation time. However, this impact is partially compensated by the robot's task allocation period. With this, if τ_{init} is the constant time to perceive task initiation, the standby period for a robot before it starts its task is evaluated from the following:

$$t_{st_i} = t_{ta_n} - t_{ta_i} + \tau_{init}(n - 1) \qquad (5)$$

With this strategy, r_1 is the first robot that completes task allocation, but it is the last robot to initiate spraying. Therefore, (5) can also be expressed as follows:

$$t_{st} = t_{st_1} = t_{ta_n} - t_{ta_1} + \tau_{init}(n - 1)$$

In here, t_{st} refers to total spraying execution time, and t_{st_1} is the first robot spraying time. Right after the task is initiated, robots start spraying. Spraying is a two-dimensional navigation task. If the velocity of a robot is denoted v, and length of a track is denoted l_p, then spraying time for a robot after task initiation can be evaluated as follows:

$$t_{s_i} = \frac{1}{v}(k_i l_p + (k_i - 1)d_f) \tag{6}$$

In here, d_f corresponds to the distance between two consecutive tracks. Bear in mind that a robot has to repeat spraying as many as k_i times, obtained from (1). In addition, as part of spraying, a robot also has to switch between tracks for $k_i - 1$ times.

The total execution time including all three steps (task allocation, task initiation and spraying) is the maximum of sum of these times for all robots. However, since task partitioning starts from the first robot, and since r_1 is the last robot that initiates its task, total task execution time can be expressed as follows:

$$t = t_{s_1} + t_{ta_1} + t_{ts_1} \tag{7}$$

The proposed strategy promises that a group of robots can spray a large field. However, robots can perceive only a limited range. In other words if the distance between two consecutive checkpoints becomes larger than the perception range of robots, the task can never be initiated. Furthermore, as number of participating robots increases, the distance between checkpoints decreases. However, there is a limit on how close two checkpoints could be placed from one another as robots require space for transiting from one track to another. In other words, there is a limit on maximum number of robots that could be applied with this strategy.

$$\lfloor \frac{W}{\Gamma} \rfloor \leq n \leq \lfloor \frac{W}{\lambda + \epsilon} \rfloor \tag{8}$$

In here, Γ is the range of perception of robots, W is the width of the field, $W = Kd_f$, λ is the length of the robots, and *epsilon* is the threshold distance between two robots.

4 Simulation Results and Discussion

In this section, we present both numerical results based on mathematical consideration and the results obtained from Stage simulation environment. Since the allowed number of participating robots depends on the dimension of the field, first the boundaries of the team sizes have to be identified. In here, few parameters about the environment have to be fixed.

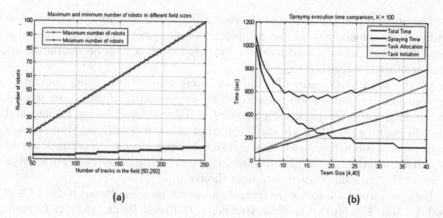

Fig. 4. Numerical results: (a) Maximum and minimum number of robots in different field sizes. (b) Time analysis results.

We assumed that there are 50 to 250 tracks are available in the field and the distance between two consecutive tracks is 20 (cm), and the length of each track (l_p) is 20 (m). Robots have equal dimensions with length (λ) equal to 50 (cm). The threshold distance ($\epsilon(epsilon)$) between two robots is set to be 30 (cm), and all robots are assumed to move in constant velocity equal to 0.5 (m/s). The checkpoint analysis duration (τ) and robot behaviour analysis duration (τ_{init}) is fixed to 5 (s).

From (8), it is possible to identify the maximum and the minimum number of robots allowed in various field sizes, since $W = Kd_f$ (see Fig. 4(a)). The definition of a team in a multi-robotic system fixes the minimum number of robots to be greater than three ($n_{min} \geq 3$). For a field with hundred tracks ($K = 100$), the number of participating robots could vary between four and forty ($n \in [4, 40]$). Consequently, task allocation and task initiation time can be predicted (see Fig. 4(b)).

As number of participating robots increases, the durations for both task allocation and task initiation increase by which the total execution time increases. In a field with 100 tracks, $K = 100$, if appropriate team sizes are applied ($n \in [4, 40]$), the total execution time opposes with what is expected from the nature of a multi-robotic system (see Fig. 4(b)). This could be concluded that the applied strategy is not efficient enough to receive positive effects from an increase in number of robots.

In (7), there are two increasing linear functions and one decreasing hyperbolic function. Therefore the resulting function will have a global minimum. This is the optimal team size that could be deployed to the given field and it can be obtained as follows:

$$n_{opt} = \sqrt{\frac{K(l_p + d_f)}{2\tau v + \tau_{init}v + \lambda + \epsilon}} \qquad (9)$$

The optimal number of robots corresponding to the minimum time from the start of time allocation and the completion of spraying depends on the parameters in the expression, and from the proposed ones, is 15.6. So the optimised number of robots is either 15 or 16.

On the other hand, series of simulations were conducted in Stage simulation environment in collaboration with ROS (Robot Operating System). In simulation, each robot is equipped with a model of Hokuyo Laser Range Finder, and a fixed camera which both are placed in front of the robot. In addition, at any given time, a robot could obtain a two dimensional coordinate corresponds to its current location in the global frame. However, during the trials no robot is aware of the position of other robots in the team.

Behaviours on each robot are controlled by collaboration between three C++ modules (see Fig. 5(a)): (1) Task Handler, (2) Reach Point, and (3) Camera Analyser. Task Handler is responsible to triggers specific behaviours in the robot: setting a new target, activating Camera Analyser, analysing the field, and executing the cooperative procedures. Reach Point is responsible to guide the robot to the requested coordinate in collision free manner. Camera Analyser is responsible to analyse the behaviour of other robots, using colour detection as robots are homogeneous, and it signals Task Handler. However, as behaviour monitoring is only necessary at particular locations and moments in the field, Task Handler will trigger the procedure whenever it is required.

Trials conducted for a field with 51 tracks (K = 51), and various team sizes were deployed ($n \in [3, 10]$). In all team sizes, robots performed task allocation and task initiation successfully. During simulation, position and time of robots are recorded. Data recording for a robot initiates when the robot passes a, and it stops as soon as the robot exits the margin of the field. The maximum execution time then is plotted and compared with the equivalent corresponding numerical prediction (see Fig. 5(b)). It can be seen that there is a slight difference between Stage simulation results and the numerical results. This difference is due to the field exiting duration which has not been considered in mathematical description.

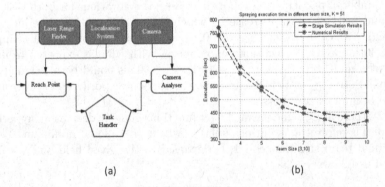

(a)   (b)

Fig. 5. Stage Simulation Results. (a) Individual behaviour on single robot. (b) Comparison between results collected during simulation and numerical visualisation.

5 Conclusion and Future Works

In this paper, we described a strategy by which a team of robots could cooperatively perform area coverage related task in a known environment. The task demands a delicate solution since each point in the field has to be sprayed only once and it has large dimensions.

In the proposed strategy, robots rely only on their local information to obtain an order and perform task allocation. The mathematical analysis of the strategy is described and numerical validation and conducted simulation in ROS and Stage suggest that this strategy can be successfully applied on real robots.

However, since local perceptions of the robots are limited, the distance between two consecutive checkpoint locations cannot exceeds beyond the robots range of detection. This makes the proposed strategy to be partially scalable. In addition, the proposed method requires that all robots participate all at once. In the future, we aim to improve the method in a way that robots can execute their share of task in more robust way.

References

1. Anil, H., Nikhil, K., Charitra, V., Gurusharan, B.: Revolutionizing farming using swarm robotics. In: 6th International Conference on Intelligent Systems, Modelling and Simulation, 19(3), pp. 141–147 (2015)
2. Blackmore, S.: Precision farming and agricultural engineering for the 21st century: How to make production agriculture more efficient. Presentation (2010)
3. Drenjanac, D., Tomic, S., Klausner, L., Kuhn, E.: Harnessing coherence of area decomposition and semantic shared spaces for task allocation. Inf. Process. Agric. 1(1), 23–33 (2014)
4. Farinelli, A., Iocchi, L., Nardi, D.: A classification focused on coordination. Mult-Agent Robot. Syst. 34(5), 2015–2028 (2004)
5. Fazeli, P., Davoodi, A., Pasquier, A.: Fault-tolerant multi-robot area coverage with limited visibility. In: Proceedings of the IEEE International Conference on Robotics & Automation (ICRA) (2010)
6. Fazli, P.: On multi-robot area coverage. In: proceedings of the 9th International Conference on Autonomous Agents and Multiagent Systems (AAMAS 2010), pp. 1669–1670 (2010)
7. Janani, A., Alboul, L., Penders, J.: Multi-agent cooperative area coverage: case study ploughing. In: 2016 International Conference of Autonomous Agents and Multi-Agent Systems (AAMAS) (2016)
8. Karla, N., Martinoli, A.: A comparative study of market-based and threshold-based task allocation. In: Gini, M., Voyles, R. (eds.) Distributed Autonomous Robotic Systems 7, pp. 91–101. Springer, Tokyo (2006)
9. Li, N., Remeikas, C., Xu, Y., Jayasuriya, S., Ehsani, R.: Task assignment and trajectory planning algorithm for a class of cooperative agricultural robots. J. Dyn. Syst. Measur. Control 137(5), 1–9 (2015)
10. Liu, C., Kroll, A.: Memetic algorithms for optimal task allocation in multi-robot systems for inspection problems with cooperative tasks. J. Soft Comput. 19(3), 567–584 (2014)
11. Liu, J., Wu, J.: Mult-Agent Robotic Systems. CRC Press, New York (2001)

12. Noguchi, N.: Robot farming system using multiple robot tractors in japan agriculture. In: Preprints of the 18th IFAC World Congress Milano, pp. 633–637 (2006)
13. Rangbar-Sahaei, B., Weiss, G., Nakisaee, A.: Stigmergic coverage algorithm for multi-robot systems. In: Proceeding of AAMAS 2012 (2012)
14. Raper, R.L., Kirby, J.M.: Soil Compaction: How to Do it, Undo it, or Avoid it. ASABE Distinguished Lecture Series. American Society of Agriculture and Biological Engineers, Michigan (2006)
15. Tsalatsanis, A., Yalcin, A., Valavanis, K.: Dynamic task allocation in cooperative robotic systems. Int. J. Adv. Robot. Syst. 6(4), 309–318 (2004)

A Step Toward Mobile Robots Autonomy: Energy Estimation Models

Lotfi Jaiem[✉], Sebastien Druon, Lionel Lapierre, and Didier Crestani

Laboratoire d'Informatique Robotique et Microélectronique
de Montpellier (LIRMM), UMR 5506, Université de Montpellier,
161 rue Ada, 34 095 Montpellier Cedex 5, France
{jaiem,druon,lapierre,crestani}@lirmm.fr
http://www.lirmm.fr

Abstract. One of the crucial questions to develop Autonomous Mobile Robotic systems is the energy consumption, its monitoring and management all along the mission. Mission complexity and fault tolerance capabilities require to exploit system redundancies, in terms of algorithms and hardware recruitment. This choice has an evident impact on energy consumption. This paper proposes an identification protocol to establish the energy consumption models for each control configuration of a Pioneer 3DX. These models are destined to be used as online predictors providing an estimation of the necessary remaining energy that mission success 'nominally' requires. At the end, the proposed energy consumption models are validated and discussed experimentally.

Keywords: Autonomy · Energy · Control architecture · Consumption model

1 Introduction

Mobile robotic systems are expected to perform long range, long term and complex missions. By contrast to industrial robotics where energy can be supposed to be infinite, the management of autonomous mobile robot missions requires to be able predicting the energy consumption of the robot with an acceptable accuracy. Moreover, the mission complexity and the environment versatility imposes to be able managing the embedded energy according to different robot configurations (sensors, actuators, control schemes,etc.), making the energy management a central issue for autonomous robotic missions. Related research addressing the energy issue can be classified into 3 main fields.

Firstly, component oriented approaches like Dynamic Power Management (DPM) in [1], want to reduce the energy consumption using hardware or software energy-aware management techniques. An estimation of the future activity of hardware and software components is used to dynamically adjust their power, while maintaining a desired performance. One of the most popular DPM techniques is the Dynamic Voltage Scaling (DVS) which dynamically adapts the frequency and the voltage of the processor for energy saving [2]. These approaches

© Springer International Publishing Switzerland 2016
L. Alboul et al. (Eds.): TAROS 2016, LNAI 9716, pp. 177–188, 2016.
DOI: 10.1007/978-3-319-40379-3_18

can be extended to robotics, for power supply management, as example, scaling the sensors acquisition frequency [1].

Secondly, some works address the robot system level. They mainly focus on an identification of the robot energy consumption and energy-aware motion for mission involving a unique control task. In [3], energy consumption predictive models, based on theoretical and experimental analysis of sensors and actuators, are proposed for Khepera III robot. [4] proposes an experimental analysis of power consumption of a Pioneer 3DX (movement energy, sensors and batteries consumption). The authors identify consumption models for DC motor in order to compare the efficiency of different motion plans to cover opened area. Energy prediction models integrating rolling resistance for unmanned ground vehicles is addressed by Sadrpour et al. in [5] to define optimal velocity profile for coverage planning. Optimal velocity for energy efficient navigation has been also addressed in [6] for car-like robots. In [7], minimum-energy trajectory for wheeled mobile robots has been studied and experimented on a Pioneer 3DX.

Few works address the energy question at the mission level involving several types of tasks. In [8], Zhang et al. develop a joint power management scheme coupling the processor frequency (image analysis) and the robot speed (traveling) to improve the whole system power performance. Authors describe in [9] the guidelines for a robot power management system adapting to tasks and environments. Unfortunately few details about its implementation are available. Power-aware tasks scheduling is addressed in [10] for mission-critical embedded systems. However these existing works related to robot energy consumption identification have some limitations:

- Multi-battery (robot and embedded computer) are not considered.
- Impact of different control schemes on energy accuracy is not studied.
- Localization error and environment impact on robot behavior are neglected.

2   Context

This work focuses on consumption models, but is part of a more general study on the question of autonomy. Autonomy means also a 'self-awareness' about the performance that the system currently exhibits, or is expected to, depending on mission requirements and environment context. We define different performance axis that has to be monitored in order to guarantee a 'nominal' mission execution from *Stability, Safety, Localization, Energy* and *Duration* viewpoints. The originality of our approach consists in the concurrent consideration of global feasibility criteria along these axes based in performance estimation of the different control schemes. In particular, an accurate estimation of the robotic system energy consumption is crucial for energy axis monitoring.

This paper presents in a first part the hardware and software architectures of the robot platform based on a Pioneer 3DX robot. Then the energy consumption of robot and laptop batteries according to the connected elements is experimentally established and modeled. Finally, before concluding, these consumption models are experimentally validated and discussed on a patrolling mission.

3   Hardware and Software Architectures

3.1   Hardware Architecture

Based on a classic Pioneer 3DX, the robot platform (Fig. 1 (a)) has been equipped with sensors and electronic devices in order to implement different motion, localization algorithms and image analysis capabilities. So, a rigorous approach of energy management is needed to manage all the control schemes configuration (hard, soft) which can be used.

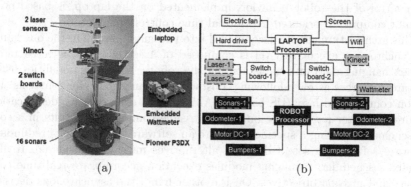

Fig. 1. (a) Robot platform; (b) Robot hardware architecture.

The Pioneer 3DX is a unicycle type, whose motion can be controlled using two independent DC motors (experimental maximum velocity is 0.8 m/s). It also integrates two arrays (front and rear) of 8 sonars (US) and 5 bumpers. Two URG-04 LX Hokuyo lasers (LAS) allow for full horizontal scanning of the environment. Their data are used for obstacle avoidance and robot localization algorithms. A Kinect© (KIN) is used to perform image analysis and robot localization task, using QR-codes. Two added switch boards allow connecting or disconnecting independently the power supply of the two lasers and the Kinect according to the chosen control schemes. The robot embeds also a battery generating up to 259 Wh of energy. An embedded watt-meter has been integrated to realize an online measurement of the energy supplied by the robot battery. The Pioneer 3DX communicates with the embedded laptop using a client-server connection and a Hitachi H8S-based micro-controller.

The embedded laptop includes an Intel-Core i5-2430M CPU 2.40 GHz. Its lithium/ion battery can generate up to 87 Wh. This battery supplies the laptop processor, screen, wifi board, hard drive and electric fan. Moreover it also supports the USB communications with the lasers, the Kinect, the switching boards, the watt-meter and the robot micro-controller. The operating system used is Linux-RTAI in order to address real-time issues. The laptop battery consumption is estimated using the linux.uevent system file. It allows an online monitoring of the battery current, voltage, and proposes an estimation of the remaining energy of the laptop battery.

Figure 1 (b) presents the hardware architecture of the platform. The elements supplied by the robot batteries are drawn with black boxes. The ones supplied by the laptop battery are drawn with white boxes. Switchable components are drawn with dotted lines. Furthermore it must be noticed that these components and the watt-meter have USB connections with the laptop for data exchange but they are supplied from the robot battery (grey boxes).

3.2   Control Architecture

The control of the robot behavior, implemented on the laptop, is based on a modular component-oriented hybrid real time control architecture.

This architecture can be decomposed into two main parts. On the one hand, the synchronous executive level is composed of a set of interconnected modules, exchanging data and implementing the robotic basic functionalities using software resources as path-following, obstacle avoidance, corridor centering, for motion control, and laser/US based SLAM and QR-code camera identification or pure odometer for localization. Each possible combination results in a control scheme, recruiting a specific hardware an software set. In case of redundant hardware/software, a set of different alternatives must be considered. It also contains a scheduler, managing modules execution according to real-time constraints and mission objectives. On the other hand, the asynchronous decision level selects the relevant control scheme according to the current mission context using a hierarchy of supervisors. Since the objective of this study is to establish energy consumption models for all possible control schemes, we focus our study on 4 specific control schemes, and their alternatives, inducing the use of different hardware components (Table 1).

Table 1. Control schemes and hardware components

Control scheme		Sensors			Actuator
		US	*LAS*	*KIN*	*DC motors*
SFM	Simple Forward Motion	-	-	-	●
PFM	Path Following Motion	○	○	○	●
QRCN	QR Code navigation	○	○	●	●

−: unused / ○: Optional / ●: Mandatory

Simple Forward Motion (*SFM*) used to measure and identify the energy needed for a straight motion without using sensors or control loop regulation. The two wheels rotation speeds are strictly the same.

Path Following Motion (*PFM*) implements the path-following algorithm SMZ [11] and exteroceptive sensors are recruited. It is used to estimate the control regulation loop impact on energy consumption with respect to the traveling distance and curvature and speed variations.

QR Code Navigation ($QRCN$) allows a global localization of the robot, using the on board camera. Note that the spatial distribution of QR-Codes in the environment influences the positioning error drift and, as a consequence, the path actually performed and the energy consumption.

4   Energy Consumption Models

As we presented previously, the studied robot has two power sources: robot battery (P_R) supplies the low-level controller, sensors, motion actuators and watt-meter, while laptop battery (P_L) supplies the control architecture execution, the data exchanged between the processor and the actuators and sensors, and the switch boards. The robot energy management requires building energy consumption models on the robot and laptop batteries.

In the following, the formulation proposed in [3] for a Khepera robot is adapted. Generally, for a given Control Scheme CS, the instantaneous power consumption P of a robot can be decomposed in dynamic and static parts. The dynamic part P_{Dyn} denotes the power consumption which can dynamically change over the time. The static part P_{Stat} denotes the constant steady state power consumption. Depending on the recruited components of a control scheme CS, the corresponding instantaneous consumption is:

$$P(CS) = \sum_{i=1}^{n_1} \alpha_i \cdot P_{Dyn\ i} + \sum_{j=1}^{n_2} \beta_j \cdot P_{Stat\ j} \ . \tag{1}$$

where n_1 is the number of dynamics components, n_2 the number of static components, α_i and β_j are equal to 1 if the considered component is involved into the current control scheme CS, 0 otherwise. For a given control scheme of Table 1 and for a given control scheme, (1) can be rephrased distinguishing optional and mandatory components. The energy consumption of a control scheme is classically obtained by multiplying the instantaneous power consumption by the duration ΔT of the active control scheme CS (2).

$$E(CS) = P(CS) \cdot \Delta T(CS) \ . \tag{2}$$

Robot and laptop battery consumption models are detailed in the sequel.

5   Power Model: Consumption on Robot Battery

We present now a detailed analysis of the robot battery consumption linked to the dynamic and static components. From these results, the power model parameters of the studied control schemes are evaluated. The experimental results are based on the embedded watt-meter data analysis.

5.1 Instantaneous Power Consumption of Actuators

DC motor motion power model for Pioneer 3DX robot has been largely studied in [4–7]. Analytically, it depends on many parameters like linear (v) and angular (r) velocities, linear (a) and angular accelerations, robot weight, slope and the type of surface. Model parameters remain difficult to evaluate experimentally and the null acceleration hypothesis is generally observed in the literature [5]. This hypothesis corresponds with the work context where mission is composed by a sequence of straight lines traveled with constant velocities. Another approach developed in [6] and adopted in [3] demonstrates that the DC motor motion power model can be defined by the following polynomial approximation:

$$P_{R\_Motion}(a, v) = C_1 \cdot a^2 + C_2 \cdot v^2 + C_3 \cdot v + C_4 + C_5 \cdot a + C_6 \cdot a \cdot v \ . \tag{3}$$

The null acceleration assumption reduces (3) to (4).

$$P_{R\_Motion}(v) = C_2 \cdot v^2 + C_3 \cdot v + C_4 \ . \tag{4}$$

From experimental measures (Fig. 2 (a)), the polynomial approximation of parameters in (4) yields to (5). This experimental data is the average of power measurements (1 s of period sampling) of 20 m straight line traveling on a flat horizontal floor using SFM control scheme.

$$P_{R\_Motion}(v) = 6.25 \cdot v^2 + 9.79 \cdot v + 3.66 \ . \tag{5}$$

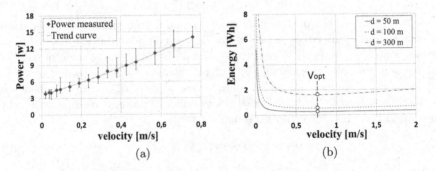

Fig. 2. (a) DC motor motion power, (b) Theoretical energy for different distances.

Some remarks can be derived from this formulation: when the robot does not move ($v = 0 \ m/s$) the power is equal to C_4 (3.66 W). This power $PR_{Cont.1}$ corresponds to the static consumption required by the different electronic boards and the micro-controller within the robot. From (2) and (5) and considering that $v = d/\Delta T$ where d represents the traveled distance at a constant velocity v during ΔT, the energy motion model is:

$$E_{R\_Motion}(d, v) = 6.25 \cdot d \cdot v + 9.79 \cdot d + 3.66 \cdot \frac{d}{v} \ . \tag{6}$$

Eq. (6) can be very useful to estimate the motion energy needed to travel over a distance d at velocity v. Figure 2 (b) represents the theoretical motion energies required to travel a given distance according to the velocity.

5.2   Instantaneous Power Consumption of US Sensors

The power consumption of the sonars is experimentally defined for different acquisition frequencies f_{US}, of the 16 sonars recruitment. Based on these data (Fig. 3), a polynomial curve trend is expressed in (7) where the presented data are the averaged consumption over 60 tests of one minute acquisition.

$$P_{R\_US}(f_{US}) = 4 \cdot 10^{-5} \cdot f_{US}^2 + 5.1 \cdot 10^{-4} \cdot f_{US} . \tag{7}$$

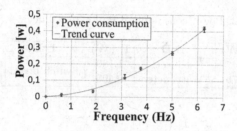

Fig. 3. Sonars consumption.

5.3   Instantaneous Power Consumption of Static Components

Power consumption of the static components has been also measured. The most consuming sensor is the Kinect with $P_{R\_Kinect} = 2.82$ W. Each of the 2 lasers sensors consumes $P_{R\_Laser} = 2.34$ W. So, using both lasers requires 4.68 W. Finally, the robot controller power consumption is $P_{R\_Cont.2} = 2.67$ W.

5.4   Control Scheme Power Consumption Model for the Robot Battery

Depending on the components involved in a control scheme, the power consumption extracted from the robot battery can be formulated as:

$$P_R(CS) = P_{R\_Motion}(v) + \alpha_1 \cdot P_{R\_US}(f_{US}) + \beta_1 \cdot P_{R\_Kinect} + k_1 \cdot \beta_2 \cdot P_{R\_Laser} . \tag{8}$$

where $k_1 \in \{0, 1, 2\}$ denotes the number of active lasers. α_1, β_1 and β_2 are boolean coefficients that indicate if sensors are used or not. It must be noticed that derivation of (6) and the consideration of (8) yields the following optimal velocity (9). $V_{opt} = 0.76\,m/s$ if no sensors used (Fig. 2 (b)). If sensors are used as expressed in (9), the theoretical optimal velocity minimizing the energy consumption cannot be practically reached.

$$V_{opt} = \sqrt{\frac{3.66 + \alpha_1 \cdot P_{R\_US}(f) + \beta_1 \cdot P_{R\_Kinect} + k_1 \cdot \beta_2 \cdot P_{R\_Laser}}{6.25}} . \tag{9}$$

6   Power Model: Consumption on Laptop Battery

The laptop battery consumption is generally ignored in the literature. However, in true autonomy context, it is essential. It supports the control architecture and data communication with the robot controller and sensors.

6.1   Instantaneous Power Consumption of External Devices

Even if the sensors are externally supplied by the robot battery, the USB communication with the laptop has also its own consumption on the laptop battery (Table 2). Moreover the switching board is also supplied by the laptop battery. That explains why three states are mentioned in the Table 2.

Table 2. Power consumption P of external devices E D

E D	P (W)	E D	P (W)	E D	P (W)		
					0	1	2
Laser	0.447	Kinect	1.180	Switch board	0.087	0.405	0.720
Controller	0.075	Watt-meter	0.490				

6.2   Instantaneous Power Consumption of Internal Devices

The Laptop integrates many internal devices consuming energy. This internal consumption depends on many factors like the executed program (processor activity), hard drive access, electric fan, wifi board consumption, screen display and of course laptop processor and frequency.

The screen power consumption can be easily measured $P_{L\_Screen} = 2.69$ W. Unfortunately, it is difficult to distinguish the impact of all other factors. However, as supposed in [1,9] all these factors can be integrated in unique power consumption $P_{L\_Processor}$ which remains constant for a given control scheme and a given constant control frequency f_c, as a first approximation. The average value of 60 measurements for different control schemes is: 10.66 W (SFM), 13.23 W (PFM) and 15.45 W ($QRCN$).

6.3   Control Scheme Power Consumption Model

The laptop processor frequency is considered as constant, then no dynamic component is involved. The laptop power consumption can also be considered constant for a given control scheme CS and a set of active external components (sensors, switch board). The laptop power consumption can be defined as:

$$P_L(CS) = P_{L\_Processor}(CS) + \beta_1 \cdot P_{L\_Kinect} + k_1 \cdot \beta_2 \cdot P_{L\_Laser}$$
$$+ \beta_3 \cdot P_{L\_Switch}(k_2) + \beta_4 \cdot P_{L\_Switch}(k_3) + \beta_5 \cdot P_{L\_Screen} \cdot \qquad (10)$$

where, β_3, β_4 and β_5 are boolean coefficient denoting if the device is used or not. k_2 and k_3 indicate whether 0, 1 or 2 relays are activated by a switch board.

7 Experimental Results

In this section, the proposed energy models applied on the Pioneer 3DX robot and its embedded laptop are validated, traveling along a distance of 50 m in a corridor from point A to B (Fig. 4). Then, a patrolling mission is considered with more complex path, stops, and different switch between control schemes.

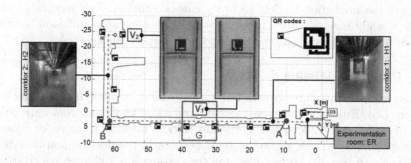

Fig. 4. Experimental context and mission description.

7.1 Energy Consumption Using Path Following Motion

A simple forward motion SFM has been used previously to define the power model of the robot motion. Now we will experimentally validate the resulting energy model using different control schemes with path following motion (control loop regulation) with different sensor configurations (supplied but not used in the control loop) and for two distances (20 and 50 m). Figure 5 (b) shows the energy measured for robot velocities from 0.1 to 0.8 m/s.

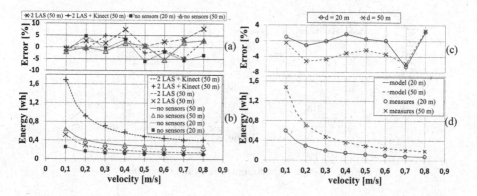

Fig. 5. Path Following Motion: (a) prediction error, (b) experimental and theoretical results for robot energy. (c) Prediction errors, (d) experimental and theoretical results for laptop energy.

Curves present the theoretical prediction model. The prediction error is computed, and displayed at Fig. 5 (a), where it varies between -7 % and +8 %. It is interesting to notice that the model accuracy of [3] for a Khepera is very similar to the one we have obtained, despite the large weight difference.

Figure 5 (c) and (d) present the laptop energy consumed and the corresponding error percentages. Energy measurements fit with model prediction with an error between -7 % and 3 %.

Since the prediction models accuracy can be considered as acceptable on a simple straight line trajectory, it becomes interesting to tackle a mission bringing ground reality like localization errors and acceleration impacts.

7.2 Patrolling Mission

A more complex mission is considered in this section. Starting from the docking station (DS) situated in the experimental room ER (Fig. 4), the robot must reach two valves V1 and V2 situated in H1 and H2 and inspects them (open/closed) by image analysis. Then, the robot must go back to the docking station (starting point). To test the correctness of energy estimation, different velocities are imposed during the mission depending on the robot location: mainly 0.46 m/s, 0.56 m/s and 0.75 m/s. Figure 6 presents the experimental measures of the robot velocity during the proposed mission and the mean velocity reached in each area is indicated in blue over the red lines (theoretical velocity). Two stops are noticed for image analysis detecting the valves status.

The total mission duration is higher than 7 min (434 s) over 190 m traveled distance. A path following control scheme QRCN (CS 1) using kinect (localization) has been used with sonars and two lasers (obstacle avoidance). On the way back, switching off one laser is planned (CS. 2). Moreover into the valves areas, a pure rotation is triggered OPR (CS. 3) and image analysis IA (CS. 4) control scheme is used to steer the robot in front of the valve and to analyze its status (open / close). The use of these schemes is shown in Fig. 6.

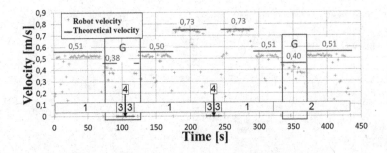

Fig. 6. Mission experimental velocity variations.

Figure 7 (a) shows that experimental robot energy consumption is a little bit higher than the expected one. However the curves are very closed. Moreover Fig. 7 (b) demonstrates that the predicted energy estimation error after an initial peak decreases to a final value of 10 % (12.1 % for laptop battery Fig. 7 (d)). The initial over-consumption is due to the system trajectory to reach the first corridor from the ER. The real trajectory differs from the expected one (odometer drift and robot oscillations) and the real consumption is higher than the predicted one. The same increase of error is noticed at the end of the mission (from 350 s) where robot returns to the ER. The same peaks are noticed for the laptop prediction error. Moreover, the over-estimation of the energy consumption (at the beginning of the mission) becomes less important with regards to the global energy consumption according to the mission progress (error decrease). The estimation error is explained also by the difference between the theoretical and the experimental velocities and the robot accelerations after the stops near the valves.

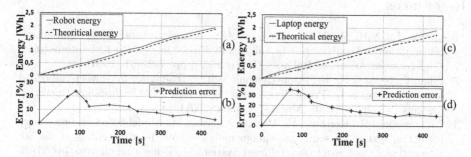

Fig. 7. Robot energy consumption: (a) Experimental and theoretical results, (b) prediction error. Laptop energy consumption: (c) Experimental and theoretical results, (d) prediction error.

These analysis demonstrate that the proposed power and energy consumption models allow for a good estimation of the real consumption for quite complex and long missions. The over-estimation observed is due to the difference between the expected and real robot trajectories. This difference is due mainly to the system transient response and the robot odometer drift for quite long distance before being corrected by QR code localization (robot oscillation and velocity variations). When robot heading changes are not frequent the consumption estimation becomes close to the reality.

8   Conclusion

This paper tackled the question of energy prediction of the Pioneer 3DX robot in connection with energy management for autonomous mobile robotics. Energy aware based on accurate energy consumption estimation becomes crucial. Then, both robot and embedded laptop where considered. Based on experimental analysis, generic energy prediction models were proposed for a given control scheme. Optimal velocity from robot energetic viewpoint is defined. Experimental results show that the measurements fit to the proposed models for path following motion with different control schemes using different sensors. Later, a more complex mission is realized and the impact of the localization accuracy and the path shape are noticed. Thus, prediction accuracy decreases but remains still acceptable.

Future works, look for identifying the gap between prediction and experimental results tacking in account localization accuracy and followed path. That will improve the prediction models accuracy. Based on this prediction, a methodology allowing energy management for autonomous robotic mission has been developed. This methodology consists on dynamic hardware and software resources allocation along the mission with regard to performance constraints and particularly the energy one.

References

1. Mei, Y., Lu, Y.H., Hu, Y.C., George Lee, C.S.: A case study of mobile robot energy consumption and conservation techniques. In: Proceedings of the 12th International Conference on Advanced Robotics (ICAR 2005), pp. 492–497 (2005)
2. Pillai, P., Shin, G.C.: Real time dynamic voltage scaling for low-power embedded operating systems. In: Proceedings of the ACM 18th ACM Symposium Operating Systems Principles, Banff, Alberta, Canada (2001)
3. Parasuraman, R., Kershaw, K., Pagala, P., Ferre, M.: Model based on-line energy prediction system for semi-autonomous mobile robots. In: Proceedings of the 5th International Conference on Intelligent Systems Modeling a Simulation, pp. 27–29 (2014)
4. Mei, Y., Lu, Y.H., Hu, Y.C., George Lee, C.S.: Energy-efficient motion planning for mobile robots. In: Proceedings of the International Conference on Robotics & Automation, pp. 4344–4349 (2004)
5. Sadrpour, A., Jin, J., Ulsoy, A.G.: Mission energy prediction for unmaned ground vehicles using real-time measurements and prior knowledge. J. Field Robot. $30(3)$, 399–414 (2013)
6. Tokekar, P., Karnad, N., Isler, V.: Energy-optimal velocity profiles for car-like robots. In: Proceedings of the IEEE International Conference on Robotics and Automation - ICRA 2011. pp. 1457–1462 (2011)
7. Kim, C.H., Kim, B.K.: Minimum-Energy Motion Planning for differential-Driven Wheeled Mobile Robots. Motion Planning INTECH Open Access Publisher, Rijeka (2008)
8. Zhang, W., Lu, Y.H., Hu, J.: Optimal solutions to a class of power management problems in mobile robots. Automatica $45(4)$, 989–996 (2009)
9. Ogawa, K., Kim, H., Mizukawa, M., Ando, Y.: Development of the robot power management system adapting to tasks and environments-the design guideline of the power control system applied to the distributed control robot. In: Proceedings of the 2006 SICE-ICASE International Joint Conference, pp. 2042–2046 (2006)
10. Liu, J., Chou, P.H., Bagherzadeh, N., Kurdahi, F.: Power-aware scheduling under timing constraints for mission-critical embedded systems. In: Proceedings of the 38th Design Automation Conference, pp. 840–845 (2001)
11. Lapierre, L., Zapata, R.: A guaranteed obstacle avoidance guidance system: the safe maneuvering zone. Auton. Robot. $32(3)$, 177–187 (2012). Springer Verlag (Germany)

Toward Performance Guarantee for Autonomous Mobile Robotic Mission: An Approach for Hardware and Software Resources Management

Lotfi Jaiem[✉], Lionel Lapierre, Karen Godary-Dejean, and Didier Crestani

Laboratoire D'Informatique Robotique et Microélectronique de Montpellier,
Montpellier, France
{jaiem,lapierre,Karen.Godary,crestani}@lirmm.fr

Abstract. Mission performance is a large concept. It is rarely addressed in the context of autonomous mobile robotics. This paper proposes a generic framework addressing the concept of performance for autonomous mobile robotic mission. Moreover it presents an approach to manage the mobile robot hardware and software resources during the mission execution according to performance objectives. Simulation results illustrate the proposed approach on a patrolling mission example.

Keywords: Performance · Resources management · Autonomy · Robot

1 Introduction

True autonomy requires the ability to decide the way to perform a mission under performance constraints. It consists on choosing the appropriate resources according to the mission current situation. The robot must be also able to react if unexpected events occur like obstacle avoidance or resource failure [1] in order to keep guaranteeing the required performances. If the mission becomes unfeasible, the robot must be able to identify this situation and modify its objectives.

However the concept of performance is not clearly defined in robotics. Some works evaluate specifically the performance of a specific single-robot task, like human following [2] or performance assessment of a group of collaborative robots [3]. Industrial robotics defined many performance criteria like speed, repeatability, accuracy, etc. International standards (ANSI/RIA R15.05, ISO/9283) are defined too. However, some papers globally consider the mobile robot context. Cabelos et al. define in [4] performance metrics for mobile robot navigation. They propose a classification according to safety, trajectory quality and duration to accomplish a task. Several performance metrics are identified in [5] depending on mobile robotic task: SLAM, obstacle avoidance, grasping, etc. The problem of robotic mission guarantee is tackled in [6] using properties (liveness, safety) formal verification but the considered mission is still simple. Moreover nothing is proposed to overcome unforeseen problems during mission execution.

© Springer International Publishing Switzerland 2016
L. Alboul et al. (Eds.): TAROS 2016, LNAI 9716, pp. 189–195, 2016.
DOI: 10.1007/978-3-319-40379-3_19

2 Experimental Context

A Pioneer 3DX© ($V_{Rmax} = 0.75\ m/s$) integrating 16 sonars and 10 bumpers is used. 2 URG-04 LX Hokuyo used for obstacle avoidance, centering motion and robot localization. Localization is also performed using a Kinect© camera and geo-referenced QR-codes. The Kinect is also used for image capture. An embedded lead/acid battery generates up to 259 Wh of energy. The robot communicates with an embedded laptop supporting a real time control architecture implementing the different algorithms. The laptop has its own battery, which is also monitored. Depending on algorithms and sensors used, 7 moving control laws, 3 localization methods and one image analysis control schemes (CS) are available.

The considered mission of 187 meters long is an autonomous patrolling to inspect the state (open/close) of two valves (V1-V2) (Fig. 1). Table 1 presents the mission decomposed into a sequence of ($n_{obj} = 9$) **objectives O**: Go from docking station DS to the valve V1 (**traveling**), the robot rotates in the direction of V1 (**turn toward**), inspects it (**image processing**), and turns back. Then the robot travels from V1 to V2 and inspects the second valve. At the end, the robot goes back to DS. These objectives are performed using n_{task} (≥ 1) concurrent **task(s) T**: Traveling objective needs both forward motion (FM) and location (L) tasks. One task (pure rotation R) is needed for turning toward a valve. Image processing is done with only valve detection (VD) task. Tasks are performed

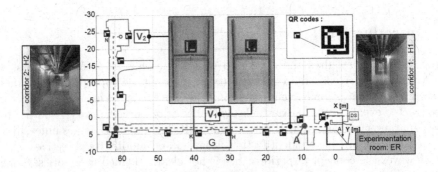

Fig. 1. Mission description

Table 1. Mission description/decomposition and complexity

$x_{i-1} - x_i$ (m)	0 - 34			37	37	37	37 - 93.5				93.5	93.5	93.5	93.5 - 187			
Objective O_i	DS → V1			↻	⋈	↺	V1 → V2				↻	⋈	↺	V2 → DS			
n_{alt_i}	21			2	1	2	21				2	1	2	21			
Task T_k	FM / L			R/L	VD	R/L	FM / L				R/L	VD	R/L	FM / L			
$A_k^{c_j}$	1	2	3	4	5	6	7	8	9	10	11	12	13	14	15	16	17
n_{alt_k}	21	9	3	2	1	2	9	21	21	7	2	1	2	21	21	9	21
$V_{max}(m/s)$																	

usually with n_{opt} (≥ 1) **options (OT)**. 7 implemented OT to perform FM task. These are path following algorithms SMZ with different sensors ($SMZ - US$, $SMZ - LAS$, etc.). Location task L can be performed with 3 OT (GOL is a Grid Oriented Localization technique based on laser data, KIN is the QR code localization method or ODOmeter). An **Alternative Implementation** AI corresponds to the selection, for an objective, a unique OT by task T. The number of alternatives n_{alt} is the product of its corresponding n_{opt}. Different areas can be identified in Fig. 1. In H1, human can be encountered, but not in H2. A glazed area G is also present where sonars are crucial.

3   Proposed Approach for Performance Management

3.1   Mission Performance: Which Relevant Viewpoints?

We distinguish *Main frame* performance viewpoints which must be respected for all autonomous robotic missions: **Safety, energy, localization** and **stability**, from *User's oriented* ones like **duration**.

The following performances constraints are defined for the studied mission: Duration axis (max $D_{perf} = 390$ s), energy axis (*robot energy*: max $E_{perf_R} = 1.9\,Wh$ and *laptop energy*: max $E_{perf_L} = 2\,Wh$), safety axis (Obstacle avoidance: $S_{perf_{OA}} = True$ and harmlessness: max $S_{perf_H} = 4J$ (maximum energy in case of impact with dynamic obstacles)).

Two **Performance Margin** classes are defined. *Boolean margin* characterizes a performance that can only be *True* or *False*. *Continuous margin* defines the gap between performance estimation (or observation) and objectives. The goals are to optimize continuous performance margins with regards to these objectives and to satisfy boolean margins.

3.2   Performance Management: The Proposed Approach

Preliminary Phase: The first step is to identify the performance inductors and to build performance estimation models allowing to predict the robot performances. To estimate the mission performance, a detailed representation of the Nominal Mission Plan (NMP) is needed. It is a sequence of n_{act} ($\geq n_{obj}$) **Activities** (Table 1). An activity $A_k^{c_j}$ is a part of a mission where an objective can be realized under a set of invariant constraints c_j. An activity can be performed using all its possible Alternative Implementations AI.

Off-line Performance Estimation: Once the NMP built, the objective is to estimate the nominal performance along the process, to determine the chosen alternative implementation AI by activity, to specify the value of performance levers in order to respect the performance constraints. So, each AI must be characterized with regard to each performance axis. The second condition is the ability to estimate the global performance of the NMP mission, by composing the local performance estimation for each eligible alternative implementation.

Following the off-line estimation phase, if performance constraints can be a priori satisfied, an alternative implementation can be selected for each activity. Initial Resources Allocation Solution is $RAS_0 = \{AI_0, ..., AI_{n_{act}}\}$.

Online Performance Evaluation and Resources Management: According to the current RAS, an estimation of the current performance behavior is available and periodically compared with evaluated performance to decide if the mission remains feasible. However, a faulty hardware component or software module can disqualify the configuration of some current or future mission activities corresponding to an AI using these faulty elements. A loss of time and energy (environment dynamism) can lead to negative performance margins. In order to consider these situations, the previous offline performance analysis is used online, on the remaining part of the mission, providing in real time a new set of alternative implementations to realize future activities (if it exits).

4   Resources Management Implementation

The mission plan being created, the mission duration M_D is estimated by adding the n_{act} duration of all A_k activities. Activities could have predefined constant duration d_k (static activities where $x_k = x_{k-1}$) or it can depend on the corresponding robot traveling length ($x_k - x_{k-1}$) and velocity V_k.

Energy estimation models are experimentally identified and expressed in [7].

The mission safety viewpoint is implemented through two safety indicators: *obstacle avoidance ability* and safe traveling (*harmlessness* of the robot movement). So, for the considered robot, the following limitations must be respected:

- Sonars are the only efficient sensors in the glazed area (G) (other sensors can be used also for obstacle avoidance). In this area $V_{max} = 0.46\ m/s$.
- Obstacle avoidance is required in the presence of human (H1 area). If lasers are used $V_{max} = 0.56\ m/s$.
- Area with no human presence (H2 area) $V_{max} = V_{Rmax} = 0.75\ m/s$.

Resources management consists on determination for each activity, the AI (algorithms and sensors) and its parameter(s) (robot velocity) that must be locally chosen for each activity to globally satisfy performance objectives.

For a mission, the Number of Global Alternatives (NGA) is equal to the product of each number alternatives implementations (n_{alt_k}) by the n_{act} activities. It becomes quickly huge ($NGA > 10^{13}$ for Table 1). It is a classical NP-hard Knapsack problem. To solve it in a real time context, the algorithm proposed in [8] has been adapted to the robotic context.

5   Simulation Results

The studied mission and robot system (cf. Sect. 2) with D, S and E axes are now considered. Table 1 summarizes the mission description (Objectives O_i) and then mission decomposition (Activities A_k). Grey color expresses constant duration objectives. Projecting the 10 m length glazed area (G) on the two ways mission is [31, 41 m] and [146, 156 m]. Zone without human presence $(H2)$ linear projection coordinates are [63.5, 123.5 m]. These areas impact the constraints (maximum velocity and eligible $\{AI\}$) from safety viewpoint. Row n_{alt_i} in Table 1 shows the number of AI by objective. $A_k^{c_j}$ row shows the number of activities by objective. n_{alt_k} shows the number of AI for the corresponding activities. It is reduced if some AI are non illegible respecting safety constraints. The mission initially composed by 9 objectives is then composed by 17 activities with different constraints. Colored boxes in the last row express maximum linear velocity V_{max} depending on activities. In blue maximum velocity is $0.46 \, m/s$, green $0.56 \, m/s$, orange $0.75 \, m/s$ and white $0 \, m/s$ (pure rotation).

Once the decomposition is done, mission feasibility is tested considering the performance axes and a first resource allocation RAS_0 is calculated (Table 2).

During the mission (Fig. 2), 8 obstacle avoidances (OA) occur and energy (laptop and robot) and duration margins decrease (Fig. 3). Robot energy margin becomes negative twice (33 m and 62 m linear coordinates) and respectively two switches RAS_1 and RAS_2 were done to overcome these perturbations and ensure mission feasibility. Table 2 shows the details of the generated solutions (AI by activity), the global number of possible alternatives implementations NGA and the number of iterations IT needed to find a new solution.

To overcome the energy loss during obstacle avoidance maneuver, the algorithm switch the selected AI (initially SMZ path following algorithm with two

Table 2. Generated resources allocation solutions

A_k	1	2	3	4	5	6	7	8	9	10	11	12	13	14	15	16	17	$NGA \geq$	IT
RAS_0	1	1	1	2	3	2	1	1	1	1	2	3	2	1	1	1	4	$2 \cdot 10^{13}$	724
RAS_1		1	1	2	3	2	1	1	1	1	2	3	2	1	1	1	5	$1 \cdot 10^{12}$	651
RAS_2								1	1	1	2	3	2	1	1	1	6	$1 \cdot 10^{9}$	499
RAS_3														7	7	7	7	$1 \cdot 10^{4}$	187

(1): SMZ-2LAS-US/KIN, (2): OPR/KIN, (3): VALVE ANALYSIS, (4): SMZ-2LAS/KIN,
(5): SMZ-US/KIN, (6): SMZ-US/NONE, (7): CENTERING-2LAS-US/GOL

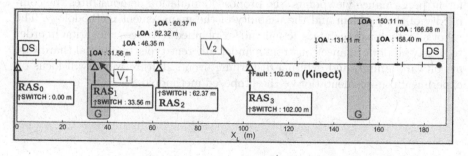

Fig. 2. Mission progress and events.

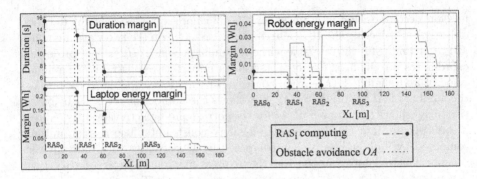

Fig. 3. Duration, laptop and robot energy margins.

lasers and Kinect for KIN localization) for the activity 17 to a less consuming OT with only sonars for RAS_1 and deactivating the Kinect for RAS_2 (localization based on odometer data).

At 102 m, the Kinect fails. Since it is planned to be used in a future activities in RAS_2 for activities 14, 15 and 16, a new RAS_3 is found after eliminating/filtering the sets of AI for the rest of activities. Robot energy margin increases because the new RAS is less consuming than the previous one. Duration margin increases too. This is due to the localization method GOL based on lasers data that allows the robot to run faster ($V_k = V_{Rmax} = 0.75\ m/s$) than with Kinect $RAS_{0,1,2}$ (blurred image beyond $V_k = 0.6\ m/s$). Margins increasing permits to tolerate the rest of the occurred obstacles avoidance. At the end, green boxes in Table 2 show the executed AI along the mission.

6   Conclusion

This paper proposes a methodology for autonomous resources management and a conceptualization of performance on mobile robotics. Based on mission description and regarding to performance constraints, this methodology determines which hardware and software must be used for each mission activity. From different performance viewpoints, the proposed simulation demonstrates the complexity of the problem and the usability of the management methodology. The robot adapts dynamically its actual and/or planned resources allocation in order to satisfy all performance constraints under different types of internal (hardware or software failure) and external disturbing events. Future works will focus on experimental implementation of the proposed methodology.

References

1. Crestani, D., Godary-Dejean, K., Lapierre, L.: Enhancing fault tolerance of autonomous mobile robots. Robot. Auton. Syst. **68**, 140–155 (2015)
2. Doisy, G., Jevtic, A., Lucet, E., Edan, Y.: Adaptive person-following algorithm based on depth images and mapping. In: IEEE/RSJ International Conference on Intelligent Robots and Systems (IROS), Vilamoura, Algarve, Portugal (2012)
3. Wellman, B.L., Erickson, B., Suriel, T., Mayo, K., Phifer, T., Acharya, K.: Effect of wireless signal attenuation on robot team performance. In: Proceedings of the 27th International Florida Artificial Intelligence Research Society Conference, pp. 412–417 (2014)
4. Cabelos, N.D.M., Valencia, J.A., Ospina, N.L.: Quantitative performance metrics for mobile robots navigation. In: Berrera, A. (ed.) Mobile Robots Navigation, pp. 485–500 (2010). ISBN 978-953-307-076-6
5. Bonsignorio, F., Hallam, J., Del Pobil, A.P.: Good Experimental Guidelines. In: European Robotics Network NoE, pp. 1–25 (2008)
6. Lyons, D.M., Arkin, R.C., Nirmal, P., Jiang, S.: Designing autonomous robot missions with performance guarantees. In: IEEE International Conference on Intelligent Robots and Systems, pp. 2583–2590 (2012)
7. Jaiem, L., Druon, S., Lapierre, L., Crestani, D.: A step toward mobile robots autonomy: energy estimation models. In: Proceedings of the 17th Towards Autonomous Robotic Systems, Sheffield, U.K. (2016)
8. Bennour, M., Crestani, D., Crespo, O., Prunet, F.: Computer aided decision for human task allocation with mono and multi performance evaluation. Int. J. Prod. Res. **43**(21), 4559–4588 (2005)

Using Google Glass in Human–Robot Swarm Interaction

Gabriel Kapellmann-Zafra[1]([✉]), Jianing Chen[2], and Roderich Groß[1]

[1] Sheffield Robotics and Department of Automatic Control
and Systems Engineering, The University of Sheffield, Sheffield, UK
{gkapellmann,r.gross}@sheffield.ac.uk
[2] School of Electrical and Electronic Engineering,
The University of Manchester, Manchester, UK
jianing.chen@manchester.ac.uk

Abstract. We study how a human operator can guide a swarm of robots when transporting a large object through an environment with obstacles. The operator controls a leader robot that influences the other robots of the swarm. Follower robots push the object only if they have no line of sight of the leader. The leader represents a way point that the object should reach. By changing its position over time, the operator effectively guides the transporting robots towards the final destination. The operator uses the Google Glass device to interact with the swarm. Communication can be achieved via either touch or voice commands and the support of a graphical user interface. Experimental results with 20 physical e-puck robots show that the human–robot interaction allows the swarm to transport the object through a complex environment.

1 Introduction

The cooperative transport of large objects by groups of comparatively small robots is a canonical task studied in collective robotics [4,5,8,10]. Algorithms for cooperative transport need to cope with individual failure and different environmental conditions.

Chen et al. [2] proposed an occlusion-based cooperative transport algorithm that does neither require the robots to communicate with each other, nor to consistently perceive the goal. Once the object has been found, a robot would push perpendicular to the surface, but only if it had no line of sight of the goal. In this situation, the robot's view of the goal is assumed to be occluded by the object. Under the assumption of quasi-static movement, it was proven that any convex object will always reach the goal when using this strategy [2].

A limitation of the occlusion-based cooperative transport algorithm is that it assumes the environment to be free of obstacles. If the line of sight between a robot and the goal is occluded by anything but the object (e.g., walls or obstacles), the strategy will not work. The challenge is to allow the robots to negotiate obstacles during cooperative transport without needing to increase their complexity [6]. Wang et al. [11] propose a leader-follower system, where

L. Alboul et al. (Eds.): TAROS 2016, LNAI 9716, pp. 196–201, 2016.
DOI: 10.1007/978-3-319-40379-3_20

a human guides the robots during the transportation task. The human exerts physical forces onto the object, causing it to displace. The robots, which are grasping the object, measure these forces and comply with them, by simulating the behavior of passive casters [9]. In this strategy, the human communicates with the robots only indirectly (through the object). However, this requires the human to physically interact with the object, and produce the lateral forces resulting in its displacement.

Rather than interacting with the object physically, we study how human operators can collaborate with a swarm of robots through portable devices. In [3], a tablet is used to interact with the robots in a swarm. The tablet shows the operator the relative positions of the robots (obtained using an overhead camera system). The operator can then modify the positions of the robots through the tablet. A similar interaction is studied in [1] where body gestures are used as command inputs through a Kinect to organize and position the swarm. The main difference of our system is that the human needs to interact with only a single robot, and yet gains control over the cooperative actions of the entire swarm. The latter comprises simple miniature robots of very low computational power. The leader is a robot of this swarm. A preliminary investigation using hand-held devices was reported in [2]. The present paper focuses on a more advanced interface—the Google Glass—which enables the operator to interact with the swarm in multiple ways, via touch or hands-free commands.

2 Cooperative Transport Using a Swarm of Robots

The experiment is conducted with the e-puck (see Fig. 1a), which is a differential wheeled miniature mobile robot [7]. The e-puck is controlled by the on-board Microchip dsPIC30F6014A. It has 8 kB RAM and 144 kB ROM. The e-puck has a directional camera mounted in its front, eight infrared (IR) proximity sensors and Bluetooth connectivity. To make its appearance more uniform, the e-puck was fitted with a black "skirt" (not shown in the figure).

The experimental setup is similar to the one used in [2]. The environment is a rectangular arena of 400×225 cm. Two obstacle walls, each of 112 cm side

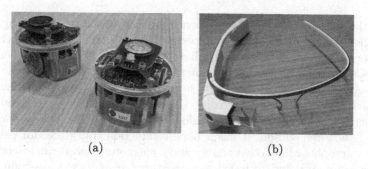

(a)   (b)

Fig. 1. (a) The e-puck robot. (b) The Google Glass.

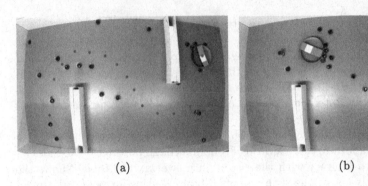

(a)   (b)

Fig. 2. A human operator, wearing the Google Glass, guides a swarm of robots that is transporting a circular object through an arena with obstacles. The swarm consists of 1 leader robot (in red) and 20 follower robots. (a) Red markers indicate the example trajectory of the object throughout a trial where the final destination was the bottom-left corner of the arena. (b) Four follower robots push the object towards the position of the leader, thereby gradually approaching the final destination. (Color figure online)

length (half of the arena width), are added as shown in Fig. 2a. The object to be transported is a blue cylinder of 42 cm diameter.

We use the occlusion-based cooperative transport controller as detailed in [2]. The robots initially move randomly through the environment, avoiding walls and each other using their proximity sensors. Once a robot detects the blue object with its camera, it approaches it. Once in contact, it performs one revolution to scan its environment for the goal—which is assumed to be of red colour. If the goal is not visible, the robot pushes the blue object for a fixed duration and then repeats scanning for the goal (the repeat scan is not necessary if two fellow robots are present to its left and right sides, respectively). Otherwise, it follows the object's perimeter and approaches it from a different angle. Full details of the controller are reported in [2].

3   Human–Robot Swarm Interaction

To give the operator the ability to influence the swarm, one robot is configured to be a leader. The leader robot is equipped with a red cylinder—to be recognized by the other robots as the (intermediate) goal. In addition, the leader robot activates the red LEDs along its perimeter.

The objective of the operator is to help the swarm transport the blue object from one corner to the other, avoiding the wall obstructions. The operator and the leader robot interact via the Google Glass (Fig. 1b). The device works as a modern hands-free add-on. It has a bone conduction transducer for delivering sound, a microphone, an accelerometer and a magnetometer. It can be controlled via voice commands and/or touch input gestures. It offers wireless connectivity through Wi-Fi and Bluetooth 4.0LE technology. A small interferometric high

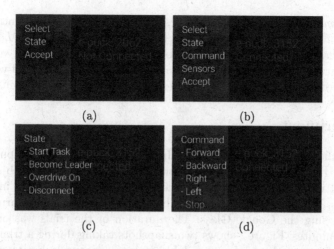

Fig. 3. Sequence of snapshots taken from the Google Glass interface. (a) Main menu when no connection to a robot is established; (b) main menu when a connection to the leader robot is established; (c) state menu when leadership is not initiated yet; (d) state menu during teleoperation.

resolution display, positioned at the right upper side of the operator's visual range, shows a graphical interface. It allows the operator to connect via Bluetooth directly to the leader robot and gain control of the robot's movement. While the robots' sight may be occluded by the obstacles in the environment, the human has the advantage of a bird's-eye view. This way the swarm can focus on the physical manipulation, while the operator focuses on the overall guidance.

Figure 3 shows the graphical interface that is presented to the operator via the Google Glass. Figure 3a shows the main menu when no connection to the leader robot has been established. When the operator chooses the *Select* instruction, they are shown a list of robots that are detected via Bluetooth. They then select the leader robot via the *Accept* instruction and connect to it via the *State* menu.

Once connected, the Google Glass automatically updates the options in the main menu (Fig. 3b) and *State* menu (Fig. 3c). The new options in the *State* menu are:

- *Start Task*: Executes the default behavior of the robot—the occlusion-based cooperative transport controller.
- *Become Leader*: Executes the leader mode—enabling the operator to teleoperate the robot.
- *Overdrive On*: Instructs the robot to ignore any commands issued by a remote control[1].
- *Disconnect*: Terminates the wireless connection to the robot.

---

[1] In the experiment, all robots get activated simultaneously by issuing a signal via an IR remote control.

The *Commands* menu (Fig. 3d) gives the operator complete control of the movement of the leader. The options are: {*Forward, Backward, Right, Left, Stop*}.

The interactions with the GUI can be performed via touch and voice commands. Instructions issued via voice command require a sequence of words; the operator needs to say *OK Glass*, followed by the name of the menu or instruction.

4 Experiment

A set of multiple trials were performed. In each trial, a total of 21 e-puck robots were used: 1 leader robot and 20 follower robots. Initially, the object to be transported was put in one corner (either top right or bottom left in Fig. 2a). In all trials, the human operator was able to lead the pushing swarm along a trajectory using the Google Glass. The duration of the trials was on average about ten minutes. Figure 2 shows two snapshots taking during a trial. A video recording of one of the trials is available at http://naturalrobotics.group.shef.ac.uk/supp/2016-002. A documentary, show-casing the experiment, featured in the *Daily Planet* program of the Discovery Channel in 2015.

The operator had direct visual contact with the robots and object throughout the trials. This feedback helped the operator to maneuver the leader robot at an appropriate pace, in response to the object's displacement. If the swarming robots were moving substantially faster, an alternative approach would be a semi-autonomous leader, which prevents obstacles by itself while getting high-level direction input by the human.

The Google Glass interface allowed the operator to influence the swarm via either touch or voice commands. The voice command option turned out to be the preferable one by the operator, as more direct and allowing a hands-free interaction with the swarm. Despite the ease-of-use of the Google Glass interface, it presented some performance issues—such as overheating—which resulted in delays of response time during prolonged use and low performance of the in-built display.

5 Conclusions

This paper proposed the use of a hands-free device (Google Glass) to gain control over a swarm of robots in a cooperative transport task. By introducing a human operator, it was possible to choose dynamically the goal to which the object was being transported, enabling the system to negotiate obstacles. The system has very low communication requirements—the robots do not need to communicate (explicitly) with each other and the operator communicates, via a portable device, with a leader robot using simple commands. Yet, the operator has enough influence over the entire swarm, and is able to direct the collective force such that the object moves in the desired direction. Through physical experiments we demonstrated that the operator's interactions resulted in a positive global feedback to the system. Future work will focus on how the operator could interact with the swarm without having a bird's-eye view of the environment.

Acknowledgment. The first author acknowledges scholarship support by CONA-CYT (Consejo Nacional de Ciencia y Tecnologia). The last author acknowledges support by a Marie Curie European Reintegration Grant within the 7-th European Community Framework Programme (grant no. PERG07-GA-2010-267354).

References

1. Alonso-Mora, J., Haegeli Lohaus, S., Leemann, P., Siegwart, R., Beardsley, P.: Gesture based human-multi-robot swarm interaction and its application to an interactive display. In: 2015 IEEE International Conference on Robotics and Automation (ICRA), pp. 5948–5953. IEEE (2015)
2. Chen, J., Gauci, M., Li, W., Kolling, A., Groß, R.: Occlusion-based cooperative transport with a swarm of miniature mobile robots. IEEE Trans. Robot. 31(2), 307–321 (2015)
3. Grieder, R., Alonso-Mora, J., Bloechlinger, C., Siegwart, R., Beardsley, P.: Multi-robot control and interaction with a hand-held tablet. In: ICRA 2014 Workshop on Multiple Robot Systems. IEEE (2014)
4. Groß, R., Dorigo, M.: Evolution of solitary and group transport behaviors for autonomous robots capable of self-assembling. Adapt. Behav. 16(5), 285–305 (2008)
5. Kube, C.R., Zhang, H.: Task modelling in collective robotics. Auton. Rob. 4(1), 53–72 (1997)
6. Miyata, N., Ota, J., Arai, T., Asama, H.: Cooperative transport by multiple mobile robots in unknown static environments associated with real-time task assignment. IEEE Trans. Robot. Autom. 18(5), 769–780 (2002)
7. Mondada, F., Bonani, M., Raemy, X., Pugh, J., Cianci, C., Klaptocz, A., Magnenat, S., Zufferey, J.C., Floreano, D., Martinoli, A.: The e-puck, a robot designed for education in engineering. In: Proceedings of the 9th Conference on Autonomous Robot Systems and Competitions, vol. 1, pp. 59–65 (2009)
8. Pereira, G.A.S., Campos, M.F.M., Kumar, V.: Decentralized algorithms for multi-robot manipulation via caging. Int. J. Robot. Res. 23(7–8), 783–795 (2004)
9. Stilwell, D.J., Bay, J.S.: Toward the development of a material transport system using swarms of ant-like robots. In: 1993 IEEE International Conference on Robotics and Automation (ICRA), vol. 1, pp. 766–771 (1993)
10. Tuci, E., Groß, R., Trianni, V., Mondada, F., Bonani, M., Dorigo, M.: Cooperation through self-assembly in multi-robot systems. ACM Trans. Auton. Adapt. Syst. 1(2), 115–150 (2006)
11. Wang, Z.D., Hirata, Y., Takano, Y., Kosuge, K.: From human to pushing leader robot: Leading a decentralized multirobot system for object handling. In: 2004 IEEE International Conference on Robotics and Biomimetics (ROBIO). pp. 441–446. IEEE (2004)

A Virtual Viscoelastic Based Aggregation Model for Self-organization of Swarm Robots System

Belkacem Khaldi[✉] and Foudil Cherif

LESIA Laboratory, Department of Computer Science,
University of Mohamed Khider, B.P. 145, R.P. 07000 Biskra, Algeria
khaldi.belkacem@gmail.com, foudil.cherif@lesialab.net

Abstract. We report a bio-inspired control model to dynamically self-organize a swarm robots system into unplanned patterns emerged through an aggregation method based upon using virtual viscoelastic links among the K-nearby robots. By varying this neighbourhood relationship, virtual viscoelastic links are dynamically created and destroyed between the robots and their sensed neighbours. Based on the equilibrium between these virtual links, the model can distribute the robots at equal angular configurations of the emergent shape being formed. A forward dependent angular motion control is designed to control at which speeds the robots are moving. The model is implemented and tested using the ARGoS simulator where many emerged self-organized configurations are formed showing the effectiveness of the model.

Keywords: Aggregation in swarm robotics · Self-organization in swarm robotics · Virtual viscoelastic model · Decentralized swarm robot system

1 Introduction

As a self-organization basic swarm behaviour that involves flocking, grouping, and herding; aggregation (or gathering together) is an important collective natural behaviour mechanism to be helpful in different tasks like survival of individuals, avoidance of predators, increase of chances in finding foods, etc. [23]. For the same reason, aggregation is a desired behaviour that is being applied in multi-agents as well as swarm robots systems. Moreover, a lot of collective behaviours that are being perceived in biological swarms and which some of them are possibly implemented in engineering of multi-agents and swarm robots systems emerge in aggregated swarms [8].

Shape (or pattern) formation, one of the behaviours which emerges from aggregation of swarms, is an interesting characteristic of swarm robotics self-organization problems that are recently driving an important interest in real world. Most of the control approaches conceived for multi-robots formation problems make use of biological inspiration. The Boids model introduced by Reynolds [16], is one of the earliest individual model that mimics birds flocking behaviour using distance metrics. The most common use of this model in swarm robotics

© Springer International Publishing Switzerland 2016
L. Alboul et al. (Eds.): TAROS 2016, LNAI 9716, pp. 202–213, 2016.
DOI: 10.1007/978-3-319-40379-3_21

is in the form of artificial physics, an approach introduced by Spears et al. [19] as a physicomimetics framework. The approach was able to drive large groups of agents moving into a desired formation such as a hexagonal lattice through using two types of physics force laws: Newtonian force law and Lennard-Jones force law [18]. Further, the framework was extended to handle moving formations through obstacle fields [10,11].

Derived from this approach, other works have been released using different virtual physics forces. For example, Howard et al. [12] used virtual electric charges to model the deployment of robots into an unknown area, Moeslinger et al. [14] applied repulsive and attractive virtual forces for swarm robots flocking behaviour. Hashimoto et al. [9] suggested a swarm robotic control algorithm based on the center of gravity of the local swarm through making use of virtual forces, local forces and an advancing force.

Moreover, virtual spring based control models have recognized a significant interest in the last years. Most of the issued works in this context are based on the graph theory. As instance: to deploy a huge number of swarm robots, a fully Distributed Robotic Macrosensor (DRM) control mechanism was proposed by Shucker and Bennett [17], the DRM control involved a set of formation algorithms based on virtual spring mesh connectivity. A fully decentralized switched spring mesh model for a multi-robots wireless communication navigation problem was investigated in the works of Bezzo and Fierro [1,2], the problem consisted of moving the robots of the system in a swarming manner with maintaining communication connectivity while searching moving targets in a two-dimensional environment populated by obstacles.

Some works combine virtual physics based spring models with other methods such as leader and follower methods. For example, in the work of Dewi et al. [6], a wedge navigation formation is created by a flock of robots using simple virtual spring-damper model between the leader and the followers. Urcola et al. [22] presents a virtual structure based on spring-damper elements to control a navigation system composed of leader and followers robots. Jeong and Lee [13] underlines a dynamic virtual spring damper model for an artificial swarm system, where a dispersion and line formation algorithms are proposed to realize attractive and repulsive forces between the artificial agents and their neighbours.

The main focus of this paper is to design a swarm robots formation control model by which unplanned shapes are emerged using an aggregation based arrangements mechanism. The mechanism uses a virtual viscoelastic links among nearby robots, the viscoelastic model adopted here is a Voigt model consisted of a Hookean elastic spring and a Newtonian damper connected in parallel. This differs from the formation control models reported in [1,2,17], where virtual dampers are added separately to decrease the oscillations caused by virtual springs. Such an idea has recently been realized in the formation control model proposed by Chen et al. [5], where a fully Virtual Spring-Damper Mesh (VSDM) is used to distribute under a gravitational potential field a system of flying spacecraft agents modelled as point zero masses.

Most of the works using a full spring-damper connectivity model robots as simple point models and generally apply fixed spring and damper coefficients to virtually connect the robots of the swarm. Our model, on the other hand, includes so much detail about the robot model and uses dynamic parameters for the spring and the damper coefficients. For that, we follow a design method based on the artificial physics framework [18] and used in [20]; this makes our model more detailed and realistic than the models used in [5,17]. The model also incorporates collision and obstacle avoidance through applying repulsive virtual forces that push away the robots from themselves or potential encountered obstacles. We propose a new robot motion control as a modification version of the variable forward speed motion control used in [20]. Our model is fully decentralized and scalable by which an emergent shape can be achieved whatever the implicated number of the robots. By using the robots range-and-bearing system and varying the robots neighbourhood relationships, virtual spring-damper links are dynamically created and destroyed. Based on the equilibrium of the forces being exerted by each virtual link, the model can distribute the robots at equal angular configurations of the emergent shape being formed. The suggested model is implemented and tested using the ARGoS simulator [15] where many individual robots are self-organized into different emerging shapes showing the effectiveness of the model.

The rest of this paper is structured as follow: first we present in Sect. 2 the method used to implement our formation control model, the section highlights the model setup of the swarm system and reports the different controls being implemented in each robot. Section 3 describes the algorithm used to implement the simulation scenario; then in Sect. 4, we present how the model is simulated with a swarm of foot-bots robots. The results obtained through executing the algorithm is then presented and discussed in Sect. 5. Finally a summary and future works are reported in the conclusion section.

2   Method

We study a swarm system consisted of a set of N robots that are dispersed in a 2D space and which are tending to self-organize into unplanned shapes emerged from the interaction of the neighbours of the robots (Fig. 1). We assume that each individual robot is able to sense its neighbouring robots within a specific range. We define a neighbouring relationship within only the closest K robots among the available neighbours; every robot-robot link is modelled using a virtual viscoelastic link represented as a Voigt model consisted of a Hookean elastic spring and a Newtonian damper connected in parallel. In addition, to avoid collision among robots, each robot of the swarm generates a virtual repulsive potential force around itself. Hence, the robots are subject to the following virtual driven force control:

$$\vec{F}_i = \vec{F}_i^{voigt} + \vec{F}_i^{rep} \tag{1}$$

We define $\vec{F}_i^{voigt}$ as the virtual viscoelastic force control (VVFC) vector and $\vec{F}_i^{rep}$ as the virtual repulsive force control (VRFC) vector. The VVFC is used to keep and arrange the robots together, the VRFC accounts for repulsion rules to keep the robots away from each other or possible detected obstacles. The robots also are subject to a motion control in charge of computing the speed of the robots.

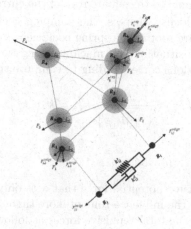

Fig. 1. A model setup of a swarm robots system based on virtual spring-damper connectivity and repulsive potential force

2.1   The VVFC

Let M_i be the neighbours of a given robot $R_i (i \in N)$ within a perception range D_r, and moreover let $M_k \in M_i$ be the K closest neighbours to the robot R_i. The VVFC control vector $\vec{F}_i^{voigt}$ is given by:

$$\vec{F}_i^{voigt} = \sum_{j \in M_k} \vec{F}_{i,j}^{voigt} \qquad (2)$$

We compute the virtual viscoelastic force vector $\vec{F}_{i,j}^{voigt}$ generated between each robot-robot link as follow:

$$\vec{F}_{i,j}^{voigt} = \vec{F}_{i,j}^s + \vec{F}_{i,j}^d = k_{i,j}^s (\vec{d}_{i,j} - d_0) + k_{i,j}^d \vec{v}_{i,j} \qquad (3)$$

With,

- $\vec{F}_{i,j}^s$ and $\vec{F}_{i,j}^d$ are the spring and the damper force vectors exerted between a robot R_i and a robot R_j,
- $k_{i,j}^s$ is the spring coefficient,
- $\vec{d}_{i,j}$ is a displacement vector that represents the current length of the spring between the two robots,
- d_0 is the equilibrium length of the spring,

– $k_{i,j}^d$ is the damping coefficient,
– $\overrightarrow{v}_{i,j}$ is the velocity vector of R_i with regards to R_j.

To calculate such a Voigt force, we use dynamic spring and coefficient parameters rather then fixed ones. Giving k_s and k_d as gain constants, the spring coefficient $k_{i,j}^s$ is inversely proportional to the square root of the current distance measured between the robots on focus; whereas, the damping coefficient $k_{i,j}^d$ is directly proportional to the square root of the spring coefficient $k_{i,j}^s$. This is to guaranty always reaching a stable situation and much reducing the oscillation caused by such systems. The equations corresponding to compute these coefficients are:

$$k_{i,j}^s = \frac{k_s}{\sqrt{\| \overrightarrow{d}_{i,j} \|}}, k_{i,j}^d = k_d \sqrt{k_{i,j}^s} \tag{4}$$

2.2   The VRFC

The VRFC uses a repulsive potential force that acts only in a zone nearby the obstacle. If a robot is in the influence zone of more than one obstacle, the robot is then pushed away by the total repulsive forces as follow:

$$\overrightarrow{F}_i^{rep} = \sum_{j \in O_i} \overrightarrow{F}_{i,j}^{rep} \tag{5}$$

Where O_i is the set of both the robots on collision or the possible detected obstacles. The repulsive potential force vector $\overrightarrow{F}_{i,j}^{rep}$ for a given obstacle is computed using the negative gradient of the potential repulsive energy $U_{i,j}^{rep}$ whose solution is given by the following equation [4]:

$$\overrightarrow{F}_{i,j}^{rep} = \begin{cases} k_{rep}(\frac{1}{L_{obs_j}} - \frac{1}{L_0})(\frac{\overrightarrow{p}_{obs_j} - \overrightarrow{p}_i}{L_{obs_j}^3}), & L_{obs_j} \le L_0 \\ 0, & \text{elsewhere} \end{cases} \tag{6}$$

Where, k_{rep} is a scaling constant, $L_{obs_j} = \| \overrightarrow{p}_{obs_j} - \overrightarrow{p}_i \|$ is the distance between the robot R_i positioned at $\overrightarrow{p}_i$ and the nearest edge of the obstacle obs_j situated in $\overrightarrow{p}_{obs_j}$, and L_0 is the obstacle influence threshold.

2.3   Motion Control

At each control step, a robot R_i updates its forward (v_i) and angular (ω_i) velocities using the immediate resulting virtual driven force control vector $\overrightarrow{F}_i$. We propose a Forward Dependent Angular Motion Control (FDAMC), a modification version of the variable forward speed motion control (VMC) proposed in [7]. In VMC, v_i and ω_i are directly proportional to the coordinate components of a given force. We differ our FDAMC from this work by scaling the angular component (α) of the force vector $\overrightarrow{F}_i$ to get the angular speed ω_i, the angle α is transformed before scaling from radians to degrees. Then the linear speed v_i

is computed as a function of the angular speed ω_i. The two equations of getting ω_i and v_i are as follow:

$$\omega_i = k_\omega(\frac{\alpha * 180}{\pi}),$$

$$v_i = \frac{v_{max}}{\sqrt{|\omega_i|+1}}$$

(7)

Where k_ω is a gain constant and v_{max} is the maximum forward speed of the robot. The equation of v_i guaranties that its value is always within the range $[0, v_{max}]$ whatever the value given to ω_i, this last should be limited within the range $[-\omega_{max}, \omega_{max}]$. Moreover, since we are controlling a two wheels differential drive mobile robot, v_i and ω_i have to be converted into the linear speeds of the left (v_{l_i}) and the right (v_{r_i}) wheels. For that we use the differential drive model applied in [21] as follow:

$$\begin{bmatrix} v_{l_i} \\ v_{r_i} \end{bmatrix} = \begin{bmatrix} 1 & \frac{b}{2} \\ 1 & \frac{-b}{2} \end{bmatrix} \begin{bmatrix} v_i \\ \omega_i \end{bmatrix}$$

(8)

Where b is the distance between the robot wheels.

3   Algorithm Description

The main Algorithm 1, executed at every time step by each individual robot, uses essentially the force resulted from the $VVFC$ function to define the target goal to be achieved by the robot. The $VVFC$ function returns the total virtual viscoelastic force exerted by the k nearby robots. The target goal might be influenced by a total repulsive force generated from detected obstacles or robots en collision, the total repulsive force is calculated in the VRFC function. The algorithm then computes the required velocities to be set to the wheels of the robot using the $FDAMC$ function. The swarm system becomes stabilized where the forces are equilibrated, and at this moment emerging shapes are automatically being created basing on the value of the aggregation parameter K.

Algorithm 1. ViscoelasticBasedSelfOrganizationAlgorithm

---

1 Initialize: $d_0, L_0, b, k_\omega, k_s, k_d, k_{rep}, v_{max}$
2 **begin**
3   **for** *every step time* **do**
4          $M_i \longleftarrow senseNeighbors()$
5          $F_i^{voigt} \longleftarrow VVFC(M_i, K)$
6          $O_i \longleftarrow senseObstacles()$
7          $F_i^{rep} \longleftarrow VRFC(O_i)$
8          $F_i \longleftarrow AddVector(F_i^{voigt}, F_i^{rep})$
9          $setRobotWheelsSpeed() \longleftarrow FDAMC(F_i)$

---

4 Self-organization with Robots

We study self-organization in swarm robots system composed of simulated versions of the foot-bot robot, an autonomous robot that was developed by Bonani et al. [3] and used in the Swarmanoid[1] project.

The foot-bot is a differentially driven mobile robot equipped by a set of sensors and actuators[2]. The ones used in our experiments are:

- A range and bearing sensing and communication device (called RAB), with which a robot is able to communicate with its neighbours and perceive their range and bearing measurements.
- Proximity sensors, by which objects are detected around the robots. The number of the sensors is 24 and they are equally distributed in a ring around the robot body.
- Two wheels actuators that are used for controlling independently the speeds of the left and the right wheels of the robot.

To achieve the VVFC with the foot-bot we use the RAB to communicate the linear speed to the neighbouring robots, and to measure the relative range and bearing (d_{ij} and ϕ_{ij}) of the jth neighbour of the robot R_i on focus. For achieving the VRFC, we use proximity sensors to get the distance L_{obs_j} of the closet detected obstacle and its coresponding angle θ_j. The linear speed of the left and right wheels of the robot is actuated through applying the FDAMC. Both the parameters related to the foot-bot robot and the constants used to compute the VVFC, the VRFC and the FDAMC, and which are chosen for the stability of the system are listed in Table 1.

Table 1. Parameters relative to the foot-bot robot, the VVFC, the VRFC and the FDAMC

Parameter	Description	Value
b	Inter-wheels distance	14 cm
r	Robot wheels radiant	3 cm
v_{max}	FDAMC max forward speed	10 cm/s
ω_{max}	FDAMC max angular speed	180°/s
D_r	Maximum perception range	185 cm
d_0	Equilibrium length of the spring	40 cm
L_0	Obstacle influence threshold	10 cm
k_{rep}	Obstacle scaling constant	1.75 force unit
k_ω	FDAMC angular speed gain	0.9°/s
k_s	Spring gain constant	1.9 force unit
k_d	Damping gain constant	1 force unit

---

[1] http://www.swarmanoid.org/.
[2] http://www.argos-sim.info.

5   Experiments Simulation

To evaluate the performance of the proposed solution; we investigated several simulation experiments using the ARGoS simulator [15]. In each ARGoS-based simulation experiment, the number of the foot-bot robots and the values of the parameter K are varied. In the subsections below we first give the regular lattices arrangements formed when a limit number of robots are aggregated. Then, based on these basic regular configurations, we illustrate the variation in shapes that are emerged when a big number of swarm robots are implicated.

5.1   Basic Regular Geometrical Arrangements

Our algorithm is able to arrange a swarm robots in basic regular geometrical configurations such as triangles, squares, pentagons, or hexagons, etc. The achievement of these configurations strongly belongs to the value of K. To demonstrate how these arrangements can be formed, we restricted the number of the simulated robots in the way that it has to be equal $(K+1)$. The illustrations in (Fig. 2) show the different possible geometrical configurations that can be achieved, as an instance: if $K=2$, which means that the corresponding viscoelastic force will be computed with regards to the 2-nearest neighbours, the resulted arrangement of three robots is then a triangle or a line (Fig. 2a); while in (Fig. 2b) the achieved configuration of four robots is a square for a value of $K=3$; whereas with $K=4$, five robots can position themselves either in a square configuration with a robot at the center or in a pentagon configuration (Fig. 2c). In the same way, other regular geometrical configurations can be also achieved such as a regular hexagon formation by six robots (Fig. 2d) or seven robots (Fig. 2e), and a Heptagon formation by eight robots (Fig. 2e) or nine robots (Fig. 2f).

The basic geometric regular arrangements obtained in these results are strongly related to the parameters used in the simulation; for example the initial length (d_0) imposes a unique distances between robots and hence it controls the size of the formed shape, the FDAMC angular speed gain (k_ω) controls the angular speed of the robots and this impose a direct influence on the linear speed of the right and the left wheels of the robots, which may affect the entire stabilization of the formed shape.

5.2   Emerging Shapes Formation Throw Regular Geometrical Configuration

Basic regular geometrical configurations achieved above can be used as a base arrangement of a very large number of swarm robots; the algorithm shows its ability to emerge shapes basing on the geometrical shape to be chosen as a configuration based arrangement. The illustrations highlighted in (Fig. 3) demonstrate the variation in shapes, which are emerged from diverse large number of swarm robots through giving different values to K. The figure shows different ARGoS simulations snapshots of how our algorithm yields to dynamically emerge various shapes by a swarm system composed of N robots throw only varying the

(a) Formation of a triangle (left) or a
line (right) ($N = 3, K = 2$)

(b) Formation of a square
($N = 4, K = 3$)

(c) Formation of a square with a
central robot (left) or a regular
Pentagon (right) ($N = 5, K = 4$)

(d) Formation of a regular Hexagon
($N = 6, K = 5$)

(e) Formation of an Hexagon with a
central robot (left) or a regular
Heptagon (right) ($N = 7, K = 6$)

(f) Formation of an Heptagon with a
central robot (left) or a regular
Octagon (right) ($N = 8, K = 7$)

Fig. 2. Basic regular shapes possibly formed by N robots running our algorithm during a maximum of 500 time simulation units each. K: aggregation based on K-nearest neighbours

method of how the neighbour robots should be aggregated. The snapshots are presented by three columns from left to right. Each column, starting from the initial positions of the N robots to diversifying the values of K (from 2 to 7), illustrates different separate experiments of the corresponding emergent final shapes achieved during running the algorithm. The right column of the figure demonstrates a situation in which obstacles can affect the bearing sensing of the robots and hence it disturbs the resulting formed shapes as a result of troubling the aggregations of the robots. The snapshots also demonstrate the ability of our K-nearest virtual viscoelastic based aggregation algorithm to self-organize robots into different shapes by different clusters of swarm robots.

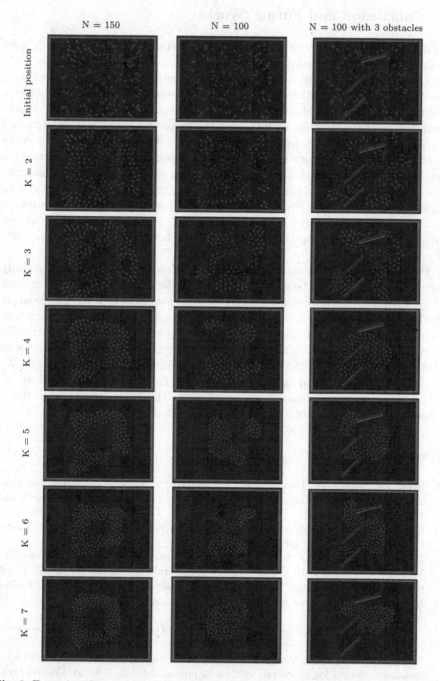

Fig. 3. Emerging shapes formed by a swarm of N robots running our algorithm during a maximum of 2000 time simulation units each. K: aggregation based on K-nearest virtual viscoelastic neighbours

6   Conclusion and Future Works

We have addressed the problem of controlling a team of swarm robots to dynamically emerge self-organized patterns basing on intra virtual physics connectivity among neighbours. We use a fully virtual viscoelastic based physics model, with incorporating virtual repulsive forces for avoiding obstacles and collision between robots. We define a neighbouring relationship within only the closest K robots among the neighbours; thus with varying the value of K, we control the virtual viscoelastic links to be created between the robots. As a result, the model shows its ability to arrange robots in basic geometric configurations such as triangles, squares, pentagons, hexagons,..etc., then founding on these geometric configurations; the model is able to emerge different unplanned shapes. The entire virtual physics based model was implemented and tested with tens of foot-bots robots using the ARGoS simulator. The achieved results report various self-organized patterns emerged from applying our model, which is very effective to be extended to a huge number of robots.

Possible directions for future work are the following: First, we plan to study decision making within the same framework of the study. Specifically, we seek for each robot to make decision on how to dynamically switch its neighbouring relationship via an automatic adaptation of the value of K. Second, we would like also to study maintaining cohesivity between clusters of robots. Specifically, we plan to study dynamic adaptation of the equilibrium length of the spring both inside the same cluster of the robots and between different clusters of robots. We think these two points can enhance our model to be applied in real applications such as area coverage or environment exploring. Third, we are willing to study self-organization of the same approach in planned shapes.

References

1. Bezzo, N., Fierro, R.: Decentralized connectivity and user localization via wireless robotic networks. In: 2011 IEEE GLOBECOM Workshops (GC Wkshps), pp. 1285–1290. IEEE (2011)
2. Bezzo, N., Fierro, R.: Swarming of mobile router networks. In: American Control Conference (ACC), pp. 4685–4690. IEEE (2011)
3. Bonani, M., Longchamp, V., Magnenat, S., Rétornaz, P., Burnier, D., Roulet, G., Vaussard, F., Bleuler, H., Mondada, F.: The marxbot, a miniature mobile robot opening new perspectives for the collective-robotic research. In: 2010 IEEE/RSJ International Conference on Intelligent Robots and Systems (IROS), pp. 4187–4193. IEEE (2010)
4. Castañeda, M.A.P., Savage, J., Hernández, A., Cosío, F.A.: Local autonomous robot navigation using potential fields. In: Motion Planning. InTech (2008)
5. Chen, Q., Veres, S.M., Wang, Y., Meng, Y.: Virtual spring-damper mesh-based formation control for spacecraft swarms in potential fields. J. Guid. Control Dyn. 38(3), 539–546 (2015)
6. Dewi, T., Risma, P., Oktarina, Y.: Wedge formation control of swarm robots. In: 14th Industrial Electronics Seminar IES (2012), Electronic Engineering Polytechnic Institute of Surabaya (EEPIS), Indonesia, pp. 294–298, 24 October 2012

7. Ferrante, E., Turgut, A.E., Huepe, C., Stranieri, A., Pinciroli, C., Dorigo, M.: Self-organized flocking with a mobile robot swarm: a novel motion control method. Adapt. Behav., 1059712312462248 (2012)
8. Gazi, V., Fidan, B., Hanay, Y.S., Koksal, L.: Aggregation, foraging, and formation control of swarms with non-holonomic agents using potential functions and sliding mode techniques. Turkish J. Electr. Eng. Comput. Sci. 15(2), 149–168 (2007)
9. Hashimoto, H., Aso, S., Yokota, S., Sasaki, A., Ohya, Y., Kobayashi, H.: Stability of swarm robot based on local forces of local swarms. In: SICE Annual Conference, pp. 1254–1257. IEEE (2008)
10. Hettiarachchi, S., Spears, W.M.: Moving swarm formations through obstacle fields. In: IC-AI, pp. 97–103 (2005)
11. Hettiarachchi, S., Spears, W.M., Hettiarachchi, S., Spears, W.M.: Distributed adaptive swarm for obstacle avoidance. Int. J. Intell. Comput. Cybern. 2(4), 644–671 (2009)
12. Howard, A., Matarić, M.J., Sukhatme, G.S.: Mobile sensor network deployment using potential fields: a distributed, scalable solution to the area coverage problem. In: Distributed Autonomous Robotic Systems 5, pp. 299–308. Springer, Heidelberg (2002)
13. Jeong, D., Lee, K.: Dispersion and line formation in artificial swarm intelligence. arXiv preprint arXiv:1407.0014 (2014)
14. Moeslinger, C., Schmickl, T., Crailsheim, K.: A minimalist flocking algorithm for swarm robots. In: Kampis, G., Karsai, I., Szathmáry, E. (eds.) ECAL 2009, Part II. LNCS, vol. 5778, pp. 375–382. Springer, Heidelberg (2011)
15. Pinciroli, C., Trianni, V., O'Grady, R., Pini, G., Brutschy, A., Brambilla, M., Mathews, N., Ferrante, E., Di Caro, G., Ducatelle, F., et al.: Argos: a modular, parallel, multi-engine simulator for multi-robot systems. Swarm Intell. 6(4), 271–295 (2012)
16. Reynolds, C.W.: Flocks, herds and schools: a distributed behavioral model. In: ACM SIGGRAPH Computer Graphics, vol. 21, pp. 25–34. ACM (1987)
17. Shucker, B., Bennett, J.K.: Virtual spring mesh algorithms for control of distributed robotic macrosensors. University of Colorado at Boulder, Technical Report CU-CS-996-05 (2005)
18. Spears, W.M., Spears, D.F., Hamann, J.C., Heil, R.: Distributed, physics-based control of swarms of vehicles. Auton. Rob. 17(2–3), 137–162 (2004)
19. Spears, W.M., Spears, D.F., Heil, R., Kerr, W., Hettiarachchi, S.: An overview of physicomimetics. In: Şahin, E., Spears, W.M. (eds.) Swarm Robotics 2004. LNCS, vol. 3342, pp. 84–97. Springer, Heidelberg (2005)
20. Stranieri, A., Ferrante, E., Turgut, A.E., Trianni, V., Pinciroli, C., Birattari, M., Dorigo, M.: Self-organized flocking with a heterogeneous mobile robot swarm. In: Advances in Artificial Life, ECAL pp. 789–796 (2011)
21. Turgut, A.E., Çelikkanat, H., Gökçe, F., Şahin, E.: Self-organized flocking in mobile robot swarms. Swarm Intell. 2(2–4), 97–120 (2008)
22. Urcola, P., Riazuelo, L., Lazaro, M., Montano, L.: Cooperative navigation using environment compliant robot formations. In: IEEE/RSJ International Conference on Intelligent Robots and Systems, IROS 2008, pp. 2789–2794. IEEE (2008)
23. Xue, Z., Zeng, J., Feng, C., Liu, Z.: Flocking motion, obstacle avoidance and formation control of range limit perceived groups based on swarm intelligence strategy. J. Softw. 6(8), 1594–1602 (2011)

A Portable Active Binocular Robot Vision Architecture for Scene Exploration

Aamir Khan, Gerardo Aragon-Camarasa$^{(\boxtimes)}$, and Jan Paul Siebert

School of Computing Science, University of Glasgow, Glasgow G12 8QQ, UK
Gerardo.AragonCamarasa@glasgow.ac.uk

Abstract. We present a portable active binocular robot vision architecture that integrates a number of visual behaviours. This vision architecture inherits the abilities of vergence, localisation, recognition and simultaneous identification of multiple target object instances. To demonstrate the portability of our vision architecture, we carry out qualitative and comparative analysis under two different hardware robotic settings, feature extraction techniques and viewpoints. Our portable active binocular robot vision architecture achieved average recognition rates of 93.5 % for frontoparallel viewpoints and, 83 % percentage for anthropomorphic viewpoints, respectively.

1 Introduction

Active robot vision systems are dynamic observers that exploit recovered information from the imaged scene to perform actions and fulfil tasks [7]. Active robot vision systems mainly comprise hard-wired, ad-hoc visual functions that are intended to be capable of robustly exploring a scene and finding objects contained in a database of pre-trained object examples [9,10]. However, current systems are limited in their visual capabilities and their software modules are crafted according to the robot's specific geometric configuration and hardware components. These limitations constrain the scope of potential applications for such vision systems.

In this paper, we present a portable active binocular robot head architecture that is able to execute *vergence, localisation, recognition and simultaneous identification of multiple target object instances.* In this paper, we focus on the development of a portable architecture while preserving visual behaviours previously reported in [2,3]. We have chosen the Sensor Fusion Effects (SFX) architecture [16] as the foundation for our portable robot head (Fig. 1). We must point out that *our robot architecture is not an attempt to model the mammalian visual pathway itself,* but it is a functional system that robustly carries out the specific high-level task of *autonomous scene exploration.* To demonstrate the portability of our system, we conducted experiments considering three important variables for any active scene exploration tasks, namely; the hardware used, visual representation and view(s) of the scene. Hence, we present experiments with three different state-of-the-art feature extraction techniques, namely SIFT [12], SURF [8] and KAZE [1] and, different hardware and scene settings.

© Springer International Publishing Switzerland 2016
L. Alboul et al. (Eds.): TAROS 2016, LNAI 9716, pp. 214–225, 2016.
DOI: 10.1007/978-3-319-40379-3_22

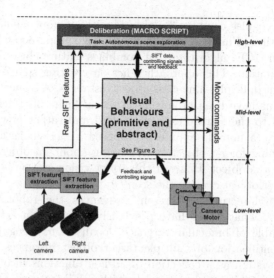

Fig. 1. Our active binocular robot vision architecture.

This paper is organised as follows: Sect. 2 presents a literature review of current robot vision technologies. Sections 3 and 4 presents our robot vision architecture. Finally, Sects. 5 and 6 details the experimental validation of the system and concluding remarks of this paper, respectively.

2   Literature Review

In robotic vision, active vision can potentially offer a sheer amount of information about the robot's environment. Should a visual task becomes ill-posed, the gaze of a robot can be shifted to perceive the scene from a different viewpoint [7]; and therefore a better understanding of the task. Current research in active robot heads has focused on the *"lost and found"* problem [15]. That is, a robot is commanded to search and locate an object in its working environment for exploration tasks [6,10], manipulation tasks [18,20] and/or navigation [15].

In an effort to replicate the nature of visual search scan paths [21], researchers have proposed a variety of visual search mechanisms according to the task at hand (e.g. [13,15,18]). These heuristic approaches are mainly driven by the outputs of available feature extraction techniques. For example, Rasolzadeh et al. [18] used depth to segment the scene according to the distance between a targeted object and the robot as part of a visual object search heuristic. Likewise, Merger et al. [15] implemented a saliency map that combines intensity, colour and depth features to drive attention, biased by a top-down feature detection based on the MSER feature extractor [14] for object recognition and navigation. Aydemir et al. [6] have recently presented a strong correlation between local 3D structure and object placement in everyday scenes. By exploiting the relationship between local 3D structure and different object classes, the authors are able to localise and recognise complex 3D objects without implementing specialized visual search routines. Finally, Collet et al. [10] have proposed an Iterative Clustering Estimation (ICE) algorithm that combines feature clustering along with robust pose estimation. This approach relies on creating sparse 3D models to localise and detect multiple same-class object instances. Advancements in visual search mechanisms have been promising in recent years of which they are not merely restricted to the feature extraction used and rather powered by cognition. For instance, a notable approach proposed in [11] looks at the problem of a robot searching for an object by reasoning about an object and possible interactions with the object. However this robot vision system is limited to one single instance per object class in the scene.

The vision architecture we present, advances the robot vision system described in [2,3]. That is, we have previously reported an active vision system that is capable of binocular vergence, localisation, recognition [2,3] and simultaneous identification of multiple target object instances [4]. We structure this initial system as a collection of ad-hoc functions in order to explore autonomously a scene by operating solely with SIFT features. Our system was also constrained to the hardware and, therefore, the limitation of its portability remained an issue. Recent developments in robotic middleware (e.g. the Robot Operating System [17]) technologies have made possible the deployment of hardware independent robotic systems. We thus propose an active binocular robot head architecture that integrates visual behaviours in a parsimonious and generic robot vision architecture based on the Robot Operating System (ROS).

While we do not make explicit use of 3D information in this paper, an explicit goal was to determine if we could reliably maintain binocular vergence of an actuated stereo-pair of cameras while actively exploring a scene. This converged binocular camera configuration supports the recovery of feature locations in 3D and also provide images for stereo-matching for dense 3D range map extraction. This feature underpins visual competences for other robotic applications as demonstrated in [19] where we presented a dual-arm robot manipulating deformable objects using the binocular system reported in this paper.

3   Robot Vision Architecture

As stated before, we have based our active vision system on the hybrid deliberative/reactive *Sensor Fusion Effector* architecture (SFX, [16]). Specifically, the SFX architecture, as implemented, relates how deliberative and reactive modules are interconnected with sensor and actuator functions. Visual behaviours in our architecture implement the configuration of the visual streams in the mid-level of the SFX architecture. This arrangement exploits sensed visual information in order to explore the environment without further reasoning (i.e. the mid-layer *senses and acts* accordingly) while the deliberative layer manages visual behaviours and, consequently, orchestrates the required set of commands to carry out a *high-level* visual task; for instance, manipulation/interaction tasks [19].

Specifically, Fig. 1 shows our architecture. The processing levels are classified in terms of their function (i.e. low-level, mid-level and high-level).

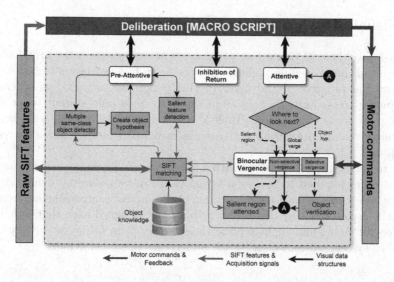

Fig. 2. Internal representation of visual behaviours (Fig. 1). White boxes denote abstract behaviours, whereas grey boxes represent primitive behaviours.

The corresponding low-level and mid-level functions consist of simple yet effective behaviours that subserve upper-level goals, whilst the high-level functions relate to the intelligence, deliberation and reasoning (out of the scope in this paper).

High-level functions (as observed in Fig. 1(a)) specify visual tasks and goals. This layer, this paper, is cast as scripted meta-behaviours (Sect. 4) that orchestrate the sequential activation of visual behaviours in order to fulfil the task of autonomous visual object exploration.

Low-level and mid-level (Figs. 1(a) and 2(b)) integrate a number of *primitive* and *abstract behaviours*. On the one hand, primitive behaviours comprise monolithic methods that only serve a single purpose; i.e. they are simple stimulus-response mappings that transform a collection of sensed information into data structures. On the other hand, abstract behaviours comprise a collection of primitive or other abstract behaviours. Figure 2(b) illustrates the **mid-level** processing architecture that comprises *pre-attentive, attentive, inhibition of return* and *binocular vergence* visual behaviours previously reported in [2,3]. Sensor and motor behaviours are decoupled from the mid- and high-level layers. This configuration allows us to maintain visual behaviours that are not constrained to the chosen feature extraction technique and hardware components.

To achieve generic and preserve a modular arrangement within the architecture, we devised an egocentric coordinate system which are not related to the real-world units of the observed environment. The egocentric coordinate map is defined as a relative pixel-based map where the frame of reference is established with respect to a *"home"* position of the robot head.

4   Visual Search Task Definition

The high-level layer is defined as a macro-script that specifies the visual search task, controls and schedules behavioural resources in lower layers (ref. [3]), and monitors the progress of the task. In this paper, we define a *pre-attentive-inhibition of return-attentive cycle* in order to allow our system to perform autonomous scene exploration (Table 1). That is, the robot acts according to the sensed visual information and reports recognised object classes stored in database.

By replacing the macro script with a cognitive/intelligent layer, the sequence of behaviours required to convey a visual task can be generated deliberatively thereby removing the fixed-task limitation of the current control scheme. Accordingly, the architecture we describe here has been designed such that a deliberative/cognitive module might replace the fixed script in future modifications of the robot system without altering the underlying visual behaviours.

Table 1. Pseudo-code of macro script in Figs. 1 and 2.

Inputs: None
Outputs: List of objects recognised and attended to.

```
 1:  Generate database
 2:  Verge cameras and extract features from the image pair
     (binocular arrangement)
 3:  Obtain pre-attentive object and salient hypotheses
 4:  Set the saccade number to 1
 5:  Loop until possible object or salient hypotheses are not empty
     or no. of saccades is less than a user-defined number
 6:     Select an object from the possible obj. hypotheses that has
        the maximum recognition score (see [3])
 7:     Verge and attend to the selected object and return features
        from both cameras after verging and the lists of the
        remaining object and salient hypotheses
 8:     Update pre-attentive object and salient hypotheses
 9:     Inhibit (inhibition of return) new pre-attentively found
        objects w.r.t previous possible object and salient hyps
10:     Saccade no. increments 1
11:  Report objects stored
```

---

5   Experiments

5.1   Robot Head Hardware and Software Interface

These experiments are designed to validate the portability of our active robot vision architecture in two different scene settings and hardware components.

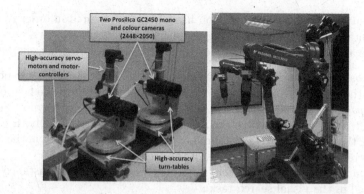

Fig. 3. Left: The Prosiclica robot head exploring the scene. Right: An image of the dual-arm robot featuring the Nikon robot head on top. Additionally, this robot features grippers specifically designed for manipulating clothing [19].

The first active binocular robot head (Fig. 3) comprise two *Prosilica* cameras (*GC2450C* and *GC2450*; colour and mono, respectively) at 5 Mega pixels of resolution fitted with *Gigabit Ethernet* interfaces and 4 high-accuracy stepper-motors and motor-controllers (Physik Instrumente). The robot vision architecture is arranged as follows for the latter robot head. Low-level components, namely, image acquisition and motor control modules (Fig. 1); are interfaced to a Pentium 4 computer with 2 GB in RAM running under Windows XP and MATLAB R2008a. Whilst, image feature extraction, mid-level and high-level components (Fig. 1) are interfaced to a 4-core Intel Xeon (model E5502) with a CPU clock speed of 2 GHz, with 24 GB in RAM running under Windows 7 and MATLAB R2009b. Both computers are interconnected through the local network by means of a collection of network socket functions for MATLAB[1].

The second active binocular robot head (Figure) consists of two Nikon DSLR cameras (D5100) at 16 Mega pixels of resolution. Cameras are mounted on two pan and tilt units (PTU-D46) with their corresponding controllers. This robot head is mounted on a dual-arm robot with anthropomorphic features. Low-level functions where implemented as ROS nodes and interfaced with Matlab 2014a with *pymatlab*[2]. The hardware is interfaced to an Intel Core i7-3930K computer at 3.20 GHz with 32 GB of RAM running Ubuntu 12.04 and ROS.

5.2 Methodology

In order to test the robustness and repeatability of our architecture, for both binocular robot heads, we performed 3 visual exploration tasks for each scene, each visual task with a random initial home position. It must be noted that we terminate the visual search task if the robot's pre-attentive behaviour does not find an object within 5 consecutive saccades; i.e. the system is only targeting

---

[1] http://code.google.com/p/msocket/ (verified on 4 March, 2016).
[2] https://pypi.python.org/pypi/pymatlab (verified on 4 March, 2016).

salient features. This halting criterion has been implemented in order to reduce the execution time while conducting these experiments.

There are three possible outcomes while actively exploring a scene:

- *True positives* comprise all correctly detected and identified object hypotheses where the system is able to centre the hypothesised object in the field of view.
- *False positives* include when the system localises an object hypothesis, but without being able to centre the object in the field of view of both cameras during the attentive cycle or, similarly, an attended object hypothesis does not correspond to the object class in the scene.
- *Not found* comprise the system's failures when an object instance is not detected in the visual search task.

For each robot head, we have arranged scenes comprising a mix of several multiple same-class and different-class object instances, arranged in different poses. We define scene complexity according to the number of similar unknown objects in the scene (i.e. a typical source of potential outliers) and by the degree of background clutter present. We detail below the experimental methodology.

Fig. 4. Left: View from the Prosilica robot head's left camera exploring a scene. Right: View of the Nikon-based robot head as viewed from the left camera.

Prosilica Robot Head. We arranged 7 different scenes[3] of differing complexity, based on combinations of 20 known object instances, of 10 different object classes. Figure 4 shows an example of a scene. Objects were placed in arbitrary poses and locations. We have also created a database of the 10 known objects by capturing stereo-pair images of an object at angular intervals of 45° and 60°. These captured images are then manually segmented in order to contain only the object of interest. We have considered two databases in order to measure the recognition performance of our system with different visual knowledge.

Nikon Binocular Robot Head. Scenes for these experiments consist of objects placed on top of a table. The goal is to investigate the response of our active vision architecture to different viewpoints, different feature extraction techniques and hardware components for the sake of portability. With this

---

[3] All 7 scenes can be accessed at http://www.gerardoaragon.com/taros2016.html.

robot head, we are also able to investigate the effects of having an anthropomorphic robot configuration as opposed to a fronto-parallel configuration as above.
Figure 5 shows examples of the scenes we created. Object databases used in
these experiments include stereo-pair images of object instances sampled randomly in order to cover the objects' view-sphere by placing the object in isolation
on the working table. Each object instance stored in the database is manually
segmented.

We therefore arranged 3 different scenes[4] of variable complexity. Each scene
is a composition of 14 known object instances observing arbitrary poses and
locations, of 9 different object classes. Scene 1 is considered to be the simplest
while scene 3, the most complex (Fig. 5). We must note that Scene 2 and Scene
3 include flat objects and objects with 3D structure while Scene 1 only comprise
objects having 3D structure. In order to effectively understand the response of
the system to different feature extraction techniques, each of the three scenes
were explored by our system with SIFT, KAZE and SURF features.

(a) Scene 1   (b) Scene 2   (c) Scene 3

Fig. 5. Scenes used for the Nikon robot head. (a) Scene 1 depicts less complexity.
(b) Scene 2, medium complexity. (c) Scene 3, most complex scene of the last two.

5.3   Analysis and Discussion

Investigating all experiments and three randomly starting position for each
scene, we can deduce that our active robot vision architecture presents stochastic
behaviours. Accordingly, neither robot vision head follows a pre-defined visual
scan path but it adapts according to the contents of the scene while exploring the
scene. Summary of the outcomes for each robot head are presented as follows.

Prosilica Robot Head. Table 2 illustrates the system's recognition rates for
all experiments. False positives emerged due to the object feature descriptors
matching with unknown objects and, in consequence, these matches were not
consistent with the reference object centre in the database while generating
object hypotheses pre-attentively (as previously reported in [2]). However, the

---

[4] All 3 scenes can be accessed at http://www.gerardoaragon.com/taros2016.html.

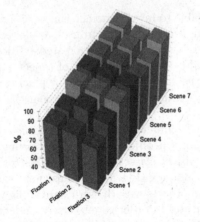

Fig. 6. Overall recognition rate for the visual tasks for the Prosilica robot head.

Table 2. Outcomes for the Prosilica robot head.

Scene no	Performance (%)	True positives	False positives	Not found	Recover from failures
1	82	56	5	7	0
2	93	57	1	3	0
3	97	60	2	0	2
4	97	60	2	0	2
5	91	59	5	1	4
6	98	59	0	1	1
7	97	59	1	1	0
Total	**93.5**	**410**	**16**	**13**	**9**

system recovered from false positives. These results further support the active vision paradigm, since the robot vision architecture is able to recover from these failures while investigating the scene from different views. Thus, the robot is able to locate almost all of the object instances, despite not noticing every object instance present during each pre-attentive cycle.

Nikon Robot Head. From Table 3, we can observe that the recognition performance is linked to the feature extraction techniques used. Average recognition rates for SURF, SIFT and KAZE are 60%, 77% and 83% percentage, respectively. SIFT and KAZE, in these experiments, achieved better recognition rates than SURF due to the inherent properties of being "almost" invariant to perspective transformations. It is also worth noting that both SIFT and KAZE techniques are less prone to false positives as opposed to SURF. As we described above, our portable active vision architecture was tested using an anthropomorphic

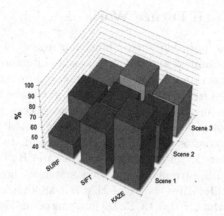

Fig. 7. Outcomes for experiments with the Nikon robot head.

Table 3. Outcomes for the Nikon robot head.

Descriptor	Scene no	Performance (%)	True positives	False positives	Not found	Recover from failure
SURF	1	53	16	5	14	54
	2	66.6	26	4	13	65
	3	59.5	25	2	17	64
Total		**60**	**67**	**11**	**44**	**184**
SIFT	1	80	24	0	6	30
	2	74	29	0	10	54
	3	76	32	0	10	59
Total		**77**	**85**	**0**	**26**	**143**
KAZE	1	100	30	0	0	30
	2	76.9	30	0	9	39
	3	71.4	30	0	12	29
Total		**83**	**90**	**0**	**21**	**98**

configuration where objects are not in similar 2D planes as it is the case from the Proscilica robot head experiments. By comparing Table 3 with Table 2, we can observe a decrease in the performance. That is, 3D structures from an anthropomorphic configuration are more difficult to recognise and, therefore, the robustness of feature descriptions decrease. We can also observe more *recoveries from failures* (last column in Table 3) in these set of experiments. We deduce that this particular configuration introduces challenging geometric transformations that state-of-the-art feature descriptions are still not able to cope with. Hence, the chosen feature extraction has a key role in the overall recognition performance. Nevertheless, our active robot head is able to explore a scene regardless of hardware configuration, different view point while maintaining acceptable recognition rates.

6   Conclusions and Future Work

We have presented a portable active binocular robot head that integrates visual behaviours in a unified and parsimonious architecture that is capable of autonomous scene exploration. That is, our robot architecture can identify and localise multiple same-class and different-class object instances while maintaining vergence and directing the system's gaze towards scene regions and objects.

Our portable robot vision architecture has been validated over challenging scenes and realistic scenarios in order to investigate and study the performance of the visual behaviours as an integrated architecture. By carrying out a qualitative comparison with current robot vision systems whose performance has been reported in the literature, we argue that our architecture clearly advances the reported state-of-the-art [3,5,13,15,18] in terms of our system's innate visual capabilities and portability to different environment settings, e.g. multiple same-class object identification and tolerated degree of visual scene complexity. Our architecture is therefore portable enough in order to be adapted to different hardware configuration, feature description and view-points.

In biological systems, it is found that a region in the scene that is sufficiently salient can capture the attention of an observer more than once during a visual task [21,22]. Our current inhibition of return behaviour, however, has been formulated explicitly to prevent the robot from visiting a previously attended location. We propose to revise this behaviour by incorporating an exponential decay criterion that dictates the mean-lifetime of inhibition of an attended location. The robot would then be able to re-visit a previously attended location, perhaps in the context of a spatial awareness model with a cognitive module.

Acknowledgements. This work was partially supported by the Programme Alβan, the European Union Programme (grant number E07D400872MX) and CONACYT-Mexico (grant number 207703).

References

1. Alcantarilla, P.F., Bartoli, A., Davison, A.J.: KAZE features. In: Fitzgibbon, A., Lazebnik, S., Perona, P., Sato, Y., Schmid, C. (eds.) ECCV 2012, Part VI. LNCS, vol. 7577, pp. 214–227. Springer, Heidelberg (2012)
2. Aragon-Camarasa, G., Siebert, J.P.: A hierarchy of visual behaviours in an active binocular robot. In: Kyriacou, T., Nehmzow, U., Melhuish, C., Witkowski, M. (eds.) Towards Autonomous Robotic Systems, TAROS 2009, pp. 88–95 (2009)
3. Aragon-Camarasa, G., Fattah, H., Siebert, J.P.: Towards a unified visual framework in a binocular active robot vision system. Robot. Auton. Syst. **58**(3), 276–286 (2010)
4. Aragon-Camarasa, G., Siebert, J.P.: Unsupervised clustering in Hough space for recognition of multiple instances of the same object in a cluttered scene. Pattern Recogn. Lett. **31**(11), 1274–1284 (2010)
5. Arbib, M., Metta, G., der Smagt, P.: Neurorobotics: from vision to action. In: Siciliano, B., Khatib, O. (eds.) Springer Handbook of Robotics, pp. 1453–1480. Springer, Heidelberg (2008)

6. Aydemir, A., Jensfelt, P.: Exploiting and modeling local 3d structure for predicting object locations. In: 2012 IEEE/RSJ International Conference on Intelligent Robots and Systems (IROS), pp. 3885–3892, October 2012
7. Ballard, D.H.: Animate vision. Artif. Intell. **48**(1), 57–86 (1991)
8. Bay, H., Ess, A., Tuytelaars, T., Van Gool, L.: Speeded-up robust features (surf). Comput. Vis. Image Underst. **110**(3), 346–359 (2008). http://dx.doi.org/10.1016/j.cviu.2007.09.014
9. Chen, S., Li, Y., Kwok, N.M.: Active vision in robotic systems: a survey of recent developments. Int. J. Robot. Res. **30**(11), 1343–1377 (2011)
10. Collet, A., Martinez, M., Srinivasa, S.S.: The moped framework: object recognition and pose estimation for manipulation. Int. J. Robot. Res. **30**(10), 1284–1306 (2011)
11. Dogar, M., Koval, M., Tallavajhula, A., Srinivasa, S.: Object search by manipulation. Auton. Rob. **36**(1–2), 153–167 (2014)
12. Lowe, D.G.: Distinctive image features from scale-invariant keypoints. Int. J. Comput. Vis. **60**(2), 91–110 (2004)
13. Ma, J., Chung, T.H., Burdick, J.: A probabilistic framework for object search with 6-DOF pose estimation. Int. J. Robot. Res. **30**(10), 1209–1228 (2011)
14. Matas, J., Chum, O., Urban, M., Pajdla, T.: Robust wide baseline stereo from maximally stable extremal regions. In. British Machine Vision Conference, vol. 1, pp. 384–393 (2002)
15. Meger, D., Gupta, A., Little, J.J.: Viewpoint detection models for sequential embodied object category recognition. In: IEEE International Conference on Robotics and Automation (ICRA), pp. 5055–5061 (2010)
16. Murphy, R., Mali, A.: Lessons learned in integrating sensing into autonomous mobile robot architectures. J. Exp. Theoret. Artif. Intell. **9**(2), 191–209 (1997)
17. Quigley, M., Conley, K., Gerkey, B., Faust, J., Foote, T., Leibs, J., Wheeler, R., Ng, A.Y.: Ros: an open-source robot operating system. In: ICRA Workshop on Open Source Software, vol. 3, p. 5 (2009)
18. Rasolzadeh, B., Bjorkman, M., Huebner, K., Kragic, D.: An active vision system for detecting, fixating and manipulating objects in the real world. Int. J. Robot. Res. **29**(2–3), 133–154 (2010)
19. Sun, L., Aragon-Camarasa, G., Rogers, S., Siebert, J.P.: Accurate garment surface analysis using an active stereo robot head with application to dual-arm flattening. In: 2015 IEEE International Conference on Robotics and Automation (ICRA), pp. 185–192, May 2015
20. Sun, L., Aragon-Camarasa, G., Rogers, S., Siebert, J.: Accurate garment surface analysis using an active stereo robot head with application to dual-arm flattening. In: 2015 IEEE International Conference on Robotics and Automation (ICRA), pp. 185–192, May 2015
21. Wolfe, B.A., Whitney, D.: Saccadic remapping of object-selective information. Attention, Percept. Psychophys. **77**(7), 2260–2269 (2015)
22. Wurtz, R.H., Joiner, W.M., Berman, R.A.: Neuronal mechanisms for visual stability: progress and problems. Philos. Trans. Royal Soc. London. Series B, Biolog. Sci. **366**(1564), 492–503 (2011)

Kinematic Analysis of the Human Thumb with Foldable Palm

Visakha Nanayakkara[1]($\boxtimes$), Ahmad Ataka[2], Demetrios Venetsanos[1],
Olga Duran[1], Nikolaos Vitzilaios[1], Thrishantha Nanayakkara[2],
and M. Necip Sahinkaya[1]

[1] School of Mechanical and Automotive Engineering, Kingston University London,
London SW15 3DW, UK
{k1454696,D.Venetsanos,O.Duran,N.Vitzilaios,M.Sahinkaya}@kingston.ac.uk
[2] Department of Informatics, King's College London, London, UK
{ahmad_ataka_awwalur.rizqi,thrish.antha}@kcl.ac.uk

Abstract. There have been numerous attempts to develop anthropomorphic robotic hands with varying levels of dexterous capabilities.
However, these robotic hands often suffer from a lack of comprehensive understanding of the musculoskeletal behavior of the human thumb
with integrated foldable palm. This paper proposes a novel kinematic
model to analyze the importance of thumb-palm embodiment in grasping objects. The model is validated using human demonstrations for five
precision grasp types across five human subjects. The model is used to
find whether there are any co-activations among the thumb joint angles
and muscuroskeletal parameters of the palm. In this paper we show that
there are certain pairs of joints that show stronger linear relationships in
the torque space than in joint angle space. These observations provide
useful design guidelines to reduce control complexity in anthropomorphic
robotic thumbs.

Keywords: Thumb kinematics · Foldable palm · Joint angle correlations · Torque correlations

1 Introduction

It is a prerequisite to understand which attributes of the human hand are the
most important to achieve functionality and agility for robotic hands that can be
used in unstructured human environments. Though most features of the human
hand have been studied and successfully replicated, the complex mechanism of
how the thumb works together with the foldable palm is not well understood.
Human motor capabilities are limited in kinematically-simplified hands [1]. However, there are many robotic hand applications in which simplified models are
not sufficient since correct anatomical movement is a necessity such as in telemanipulation.

It is undoubted that the thumb plays a vital role in improving grasp performance and manipulation tasks. Human thumb has three joints, Carpometacarpal

© Springer International Publishing Switzerland 2016
L. Alboul et al. (Eds.): TAROS 2016, LNAI 9716, pp. 226–238, 2016.
DOI: 10.1007/978-3-319-40379-3_23

(CMC), Metacarpophalangeal (MCP), and Interphalangeal (IP). According to the studies in [2], a few postural synergies are sufficient to explain hand shape in pre-grasp, in which thumb adduction and internal rotation account for a considerable movement compared to that of MCP and IP.

There is less agreement about the kinematic thumb models that can be found in the literature. This is due to the different ways of defining thumb motion and the difficult nature of assigning classic planes to thumb movements and postures [3]. The proposed thumb kinematic description in [4], which has five Degrees of Freedom (DOF) along with a fixed carpal bone at the base of the thumb (trapezium), gives unrealistic torque/force values at the thumb tip. They suggest that abstracting bone architecture merely as invariant hinge type joints [5] does not represent the true transformation of muscle forces to thumb tip output. Authors in [6,7] propose that including trapezium movement in thumb modeling and joint axes location movement could enhance accuracy. Since muscle forces affect thumb kinematics, a detailed kinematic description of the thumb is vital for robotic researchers and clinicians alike [4]. Thumb kinematics were studied *in vitro* based on axes of rotation of CMC, MCP, and IP joints in [8,9]. According to them, abduction-adduction (A-A) axis of CMC is in the first metacarpal and flexion-extension (F-E) axis is in the trapezium. They suggest that the A-A and F-E motions that occur about these two non-orthogonal non-intersecting CMC axes, are not in the anatomic planes.

The number of DOFs required to fully describe the thumb mechanism and which DOFs are sufficient to grasp and manipulate objects designed for the human hand are contended. Even though IP joint was considered as 1-DOF in most of the kinematic thumb models [10,11], MCP and CMC joints' DOFs were less agreed [5,8,12]. Active pronation-supination (P-S) movement at the CMC joint is 23^0 on average according to [12] indicating its significance in overall movement. Kinematic hand model developed in [7] for data glove calibration, used an unsensed axis along the metacarpal for P-S motion. According to [6], the widely accepted thumb kinematic model is the virtual five-link model presented in [5]. Moreover, the anatomical and functional characteristics of the human thumb can be modeled using the kinematic models in [13]. In any of these approaches, either trapezium is adopted as fixed or merely rigid bone rotations around joints are considered. Contribution of the foldable palm as a musculoskeletal structure is not considered in creating a kinematic model of the thumb.

In vivo studies in [14] indicate that muscles add more stability and passive guidance that merely a bone model cannot achieve. We introduce a 7-DOF thumb kinematic model with variable virtual link connections on the palm to analyze the role of thumb and foldable palm morphology in grasping. The model has 12 kinematic variables and parameters. It gives promising validation results using experimental data in thumb motion tracking for five subjects in precision grasping of five objects. It is a prerequisite to understand any available joint or link correlation patterns that contribute to reduce the complexity of these higher number of DOF (seven in this case). Therefore, we analyze any existing thumb

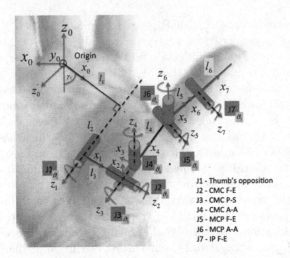

Fig. 1. 7-DOF thumb kinematic model. J1-J7 joint rotational movements are represented by $\theta_1 - \theta_7$. Thumb link lengths $l_5 - l_6$ are measured values from the subjects whereas $l_1 - l_4$ and γ_1 are model parameters.

Table 1. D-H parameters of the 7-DOF thumb kinematic model

Frame number	Link twist (deg.)	Link length (cm)	Link offset (cm)	Joint angle (deg.)
0	0	0	0	0
1	0	l_1	l_2	θ_1
2	90	0	l_3	θ_2
3	-90	0 .	0	θ_3
4	90	0	0	θ_4
5	90	l_4	0	θ_5
6	-90	0	0	θ_6
7	90	l_5	0	θ_7
8	0	l_6	0	0

joint angle and joint torque correlations using the proposed model that can be adopted in tendon transfer surgeries as well as in robotic hand design.

The rest of the paper is arranged in the following order. In Sect. 2, the proposed kinematic model methodology is discussed in detail. Kinematic model validation results and joint angle/torque correlations are elaborated in Sect. 3. Section 4 provides discussion followed by the conclusion.

2 Methodology

2.1 Kinematic Thumb Design

The proposed kinematic thumb model with integrated palm musculoskeletal behavior, has 7-DOFs (J1-J7) represented by revolute joints for each rotational movement (Fig. 1). In Fig. 1, coordinate frames for each joint, link connections and design parameters of the kinematic chain are defined according to the Denavit-Hartenberg (D-H) notation [15] (Table 1). The origin of the reference frame is taken to align with the human data reference, which is the CMC joint of the middle finger. The origin x and z axes, which lie on the palm plane, are rotated $(90+\gamma_1)$ around y axis to align x axis along the link length l_1. The virtual link lengths $l_1 - l_4$ and γ_1 on $x - z$ space tend to vary due to the muscles and ligaments that act on the thumb.

Among seven muscles and the ligaments, that maintain the stability of the CMC joint, adductor pollicis (oblique head) muscle inserts the largest torque across the palm [16,17]. We abstract this musculoskeletal behavior into the virtual joint, J1 to represent thumb's opposability with the foldable palm. J1's rotational axis lies along virtual link l_2, which is the foldabe palm crease. This joint provides interaction between thumb and the fingers allowing it to grasp various sizes and shapes. Virtual link lengths l_1, l_2, and l_3 are orthogonal in corresponding order.

Thumb's CMC joint is approximated into three revolute joints with orthogonal and intersecting rotational axes for F-E (J2), P-S (J3), and A-A (J4) with their intersecting point on the palm plane to kinematically analyze their individual contribution in CMC joint motion as a whole (in Fig. 1, these three joint axes are shown apart for clarity). P-S axis lies along link l_4 which is more or less the thumb's 1st metacarpal [12]. Merely solid bone structure will not represent kinematics properly, unless the enveloping muscles' contribution in grasping is included [18]. Hence, link lengths $l_1 - l_4$ cannot be fixed when MCP-IP-thumb tip-linkage movement occurs in coordination with the CMC joint movement.

Authors in [8] prove that MCP joint has A-A and F-E axes and the angle between them is $85^0 \pm 12^0$. We approximate 2-DOF MCP joint (J5-J6) A-A axis as orthogonal to F-E axis and IP joint (J7) has only single DOF (F-E). F-E axes of MCP and IP are taken to be parallel to each other and align along the biological joint axes.

Relative joint angle ranges and rotation directions of the revolute joints are determined based on the D-H convention [15]. Parameter/variable boundaries and initial values are tuned using forward kinematics that produce anatomically feasible motions as possible.

Each consecutive joint position and orientation (Fig. 1) can be evaluated using link parameters assigned in Table 1 and transforming frame N corresponding to the Nth joint to reference frame 0 using the transformation,

$$\substack{0 \\ N}T = \prod_{i=0}^{N-1} \substack{i \\ i+1}T \tag{1}$$

where the matrices ${}^{i}_{i+1}T, i = 0, 1, 2, \cdots, 8$ are shown in the Appendix. Taking the product of N transforms for each joint gives the standard homogeneous transformation matrix for the Nth joint relative to the reference frame [15]. Since N depends on the configuration of the model, $N = 7$ gives the standard homogeneous transformation matrix for the MCP joint with respect to the reference and $N = 9$ gives that of the tip. These kinematic model position values are compared against human thumb grasp data for the MCP joint and thumb tip for validation.

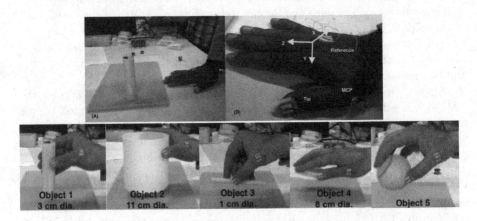

Fig. 2. (*top-A*) Initial experimental setting, (*top-B*) Sensor positions on the hand, (*bottom*) Grasping objects 1–5 for data collection.

2.2   Human Grasp Data Acquisition

The NDI Aurora electromagnetic tracking system is used to capture the thumb motion using three magnetic sensors. The 6-DOF sensors are placed on the dorsal side of the prominent hand's thumb fingernail, thumb MCP joint and one as the reference on the MCP of the middle finger as shown in Fig. 2. We measure thumb tip and MCP joint movement since these two locations are the most identifiable kinematic landmarks. The position and orientation of each sensor is measured with respect to the reference sensor. The measurement rate is 40 Hz and the sensor accuracy is 0.48 mm for position and 0.30^{0} for orientation inside the $(0.5 \times 0.5 \times 0.5)$ m cube volume region next to the field generator. Three male and two female subjects participate in grasping five objects (Fig. 2). Object dimensions and reference sensor positions (Fig. 2) are adopted from [19]. Each subject is instructed to move the flat hand from the start position, grasp the object in precision grasp strategy without squeezing the object as much as possible, lift it, place it in a new marked position (25 cm apart) and move the hand back to the original position (Fig. 2). Each grasp type is done four times for each object. Data recording is started when the hand moves from the starting position and finished when it comes back to the initial position.

Data: Human grasp data (for MCP joint and thumb tip) and thumb kinematics
Result: Optimum joint movements
Initialize joint angles to $\theta_{0,1-4}$, link inclination to γ_0 and link lengths to l_0 ;
while $C_1 >$ *threshold A* **do**
> Run Matlab $\theta_{1-4}, \gamma_1, l_{1-4} =$ gamultiobj$(E_p, \theta_{0,1-4}, \gamma_0, l_0)$;
> **if** $C_1 <$ *threshold B* **then**
>> Exit with the optimum joint angles, θ_{1-4}^*, link inclination, γ_1^* and link lengths, l^*;
>
> **else**
>> Loop again;
>
> **end**

end
Initialize joint angles to optimized θ_{1-4}^*, initialize other joint angles, $\theta_{0,5-7}$, link inclination, γ_1^* and link lengths, l^*;
while $C_1, C_2 >$ *threshold C* **do**
> Run Matlab $\theta_{1-7}, \gamma_1, l_{1-4} =$ gamultiobj$(E_p, \theta_{1-4}^*, \theta_{0,5-7}, \gamma_0, l_0)$;
> **if** $C_1, C_2 <$ *threshold D* **then**
>> Exit with the optimum joint angles, θ_{1-7}^*, link inclination, γ_1^* and link lengths, l^*;
>
> **else**
>> Loop again;
>
> **end**

end

Algorithm 1. Finds optimum kinematic model joint parameters/variables for both MCP and thumb tip human trajectories. C_1: Euclidean error at the MCP joint, C_2: Euclidean error at the thumb tip. θ and l are the angles and links vectors respectively.

2.3   Kinematic Model Variable and Parameter Estimation

The MATLAB Global Optimization Toolbox is utilized to optimize the 12 parameters of the inverse kinematic model in Fig. 1. Since the error at the thumb tip is contributed by seven DOFs and that at the MCP is contributed by only four DOFs, we introduce two cost functions: C_1 and C_2 (**Algorithm 1**). Then solutions for C_1 and C_2 are obtained using a multi-objective optimization Genetic Algorithm (MATLAB function *gamultiobj*) [20] with population size 100 and maximum number of generations 150. The cost functions were introduced in two steps following [21,22].

Let $\boldsymbol{P}_h = [H_x \ \ H_y \ \ H_z]$ and $\boldsymbol{P}_r = [R_x \ \ R_y \ \ R_z]$ denote the position vector of each human data point and that of the kinematic model respectively. Then the Euclidean distance of the two position vectors is given by,

$$E_p = \left[(\boldsymbol{P}_h - \boldsymbol{P}_r)^T(\boldsymbol{P}_h - \boldsymbol{P}_r)\right]^{1/2} \tag{2}$$

where C_1 and C_2 in **Algorithm 1** denote E_p at MCP joint and thumb tip positions respectively.

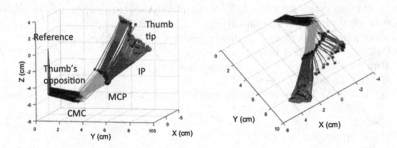

Fig. 3. Human thumb MCP joint and tip trajectories for a selected grasp type (tennis ball) with the corresponding fitted thumb kinematic model. Each link is drawn in a different color. (*left*) Overall trajectory in $x - y - z$ space, (*right*) rotated version of the same trajectory in $x - y$ space.

2.4   Joint Angle and Emulated Torque Correlations

In order to identify strong correlations between pairs of joint angles and joint torques in the kinematic model, we look at the R^2 values of linear regression during the pre-grasp and grasp stages. The R^2 value represents the degree to which the model explains the variability of the two variables considered in any given pair. Therefore relatively high R^2 values indicate that there is a high linear relationship between the pair of variables concerned than the others. In all the trials, 100 data points are used from the transient period so that 90 data samples are before and 10 data samples after the point at which the thumb tip touches the object.

Moreover, joint torques due to thumb tip-object interaction forces are examined to see whether there are any correlated joint torques which facilitate precision grasping. Equation (3) is used to calculate joint torques τ for 0.5^0 increments in all joints for an isometric 3D virtual unit vector force F (in N) at the thumb tip.

$$\tau = J^T F \tag{3}$$

where τ is the 7×1 torque vector, J is the 3×7 Jacobian matrix, and F is the 3×1 unit vector force.

These calculations are based on subject-wise virtual palm and real thumb link lengths (measured in cm). The virtual force vector applied in the overall transient thumb trajectory consists of 100 data points. To make it fair for all the trials, the transient period is evaluated 50 data points in either direction from the point where thumb tip touches the object. This approach shows how the thumb prepares to take the force and how it apportions the torque τ among the joint configuration once the contact is established with the object.

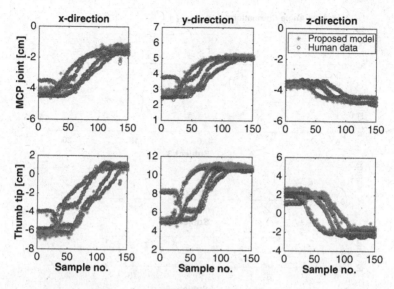

Fig. 4. A representative thumb MCP joint and tip trajectories (human and kinematic) in $x - y - z$ space plotted against the sample number.

3   Results

3.1   Kinematic Model Validation

The validated kinematic model gives Root Mean Square Error (RMSE) for MCP position < 0.5 cm and that for thumb tip position < 0.9 cm across all the five subjects, four trials and five grasp types. Figure 3 illustrates a representative kinematic model fit with joint and link variations in overall thumb trajectory in grasping object 2 (11 cm dia. cylinder) with experimental data in $x - y - z$ space (defined in Fig. 1). The figure indicates the individual joint Range of Motion (ROM) at the foldable palm, CMC, MCP and IP joints. Each kinematic linkage of this trajectory is caused by 12 kinematic model variables and parameters. Corresponding thumb MCP and tip movements in $x, y,$ and z space for four trials are shown in Fig. 4.

3.2   Joint Angle and Torque Correlations

According to joint angle correlation analysis using optimized thumb joint rotations ($\theta_1 - \theta_7$) during pre-grasp and grasp stage, we can observe that $R^2 > 0.8$ in joint pair number 20 (θ_5 and θ_7) associated with MCP and IP joint flexions in Fig. 5 (blue plots).

In torque correlations (red plots), joint pair number 8 (τ_2 and τ_4 represented by CMC flexion and abduction) also show $R^2 > 0.8$ along with number 20 (τ_5 and τ_7). These two highly correlated torque pairs are plotted in Fig. 6 across subjects, objects and trials. It shows higher torque values in the proximal pair

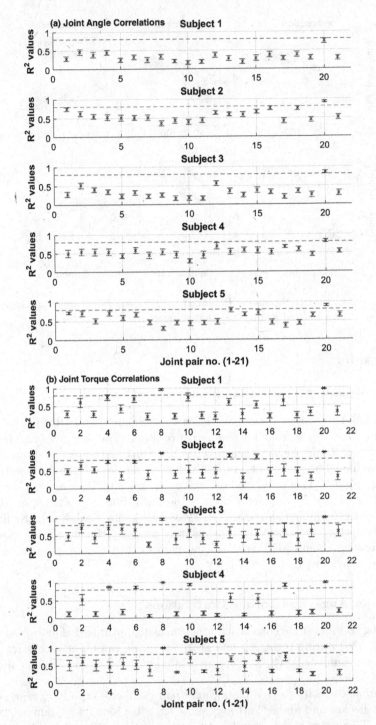

Fig. 5. R^2 values for (a) joint angle and (b) joint torque correlations plotted against 21 pairs of 7-joints for five subjects. The vertical error bars represent the standard error across five objects and four trials. Green lines mark $R^2 = 0.8$. The order of joint pairs (1-21) are given in Table 2. (Color figure online)

(associated with CMC abduction and flexion) than the distal pair (MP and IP flexion).

4   Discussion

Experimental validation of the proposed kinematic model based on thumb and integrated foldable palm gives promising results with RMSE for MCP position < 0.5 cm and that for thumb tip position < 0.9 cm across all the five subjects, four trials and five grasp types. Authors in [4] point out the lack of experimentally validated thumb kinematic descriptions and the absence of inter-subject variability in so far adopted models. Since the proposed kinematic model incorporates inter-subject differences in palm musculostructure with variable links, it can be adopted in calculating inter-subject grip forces and torques.

We observe $R^2 > 0.8$ for thumb MCP and IP joint flexion pair in both pre-grasping and grasping the objects. According to [23], ROM among thumb joint angles is different due to enveloping muscle and ligament structure constraints. Angular ROMs of CMC and MCP joints during functional motion are less than 65 % of their maximum ranges [24]. In [25], CMC and MCP A-A motions are correlated only during initial stage of thumb's opposition motion and not in the whole trajectory. Based on these findings, each thumb joint shows limited movement depending on the tasks. This behaviour could be the reason why we cannot find any other strong joint angle correlations in thumb joints while grasping.

The two rigid link thumb models adopted in [26] to compare tendon tension in thumb joints show that CMC joint flexion and abduction muscle relationships are poorly estimated. They point out the non-independent modeling of CMC P-S could be one factor. Whereas in our results, torque associated with CMC flexion and abduction axes shows strong correlation. According to the biomechanical studies in [14], Opponens Pollicis (OPP) muscle activity contributes 80 % for flexion and 100 % for abduction at CMC joint. Our results (Fig. 5) are consistent with them showing a strong joint torque correlation ($R^2 = 0.9808$) in CMC flexion and abduction for precision grasping. This reliable prediction could come from the following factors considered in the kinematic model: (1) CMC joint is modeled as three separate DOFs. Hence P-S is a kinematically independent motion within CMC. (2) A virtual joint J1 (Fig. 1) is introduced to abstract thumb's opposition. (Relatively higher torque correlation can be observed in joint axes pair J1 and CMC P-S than that of joint axes pair J1 and CMC F-E.) (3) Subject-wise link length variations are accounted.

However, the authors in [27] point out that CMC and MCP A-A are correlated. Our results do not illustrate this behaviour. It could also be due to the model CMC A-A and MCP A-A axes lie on the same $y - z$ plane. In addition, MCP flexion could be the dominant motion compared to abduction in precision grasping.

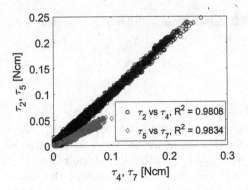

Fig. 6. Linearly correlated joint torque pairs with $R^2 > 0.8$.

Table 2. Joint pair numbers

Joint pair no	Joint angles	Joint pair no	Joint angles	Joint pair no	Joint angles
1	θ_1, θ_2	8	θ_2, θ_4	15	θ_3, θ_7
2	θ_1, θ_3	9	θ_2, θ_5	16	θ_4, θ_5
3	θ_1, θ_4	10	θ_2, θ_6	17	θ_4, θ_6
4	θ_1, θ_5	11	θ_2, θ_7	18	θ_4, θ_7
5	θ_1, θ_6	12	θ_3, θ_4	19	θ_5, θ_6
6	θ_1, θ_7	13	θ_3, θ_5	20	θ_5, θ_7
7	θ_2, θ_3	14	θ_3, θ_6	21	θ_6, θ_7

5 Conclusion

The proposed novel kinematic model for the thumb and the foldable palm explains the essential musculoskeletal behaviors in grasping. The model shows a strong joint angle correlation in MCP and IP flexion while having strong torque relationships in CMC abduction and flexion pair and MP and IP flexion pair. The proposed modeling approach provides a model to determine tendon routing and pulling forces in anthropomorphic robotic thumbs.

A APPENDIX: The Transformation Matrices for the Kinematic Model

$$
{}^0_1T = \begin{bmatrix} cos(90 + \gamma_1) & 0 & sin(90 + \gamma_1) & 0 \\ 0 & 1 & 0 & 0 \\ -sin(90 + \gamma_1) & 0 & cos(90 + \gamma_1) & 0 \\ 0 & 0 & 0 & 1 \end{bmatrix} \quad {}^1_2T = \begin{bmatrix} cos(\theta_1) & -sin(\theta_1) & 0 & l_1 \\ sin(\theta_1) & cos(\theta_1) & 0 & 0 \\ 0 & 0 & 1 & l_2 \\ 0 & 0 & 0 & 1 \end{bmatrix}
$$

$$
{}^{2}_{3}T = \begin{bmatrix} cos(\theta_2) & -sin(\theta_2) & 0 & 0 \\ 0 & 0 & -1 & -l_3 \\ sin(\theta_2) & cos(\theta_2) & 0 & 0 \\ 0 & 0 & 0 & 1 \end{bmatrix} \quad {}^{3}_{4}T = \begin{bmatrix} cos(\theta_3) & -sin(\theta_3) & 0 & 0 \\ 0 & 0 & 1 & 0 \\ -sin(\theta_3) & -cos(\theta_3) & 0 & 0 \\ 0 & 0 & 0 & 1 \end{bmatrix}
$$

$$
{}^{4}_{5}T = \begin{bmatrix} cos(\theta_4) & -sin(\theta_4) & 0 & 0 \\ 0 & 0 & -1 & 0 \\ sin(\theta_4) & cos(\theta_4) & 0 & 0 \\ 0 & 0 & 0 & 1 \end{bmatrix} \quad {}^{5}_{6}T = \begin{bmatrix} cos(\theta_5) & -sin(\theta_5) & 0 & l_4 \\ 0 & 0 & -1 & 0 \\ sin(\theta_5) & cos(\theta_5) & 0 & 0 \\ 0 & 0 & 0 & 1 \end{bmatrix}
$$

$$
{}^{6}_{7}T = \begin{bmatrix} cos(\theta_6) & -sin(\theta_6) & 0 & 0 \\ 0 & 0 & 1 & 0 \\ -sin(\theta_6) & -cos(\theta_6) & 0 & 0 \\ 0 & 0 & 0 & 1 \end{bmatrix} \quad {}^{7}_{8}T = \begin{bmatrix} cos(\theta_7) & -sin(\theta_7) & 0 & l_5 \\ 0 & 0 & -1 & 0 \\ sin(\theta_7) & cos(\theta_7) & 0 & 0 \\ 0 & 0 & 0 & 1 \end{bmatrix} \quad {}^{8}_{9}T = \begin{bmatrix} 1 & 0 & 0 & l_6 \\ 0 & 1 & 0 & 0 \\ 0 & 0 & 1 & 0 \\ 0 & 0 & 0 & 1 \end{bmatrix}
$$

References

1. Martell, J.S., Gini, G.: Robotic hands: design review and proposal of new design process. World Acad. Sci. Eng. Technol. **26**, 85–90 (2007)
2. Santello, M., Flanders, M., Soechting, J.F.: Postural hand synergies for tool use. J. Neurosci. **18**(23), 10105–10115 (1998)
3. Grinyagin, I.V., Biryukova, E.V., Maier, M.A.: Kinematic and dynamic synergies of human precision-grip movements. J. Neurophysiol. **94**(4), 2284–2294 (2005)
4. Valero-Cuevas, F.J., Johanson, M.E., Towles, J.D.: Towards a realistic biomechanical model of the thumb: the choice of kinematic description may be more critical than the solution method or the variability/uncertainty of musculoskeletal parameters. J. Biomech. **36**(7), 1019–1030 (2003)
5. Giurintano, D., Hollister, A., Buford, W., Thompson, D., Myers, L.: A virtual five-link model of the thumb. Med. Eng. Phys. **17**(4), 297–303 (1995)
6. Bullock, I.M., Borràs, J., Dollar, A.M.: Assessing assumptions in kinematic hand models: a review. In: 4th IEEE RAS & EMBS International Conference on Biomedical Robotics and Biomechatronics (BioRob), pp. 139–146 (2012)
7. Griffin, W.B., Findley, R.P., Turner, M.L., Cutkosky, M.R.: Calibration and mapping of a human hand for dexterous telemanipulation. In: ASME IMECE 2000 Symposium on Haptic Interfaces for Virtual Environments and Teleoperator Systems, pp. 1–8 (2000)
8. Hollister, A., Giurintano, D.J., Buford, W.L., Myers, L.M., Novick, A.: The axes of rotation of the thumb interphalangeal and metacarpophalangeal joints. Clin. Orthop. Relat. Res. **320**, 188–193 (1995)
9. Hollister, A., Buford, W.L., Myers, L.M., Giurintano, D.J., Novick, A.: The axes of rotation of the thumb carpometacarpal joint. J. Orthop. Res. **10**(3), 454–460 (1992)
10. Chang, L.Y., Matsuoka, Y.: A kinematic thumb model for the act hand. In: IEEE International Conference on Robotics and Automation (ICRA), pp. 1000–1005. IEEE (2006)
11. Chalon, M., Grebenstein, M., Wimböck, T., Hirzinger, G.: The thumb: guidelines for a robotic design. In: IEEE/RSJ International Conference on Intelligent Robots and Systems (IROS), pp. 5886–5893 (2010)

12. Chang, L.Y., Pollard, N.S.: Method for determining kinematic parameters of the in vivo thumb carpometacarpal joint. IEEE Trans. Biomed. Eng. **55**(7), 1897–1906 (2008)

13. Santos, V.J., Valero-Cuevas, F.J.: Reported anatomical variability naturally leads to multimodal distributions of Denavit-Hartenberg parameters for the human thumb. IEEE Trans. Biomed. Eng. **53**(2), 155–163 (2006)

14. Kaufman, K.R., An, K.N., Litchy, W.J., Cooney, W.P., Chao, E.Y.: In-vivo function of the thumb muscles. Clin. Biomech. **14**(2), 141–150 (1999)

15. Craig, J.J.: Introduction to Robotics: Mechanics and Control, vol. 3. Pearson Prentice Hall, Upper Saddle River (2005)

16. Smutz, W.P., Kongsayreepong, A., Hughes, R.E., Niebur, G., Cooney, W.P., An, K.N.: Mechanical advantage of the thumb muscles. J. Biomech. **31**(6), 565–570 (1998)

17. Neumann, D.A., Bielefeld, T.: The carpometacarpal joint of the thumb: stability, deformity, and therapeutic intervention. J. Orthop. Sports Phys. Ther. **33**(7), 386–399 (2003)

18. Ladd, A.L., Crisco, J.J., Hagert, E., Rose, J., Weiss, A.P.C.: The 2014 ABJS nicolas andry award: The puzzle of the thumb: mobility, stability, and demands in opposition. Clin. Orthop. Relat. Res. **472**(12), 3605–3622 (2014)

19. Feix, T., Romero, J., Ek, C.H., Schmiedmayer, H.B., Kragic, D.: A metric for comparing the anthropomorphic motion capability of artificial hands. IEEE Trans. Robot. **29**(1), 82–93 (2013)

20. Deb, K.: Multi-Objective Optimization Using Evolutionary Algorithms, vol. 16. John Wiley & Sons, Hoboken (2001)

21. Nanayakkara, T., Watanabe, K., Kiguchi, K., Izumi, K.: Evolving a multiobjective obstacle avoidance skill of a seven-link manipulator subject to constraints. Int. J. Syst. Sci. **35**(3), 167–178 (2004)

22. Nanayakkara, T., Kiguchi, K., Murakami, T., Watanabe, K., Izumi, K.: Skillful adaptation of a 7-dof manipulator to avoid moving obstacles in a teleoperated force control task. In: IEEE International Symposium on Industrial Electronics (ISIE), vol. 3, pp. 1982–1987 (2001)

23. Cooney, W.P., Lucca, M.J., Chao, E., Linscheid, R.: The kinesiology of the thumb trapeziometacarpal joint. J. Bone Joint Surg. **63**(9), 1371–1381 (1981)

24. Tang, J., Zhang, X., Li, Z.M.: Operational and maximal workspace of the thumb. Ergonomics **51**(7), 1109–1118 (2008)

25. Li, Z.M., Tang, J.: Coordination of thumb joints during opposition. J. Biomech. **40**(3), 502–510 (2007)

26. Vigouroux, L., Domalain, M., Berton, E.: Comparison of tendon tensions estimated from two biomechanical models of the thumb. J. Biomech. **42**(11), 1772–1777 (2009)

27. Hollister, A., Giurintano, D.J.: Thumb movements, motions, and moments. J. Hand Ther. **8**(2), 106–114 (1995)

Private Cloud Deployment Model in Open-Source Mobile Robots Ecosystem

Petri Oksa[✉] and Pekka Loula

Telecommunications Research Center, Tampere University of Technology,
Pohjoisranta 11 A, Pori, Finland
{petri.oksa,pekka.loula}@tut.fi

Abstract. The focus of this paper is on secure cloud service platform for mobile robots ecosystem. Especially the emphasis is based on the scope of open-source software frameworks such as Apache Hadoop which offers numerous possibilities to employ open-source designing tools and deployment models for private cloud computing planning. This paper presents implementation of the OpenCRP (Open CloudRobotic Platform) locally-operated private cloud infrastructure and configuration methods by using Hadoop distributed file system (HDFS) for easing the ecosystem communications set-up in its entirety. For robot teleoperation, ROS (Robot Operating System) is used. The presented ecosystem utilizes security features for autonomous cloud robotic platform, software tools to manage user authentication and methods for large-scale robot-based data management and analysis. In addition to robot trial set-up of robot data storage and sharing, an ecosystem built with two low-cost mobile robots is presented.

Keywords: Private cloud · Cloud robotics · Open-source · Ecosystem

1 Introduction

The past five years have been seen an emerging period of mobile robots e.g. in the growth of robot lawnmowers, vacuum cleaners, automated guided vehicles and healthcare robots. Today, when new web-based technologies constantly evolve—along with cloud computing infrastructures—which enables mobile cloud robots to benefit existing internet resources. Web can be used as a powerful computational resource, communication medium and a source of shared information allowing developers to overcome current limitations by building cloud robotics applications. Both Wireless– and Web–based technologies and availability of data centers have made possible for mobile robots to tap into the cloud [1–3, 36].

Cloud platforms capability in computation, communication and information processing are realized both in robot-wise solutions and massive data set storages in the distributed cloud frameworks. This allows fast access to a dedicated data storages and integrates together separate global libraries, parallel computation and robot data sharing. A cloud robotics platform helps robots to offload heavy computation by providing secured customizable computing environments in the cloud. One fundamental demonstrator is that robots can use private cloud platform to share the knowledge and

© Springer International Publishing Switzerland 2016
L. Alboul et al. (Eds.): TAROS 2016, LNAI 9716, pp. 239–248, 2016.
DOI: 10.1007/978-3-319-40379-3_24

benefit the experience collected of other robots. This has become the base of the many robotic team– and multi-robot deployments [1–5].

One of the most popular cloud services is Apache Hadoop presenting open-source software composed with scalability, service delivery and capacity planning methods. Hadoop is a software framework for applications running on large clusters built of commodity hardware. Hadoop provides a distributed file system (HDFS) and a parallel processing framework based on the MapReduce programming paradigm [6–9].

In the matter of open-source operating systems for mobile robots, ROS introduces Linux-based system providing software libraries and tools to build or develop robot applications. Together with basic robot control functions, ROS can be employed with Java-based repositories e.g. for robot simulation, visualization and freely downloadable applications for Android mobile operating system. Currently there are available over 80 robot hardware platforms that use or can be used with ROS software framework and 2000 + libraries for different ROS applications [10].

In the past few years mobile cloud robotics initiatives have emerged such as A'STAR Data Storage institute of Singapore proposing a software framework (ASORO) [37], Google (Google Object Recognition Engine) [38, 39], DAvinCi (Distributed Agents with Collective Intelligence) [42] and Researchers at the Laboratory of Analysis and Architecture of Systems (LAAS) in Toulouse [40]. One of the pioneering initiatives is European project RoboEarth proposing the implementation of Rapyuta [36], Web-like open source network for robots using a PaaS (Platform-as-a-Service). Rapyuta is based on an elastic computing model that dynamically allocates secured computing environments for robots [41]. The DAvinCi framework combines the distributed ROS architecture, Hadoop's HDFS and the Map/Reduce framework [42]. The DAvinCi is a cloud computing architecture for service robots offloading data intensive and computationally intensive workloads from the onboard resources on the robots to a backend cluster system [42].

Aforementioned initiatives are among the first paving the way for the research of cloud robotics era. The OpenCRP ecosystem initiative continues onward this significant work by adopting information security features for open-source private cloud platform, robot teleoperation, user authentication and methods for large-scale robot-based data management and analysis. The presented platform uses ROS-based computing environment as well as Apache Hadoop software tools to build an ecosystem for multi-robot infrastructure. It is worth of noting that the presented design choice focus on locally-operated robot data analyzing, planning and reasoning by having a corresponding software clone in the cloud.

The rest of the paper is organized as follows. Section 2 clarifies functional determination and objectives of the OpenCRP ecosystem. Environment for secure robot trials to ROS network communications set-up and Hadoop private cloud tasks are presented in Sect. 3. Finally, Sect. 4 concludes the paper with a brief discussion on future developments and work.

2   OpenCRP Ecosystem

An autonomous robotic ecosystem consists of mobile robots, software frameworks/toolsets and charging stations at least. These primary issues are taken into account in the planning of ecosystem. OpenCRP ecosystem is, in essence, based on service-oriented robotic (SOR) cloud enabling PaaS mobile multi-robot architecture. Moreover, the main focus is based on cloud-communicating service robots with lightweight workload requirements for the purpose of offloading heavy computation into a cloud. Thus, to provide fast and flexible access to data repository, most of the data-processing functions are processed in the cloud; see Fig. 1 for the simplified architecture model. The main contributions of the OpenCRP ecosystem demonstration platform include security features for open-source cloud robotic ecosystem, software tools to manage user authentication and methods for large-scale robot-based data management and analysis. Emphasis is also aimed to further develop scalable data storage functions, data analysis and metadata processing tools on top of the Hadoop HDFS storage infrastructure. The main idea of the presented cloud robotic system is that robots offload their heavy computational load into a private cloud and share their information. Private cloud then shares the knowledge and experience with other robots connected to the network. This leads to a less complicated situation where extensive computational power and hardware on the robot side is not necessary needed anymore. Ecosystem users can also benefit by utilizing the ROS network set-up configuration steps in their experimentations as presented in Subsect. 3.1.

Figure 1 gives a simplified overview of the OpenCRP multi-robot ecosystem. The ecosystem consists of two low-cost and tele-operated mobile robots TurtleBot and Groma, depicted also in Fig. 2 showing how the robots are connected to ROS network. We have used ROS Indigo Igloo platform for the trial set-up environment. Two PCs have Hadoop installed on Ubuntu Linux 14.04 LTS; TurtleBot laptop PC runs a master node (NameNode) and desktop PC runs a client node (DataNode). Master side Hadoop release version is 2.7.0 and client is 2.5.0 [6]. Robots with ROS-based hardware or virtual machines (VM) can use the database to share their information.

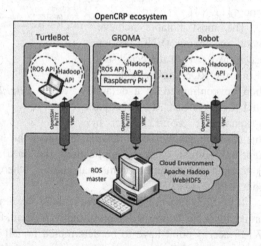

Fig. 1. A simplified overview of the OpenCRP ecosystem.

3 Environment for Secure Robot Trials

3.1 ROS Network Set-Up

In this section, ROS network set-up variables and practical trial set-up are clarified. One of the most significant methods we discovered is useful procedure adopted with VNC (Virtual Network Communicating) in managing the ROS platform communications. In our implementation set-up, we have installed RealVNC [11] for the purpose of robot teleoperation from desktop PC (ROS client node). This is due to the IPv4 configuration problem which can be occurred when connecting ROS master to ROS slaves. Solution for this problem is presented later in this section. Detailed ROS network is described in Fig. 2.

RealVNC enables graphical remote access and control computers from another computer or mobile device. RealVNC clients using vncviewer can run in full-screen mode and vncserver allows a computer to be remotely controlled by another computer. Many security features are available for registered users: authentication schemes, encrypted connections, blacklisting, IP filtering and gatekeeping among others [11, 14].

The most notable achievement of taking VNC remote access as a part of the ecosystem is that the ROS master node can then be configured as a default URI `localhost:11311` and hostname as `localhost`. In

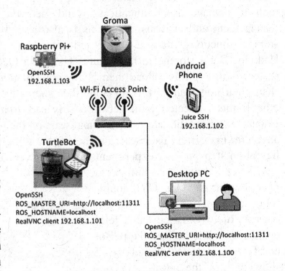

Fig. 2. OpenCRP ROS network.

this case the robotic network operates under the local mode and the `.bashrc` file is kept unmodified. Once all applications run on a single entity, private cloud is then better suited for some commercial entities where trust and security is the highest concern [12–14, 17].

If the ROS network is configured without any remote control system (VPN tunneling) unlike the VNC connected in proposed ecosystem, fixed IPv4 address assigned from the Wi-Fi access point have to be determined for all machines to establish the network. Because the ROS master node checks communication channels before attaching computers to the network, the network connection will fail if one or other IP address is incorrect. In OpenCRP network configurations we encountered a problem causing the ROS slave node unable to establish network connection to the master computer (TurtleBot laptop) even though both IP addresses were correctly set. As clarified in the previous section and also depicted in Fig. 2, we came upon a solution to the problem by using VNC remote control and authentication for relaunching the network.

To establish the ecosystem virtual machines interconnection, VNC server is installed into the TurtleBot laptop PC and VNC viewer for desktop PC. When all traffic is now unencrypted, we conducted OpenSSH [15] and PuTTY [16] to provide secure tunneling and authentication methods.

OpenSSH is a free security-related middleware based on the SSH protocol. It provides encryption for network services like remote login and remote file transfer. OpenSSH help to secure network communications via the encryption of network traffic over multiple authentication methods and by providing secure tunneling capabilities [15].

PuTTY is an SSH and telnet client, developed originally for the Windows platform. Official ports are available for some Unix-like platforms such as Linux. PuTTY is an open source terminal emulator, serial console and network file transfer application that is available with source code and is developed and supported by a group of volunteers [16].

For Android device users, Google play store offers freely downloadable applications for controlling and coordinating the Turtlebot, including TurtleBot follower and panorama algorithms, teleoperation of ROS-enabled robots and ROS make a map application. We experimented the remote access of Turtlebot by using ZTE Blade III Android phone connected through the Juice SSH as illustrated in Fig. 2. Supported ROS releases are Indigo Igloo and Hydro Medusa [18].

Figure 3 illustrates the browse directory of WebHDFS REST API (Representational State Transfer Application Programming Interface) running on the TurtleBot laptop. The WebHDFS supports all HDFS user operations including reading files, writing to files, making directories, changing permissions and renaming [6, 7, 31, 33]. WebHDFS defines a public HTTP REST API, which permits clients to access Hadoop from multiple languages without installing Hadoop [31]. In addition, WebHDFS clients can talk to clusters with different Hadoop versions. WebHDFS retains the security that the native Hadoop protocol offers and is suitable into the overall strategy of providing web services access to all Hadoop components [7, 31–35].

Fig. 3. WebHDFS REST API directory.

3.2 Hadoop Private Cloud

As the ecosystem architecture is PaaS-based, Hadoop offers flexibility on pre-built patterns for deploying clusters more easily. The very primary reason for private cloud deployment is a data security and access control for robotic data storage [19–21].

In this section, secure file transfer and user authentication schemes for OpenCRP Hadoop cluster set-up is described. For installation of cluster node, SSH access is required. We propose OpenSSH remote access for its suitability in secure network communications via the encryption of network traffic over multiple authentication methods and by providing secure tunneling capabilities. Furthermore, OpenSSH can be installed on many other computing platforms and VMs giving the possibility of different platforms to interconnect with wide authentication schemes.

To strengthen the robotic network file transfer security, PuTTY is used to ensure the user control over SSH encryption key and several protocol versions, especially when using VNC in robots remote control. That is because of the VNC's low data encryption level. When tunnelling the data communication, PuTTY encrypted public key is more secure than a password for instance SSH-2 RSA public key used in our trial-set-up. To accommodate several robots to the private cloud, RSA keys have to generate for every new robot when connecting them into the network. Following the SSH configurations we start the HDFS daemon and then NameNode is started on master and DataNode daemons are started on all slaves. Figure 4 shows cluster node information in a single node set-up on a desktop PC including one directory dedicated for robotic data.

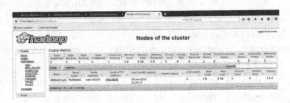

Fig. 5. OpenCRP data management.

As the system set-up originates at open-source basis, some convenient Apache software tools were exploited in the platform as illustrated in Fig. 5. The ecosystem architecture further utilizes the connectivity of Apache Hive data warehouse and data processing facilities of Apache Spark [27, 28, 30]. These software tools are perceived very useful in massive data-set operations and obviously support the data processing actions.

Software tools in the data management framework run in desktop PC as other applications rely on the certain architecture model component. The ecosystem data management is a top-level framework utilizing the Hadoop HDFS background processing storage infrastructure [22–25].

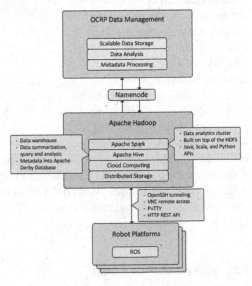

Fig. 4. Hadoop cluster information.

Apache Spark is an open-source cluster computing framework providing faster application performance than MapReduce paradigm [9]. Spark requires a cluster manager and a distributed storage system accessing diverse data sources including HDFS. In large-scale data processing, Spark extends software tools for data analytics and API for writing applications in Java, Scala, Python, and R. Spark supports SQL queries, streaming data and real-time data processing, latter being important for remote-controlled robot data management [26, 27].

The OpenCRP ecosystem data management platform uses Spark algorithms of data querying, sorting and cluster management for multi-robot data analysis to force the dataflow of environmental data collected by mobile robots. This approach is a major part of our work towards the healthcare case where the research focus is on development an automation platform for the support of elderly persons and disabled veteran nursing operations. In this case we plan a scalable data storage functions, data analysis and metadata processing tools on top of the Hadoop platform by using Spark in data processing of collected surroundings data in hospice environment.

The Apache Hive is built on top of Hadoop for providing data summarization, query, and analysis. Hive data warehouse software facilitates querying and managing large datasets residing in distributed storage. Hive provides a mechanism to project structure onto this data and query the data using a SQL-like language called HiveQL. At the same time this language also allows traditional MapReduce programmers to plug in their custom mappers and reducers when it is inconvenient or inefficient to express this logic in HiveQL. Also Hive supports analysis of large data-sets stored in Hadoop's HDFS [29, 30].

As explained earlier in this section, the private cloud platform is implemented by using Apache Hadoop framework tools. Other cloud types that have succeeded in robotic systems so far are community, public and hybrid clouds [3]. Hadoop is not yet being so well established in robotics field and for that reason OpenCRP implementation would give directions and instructions of how multiple robot and cloud platforms can tightly be interacted. OpenCRP data management is scalable when using HDFS and a parallel processing framework based on the MapReduce programming paradigm.

4 Conclusions and Discussion

In this paper, we have presented designing methods and deployments for open-source private cloud for mobile robots. The novel OpenCRP ecosystem provides useful benefits for robot cloud service establishment as follows:

- purpose to ease the ROS network configuration with security features;
- robots offload heavy computational load into a private cloud;
- no need to extensive computational power or hardware on the robot side;
- mobile robot teleoperation to provide secure tunneling and authentication methods;
- environment for secure robot trials;
- scalable data storage functions, data analysis and metadata processing for data collected by mobile robots.

We showed how to take advantage of cloud-driven pilot- and demonstration environment for testing automation processes. To ensure the system functions, we have made some practical provisions for the network configurations. Also we have proposed improvements of the existing ecosystem architecture adding some convenient applications to accommodate security solutions.

During the measurement scenario set-up executions and experiments we encountered some technical problems with Ubuntu Linux. One of the crucial exceptions came up with the `.Xauthority` file. When entering the right password the screen only flashes and then goes back to login screen. Fortunately, we had added another user account `hduser` and password so it was possible to log on to the computer. We solved the problem by running the command `sudo mv.Xauthority.XauthorityBak` in the command line terminal.

The insight of this research work is kept practical and suggestive for researchers and system specialists who are interested to develop and improve open-source robotic implementations, programming tools and measurement testbeds.

Our future efforts will focus on the private cloud infrastructure information security, management and task execution in the healthcare case which presents an automation platform for the support functions of elderly persons and disabled veteran nursing operations. The plan of this case is to implement a scalable data storage functions, data analysis and metadata processing tools on top of the Hadoop platform by using Spark in data processing of collected surroundings data in hospice environment. The aim of the OpenCRP ecosystem is to be autonomously controllable, coordinable and self-adaptable. Next, we develop scalable data storage functions, data analysis and metadata processing tools on top of the Hadoop platform as pictured in Fig. 5. This novel platform would obviously help in processing and reusing massive data-sets collected by mobile robots. Though the ecosystem concept is virtuous, it would be the direction of future state of art systems where mobile robots automatically do their tasks, communicate with each other (machine-to-machine M2 M), share and reuse/share their collected information via a cloud (machine-to-cloud, M2C) allocated for real time demand.

References

1. Mohanarajah, G., Usenko, V., Singh, M., D'Andrea, R., Waibel, M.: Cloud-based collaborative 3D mapping in real-time with low-cost robots. IEEE Trans. Autom. Sci. Eng. 12(2), 423–431 (2015)
2. Wang, L., Liu, M., Meng, M.Q.H. Towards cloud robotic system: a case study of online co-localization for fair resource competence. In: IEEE 2012 International Conference on Robotics and Biomimetics (ROBIO), pp. 2132–2137 (2012)
3. Liu, B., Chen, Y., Hadiks, A., Blasch, E., Aved, A., Shen, D., Chen, G.: Information fusion in a cloud computing era: a systems-level perspective. IEEE Aerosp. Electron. Syst. Mag. 29(10), 16–24 (2014)
4. Wen, Y., Zhu, X., Rodrigues, J.J., Chen, C.W.: Cloud mobile media: reflections and outlook. IEEE Trans. Multimedia 16(4), 885–902 (2014)
5. Akella, A.: Experimenting with Next-Generation Cloud Architectures Using CloudLab. IEEE Internet Comput. 19(5), 77–81 (2015)
6. The Apache Software Foundation (2016). https://hadoop.apache.org
7. Kiencke, T.: Hadoop distributed file system (HDFS) (2013)
8. Rao, B.T., Reddy, L.S.S.: Survey on improved scheduling in Hadoop MapReduce in cloud environments. Int. J. Comput. Appl. (IJCA) 34(9), 29–33 (2011). arXiv:1207.0780

9. Ibrahim, S., Jin, H., Lu, L., Qi, L., Wu, S., Shi, X.: Evaluating mapreduce on virtual machines: the Hadoop case. In: Jaatun, M.G., Zhao, G., Rong, C. (eds.) Cloud Computing. LNCS, vol. 5931, pp. 519–528. Springer, Heidelberg (2009)
10. ROS.org (2016). http://wiki.ros.org
11. RealVNC (2016). https://www.realvnc.com
12. Zhang, K., Yang, K., Liang, X., Su, Z., Shen, X.S., Luo, H.H.: Security and privacy for mobile healthcare networks: from a quality of protection perspective. IEEE Wirel. Commun. 22(4), 104–112 (2015)
13. Visalakshi, M.P., Deepak, M.: Remote desktop access using remote frame buffer in mobilecloud environment. Int. J. Mod. Eng. Res. (IJMER), 30–34 (2013)
14. Jadhav, A., Oswal, V., Madane, S., Zope, H., Hatmode, V.: Vnc architecture based remote desktop access through android mobile phones. Int. J. Adv. Res. Comput. Commun. Eng. 1(2) (2012)
15. OpenSSH (2016). http://www.openssh.com
16. PuTTY (2015). http://www.putty.org
17. Mohanarajah, G., Hunziker, D., D'Andrea, R., Waibel, M.: Rapyuta: A cloud robotics platform. IEEE Trans. Autom. Sci. Eng. 12(2), 481–493 (2015)
18. Google Play (2016). https://play.google.com/store/apps/developer?id=OSRF
19. Kaufman, L.M.: Data security in the world of cloud computing. IEEE Secur. Priv. 7(4), 61–64 (2009)
20. Shvachko, K., Kuang, H., Radia, S., Chansler, R.: The Hadoop distributed file system. In: IEEE 26th Symposium on Mass Storage Systems and Technologies (MSST), pp. 1–10, May 2010
21. Leverich, J., Kozyrakis, C.: On the energy (in) efficiency of Hadoop clusters. ACM SIGOPS Operating Systems Review 44(1), 61–65 (2010)
22. Lu, H., Hai-Shan, C., Ting-Ting, H.: Research on Hadoop cloud computing model and its applications. In: IEEE 2012 Third International Conference on Networking and Distributed Computing (ICNDC), pp. 59–63, October 2012
23. Kala Karun, A., Chitharanjan, K.: A review on Hadoop—HDFS infrastructure extensions. In: IEEE Conference on Information and Communication Technologies (ICT), pp. 132–137, April 2013
24. Cohen, J., Acharya, S.: Towards a more secure apache Hadoop HDFS infrastructure. In: Lopez, J., Huang, X., Sandhu, R. (eds.) NSS 2013. LNCS, vol. 7873, pp. 735–741. Springer, Heidelberg (2013)
25. Ko, S.Y., Hoque, I., Cho, B., Gupta, I.: Making cloud intermediate data fault-tolerant. In: ACM Proceedings of the 1st symposium on Cloud computing, pp. 181–192, June 2010
26. Apache Spark™ (2016). https://spark.apache.org
27. Shanahan, J.G., Dai, L.: Large scale distributed data science using apache spark. In: Proceedings of the 21th ACM SIGKDD International Conference on Knowledge Discovery and Data Mining, pp. 2323–2324, August 2015
28. You, S., Zhang, J., Gruenwald, L.: Large-scale spatial join query processing in cloud. In: IEEE CloudDM Workshop (2015, to appear). http://www-cs.ccny.cuny.edu/~jzhang/papers/spatial_cc_tr.pdf
29. Apache Hive™ (2016). https://hive.apache.org
30. Thusoo, A., Sarma, J.S., Jain, N., Shao, Z., Chakka, P., Zhang, N., Murthy, R.: Hive-a petabyte scale data warehouse using Hadoop. In: IEEE 26th International Conference on Data Engineering (ICDE), pp. 996–1005, March 2010
31. Hortonworks Inc. (2016). http://hortonworks.com/blog/webhdfs-http-rest-access-to-hdfs

32. Li, Y., Zeng, W.H., Yang, L.Q., Wu, Z.L., Wang, M.H.: Architecture of Campus Security Management System Based on WCF and RFID Technology. Appl. Mech. Mater. **411**, 462–466 (2013)

33. Saibharath, S., Geethakumari, G.: Cloud forensics: evidence collection and preliminary analysis. In: IEEE International Advance Computing Conference (IACC), pp. 464–467, June 2015

34. Kumar, P.K.R., Aparna, R.: Storage and access in product review system using Hadoop. Int. J. Recent Adv. Eng. Technol. (IJRAET) **2**(6-7), 34–38 (2014)

35. Scavuzzo, M.: A distributed file system over heterogeneous SaaS storage platforms. In: IEEE 16th International Symposium on Symbolic and Numeric Algorithms for Scientific Computing (SYNASC), pp. 417–421, September 2014

36. Mohanarajah, G., Hunziker, D., D'Andrea, R., Waibel, M.: Rapyuta: A cloud robotics platform. IEEE Trans. Autom. Sci. Eng. **12**(2), 481–493 (2015)

37. A*STAR Social Robotics Laboratory (ASORO) (2016). http://www.a-star.edu.sg/asoro/

38. Kehoe, B., Matsukawa, A., Candido, S., Kuffner, J., Goldberg, K.: Cloud-based robot grasping with the Google object recognition engine. In: 2013 IEEE International Conference on Robotics and Automation (ICRA), pp. 4263–4270, May 2013

39. Kehoe, B., Matsukawa, A., Candido, S., Kuffner, J., Goldberg, K.: Grasping with Google Goggles. In: Robotics and Automation (ICRA) (2013)

40. Laboratory of Analysis and Architecture of Systems (LAAS) (2016). https://www.laas.fr/public/en/node/139

41. Hunziker, D., Gajamohan, M., Waibel, M., D'Andrea, R.: Rapyuta: the roboearth cloud engine. In: 2013 IEEE International Conference on Robotics and Automation (ICRA), pp. 438–444, May 2013

42. Arumugam, R., Enti, V.R., Bingbing, L., Xiaojun, W., Baskaran, K., Kong, F.F., Kit, G.W.: DAvinCi: a cloud computing framework for service robots. In: 2010 IEEE International Conference on Robotics and Automation (ICRA), pp. 3084–3089, May 2010

Efficient People-Searching Robot

Anh Pham[(⊠)], Mayang Parahita[(⊠)], Andy Tsang, Mathias Chaouche,
and James Rees

University of Birmingham, Birmingham, UK
{bdp414,mdp358,cyt493,mxc526,
jxr227}@student.bham.ac.uk

Abstract. This paper describes a robot that performs person search in an office environment. The system brings together a number of technologies to ensure efficient and robust search in the face of varying conditions. These include face detection, face recognition, speech recognition, localization, path planning and re-planning in the presence of dynamic obstacles, and a decision theoretic search strategy. The main novel contribution of the system is an ensemble method for combining the outputs of multiple classifiers for face recognition. A second contribution is the use of a strong priority over the location of individuals in the database to guide search. The system is also able to use speech generation and recognition to interact with individuals to achieve its goals, as well as to receive goals via a mobile interface.

Keywords: Person search · Navigation · Face recognition · Ensembles

1 Introduction

Mobile robots are often required to interact with people. An important skill is the ability to efficiently and robustly search for specific people. To be efficient, the robot should search using strong prior knowledge over the likely locations of a particular person. To be robust, person recognition must be reliable in the face of varying lighting conditions, dynamic obstacles and unreliable modes of interaction (speech recognition, face recognition). In this paper we present a system that has been engineered to be robust to these challenges. In particular, the system uses strong prior knowledge over person locations and multiple face recognition methods to improve performance.

The implementation is on a P3-DX robot, controlled using speech and keyboard. Given a command, the robot guides a user through the most likely sequence of locations for the target person. The robot drives smoothly at a suitable pace and avoids dynamic obstacles. The robot can detect whether doors are closed, and request help, identifying when each door has been opened. The target person is found using state-of-the-art face recognition methods that are combined in an ensemble using plurality vote. If the target person is not found, the robot will search other possible rooms in descending order of probability.

© Springer International Publishing Switzerland 2016
L. Alboul et al. (Eds.): TAROS 2016, LNAI 9716, pp. 249–254, 2016.
DOI: 10.1007/978-3-319-40379-3_25

2 Background

2.1 Navigation

Before navigation takes place, often a map must be generated in order to guide planning and movement. This involves using sensors that allow the robot to perceive (such as cameras, sonar, and laser) to form a model of the environment. Since sensors are subject to noise and the environment can change over time, most state-of-the-art mapping techniques are probabilistic and simultaneous [1].

Existing navigation techniques employ two types of internal map representation: Metric-based and topological-based. Metric-based representations use a 2D coordinate systems in which each object is assigned a global position. Topological approaches usually define a space using places and the connection between them, essentially a graph with vertices as gateways, places of interest, or intermediate points, and a set of edges as paths between them. Having obtained the appropriate navigational map, a search is performed to find a path given starting and goal positions. The path is then used to navigate, complemented by localization to confirm the robot's position.

2.2 Face Recognition

There are three popular face recognizers: Fisher, Eigen and Local Binary Pattern Histograms, each operating differently. A Fisher recognizer maximizes the mean distance of different classes while minimizing the variance within the class, in order to perform better on linear discriminant analysis. An Eigen recognizer finds a linear combination of features that maximize the total variance in data and produce Eigen-vectors for the recognition process. An LBPH recognizer compares the 8 neighbors of each pixels and used the result as a texture descriptor.

It is clear that each recognizer has its own advantages. For instance, Fisher has better accuracy with various facial expression, Eigen provides high efficiency in computation resource but very sensitive to light variation [2], while LBPH is invariant to monotonic gray-level changes and computational efficient [3]. However, to our knowledge, few attempts were made to combine different face recognizers [4].

3 Design

The physical set up consists of a P3-DX mobile robot connected to a laptop with an external camera and microphone mounted on a tripod (height of 150 cm). The system is made up of separate constituent components with a single node to control and manage the behavior of the robot as a whole. The software is written in Python using Rospy, a popular library providing interface to mobile robots.

3.1   Database

During this study, topological data for 9 rooms and 10 people were used. The initial probability of finding each person in each room is stored in an SQLite table, which can be queried to obtain possible locations of target person in order of descending probabilities. If different rooms have the same probability, the robot go to the closest first to ensure efficient navigation. Furthermore, during and after a task, the knowledge about the target person is updated, allowing the system to keep an accurate record of the targets, hence maximizing the efficiency of the task.

3.2   Input and Output

Speech recognition was implemented to provide a natural control method. The acoustic signal is first converted to an actual sentence using Google's REST API, which is then parsed using a set of regular expressions. For instance, if the parser receives "Find Andy", the task will start with "Andy" as the target person. As the module involves sending the voice recording via internet to Google, the system is prone to wireless disconnectivity. Hence, the module emits regular beeps between which the user could speak, and emitting a double beep to signal a connection error.

Despite many benefits, speech recognition was found to be unreliable. Therefore a keyboard input, which uses the same parser - was implemented for a more reliable and robust approach which also caters for users with speech impairment. A mobile interface to send commands to the robot was also developed.

A speech synthesizer was implemented to provide user friendly feedback by vocalizing the current progress. With Ubuntu, a text-to-speech application Espeak is provided natively, and a Python wrapper module Pyttsx has been used to utilize it with a speech rate of 200 words per minute.

3.3   Localization and Navigation

The implementation uses Adaptive Monte Carlo Localization [5] and Grid-based navigation. At initialisation, an overlay grid is generated, with cell marked as either an obstacle or a free space based on the map data. After that, A* search is run using the grid to find the shortest path to goal. To avoid dynamic obstacles, each free cell is given an occupancy probability, which is then increased or decreased based on the laser scan. Cells exceeding 50 % will be marked as occupied. If needed, a new path will be computed and followed, allowing the robot to avoid unexpected obstacles such as bags and people very well.

Initially, the robot moved points to points, pausing to rotate when needed. Despite guaranteeing the shortest distance travelled, a large amount of time was spent to pause and turn. Hence, the angular and linear speeds of the robot are changed smoothly according to different factors, allowing smooth and natural movement. In addition, a feedforward vector representing momentum is updated based on the robot's movement – the faster the robot moves, the bigger its momentum becomes. This feedforward control is used to predict the robot's future pose and steer the robot according to the extrapolation of its position.

To address localization failures, a relocalize mode is activated if the average weighting of particles remains too low for too long. In such cases, the robot pauses its current task and starts spinning around while adding a number of random particles on the map every frame. After a period of time, the right position will be found and the average weighting will increase enough to stop the relocalize mode, allowing the robot to return to its current task.

3.4   Face Detection and Recognition

Using OpenCV library, the face detector extract faces from a video feed and creates grayscale images of 256×256 pixels, which are then fed into face recognition, avoiding unnecessary computation spent on non-face pixels.

The face recognition module controls the recognizers provided by OpenCV including LBPH, Fisher and Eigen. The training set includes 10 faces, 7 of which from the AT&T Face Database [5]. Training data of the remaining 3 faces was taken in different places within the testing area, some of which were lit by artificial sources from various directions. Whenever new data is added, the whole training set is reprocessed. Otherwise the recognizer will load a saved training file (.yml) to minimize computation time.

After the training process, whenever a face is given to the recognizer, it will return the most likely person, along with the Euclidean distance between the testing image and the closest found image. Therefore, the lower the returned value is, the higher the chance for the prediction to be correct. A threshold is set for every recognizer in order to ignore faces with low probability.

Each recognizer has different performance in reference to the size of training set, light variance of testing image and the variation of facial expressions. Therefore, instead of using a single recognizer, our design uses all three recognizers to improve the result. The combined method will recognize a face only when all three recognizers agree on the same ID (comparable to using AND operators). The combination was found to be more robust with lower false positive rate, justifying the extra computational resource required.

4   Preliminary Results

4.1   Speech Recognition

The success frequency of speech commands was recorded. Using 5 different voices, command sentences were pronounced a total of 100 times. The results shows a very low True Positive Rate of 0.561, indicating how the speech command is only successfully processed after several attempts.

4.2   Path Planning and Driving

Choosing different starts and goals, the performance of pathfinding is found to be 0.001177 s on average, with the maximum time recorded being 0.03298 s, which means the robot can quickly re-plan to avoid obstacles without using up resource and affecting the system. To show the improvements of the driving code, tests with each feature turned off were compared against the complete driving code, using the same 31-m path. In the first set of tests, the current position given by localisation is used without extrapolation. In the second set, the robot can only do one type of movement at a time, either rotating or moving forward. In the last set, the complete code was used. Moreover, to keep the results deterministic and reliable, a static environment was used with no dynamic obstacles and detection (Fig. 1).

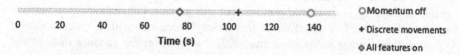

Fig. 1. Average performance of each set of tests in seconds

The difference was tremendous: feedforward vector and continuous movement improve driving time by roughly 45 % and 27 % respectively. In addition, the code performed similarly when tested with different routes, showing its consistency. Furthermore, Fig. 2 shows how the robot can avoid dynamic obstacles using very narrow pathways in between.

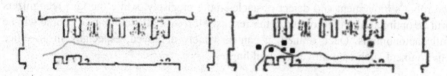

Fig. 2. Navigation path through empty corridor (left) and obstructed by obstacles (right)

4.3 Face Recognition

For face recognition, we conducted several tests and found that 800, 95 and 10000 were the optimal distance thresholds for Eigen, LBPH and Fisher respectively.

Figure 3 shows that the ensemble of three recognizers performs better than single ones, giving an accuracy of 0.85. Moreover, it has a False Positive Rate of 0 which is extremely desirable in our system since even with a large database and a constant video feed, it will never wrongly recognize the target person.

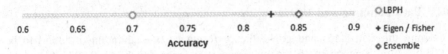

| 0.6 | 0.65 | 0.7 | 0.75 | 0.8 | 0.85 | 0.9 |

Accuracy

○ LBPH
◆ Eigen / Fisher
◇ Ensemble

Fig. 3. Accuracy of individual face-recognizers and the ensemble of all three

4.4 Systems Tests

The system was tested as an integrated unit at multiple stages throughout the development process to ensure that all components work well together to fulfill the overall task, and to give guidance to necessary changes. In the final iteration, the system was able to receive keyboard commands, while speech recognition was not reliable. The robot could then navigate to the destination with a 95 % success rate. The few trial failures were due to obstacles that could not be navigated around. The robot proved capable of stopping in a suitable position that will provide sufficient space to perform face detection. The face recognition was mostly successful with a success rate of 85.7 % and an average recognition time of 22.8 s, however the variance of light and facial expression relative to the quality and quantity of the training set had a strong negative effect on the result. Overall, the system takes around 2-5 min to find a person, depending on where he or she is.

5 Conclusion

In conclusion, although it is still work in progress, the project has successfully combined various techniques, including both standard and novel methods. In future work, components will be improved in various ways: a denser grid will be used to smoothen the robot's movement and detect obstacles more precisely, while the face recognizers will be optimized further to give better results. Moreover, the tracking of people will be fully autonomous. Once refined, it can be widely deployed, especially in an office environment such as a university building.

References

1. Thrun, S.: Robotic mapping: a survey. In: Lakemeyer, G., Nebel, B. (eds.) Exploring Artificial Intelligence in the New Millennium, pp. 1–35. Morgan Kaufmann Publishers Inc., San Francisco (2003)
2. Jaiswal, S., Bhadauria, S., Jadon, R.: Comparison between face recognition algorithms - eigenfaces, fisherfaces, and elastic bunch graph mapping. JGRCS 2(7), 187–193 (2011)
3. Ahonen, T., Hadid, A., Pietikainen, M.: Face description with local binary patterns: application to face recognition. IEEE Trans. Pattern Anal. Mach. Intell. 28(12), 2037–2041 (2006). doi:10.1109/TPAMI.2006.244
4. Zhao, W., Chellappa, R., Phillips, P.J., Rosenfeld, A.: Face recognition: a literature survey. ACM Comput. Surv. 35(4), 399–458 (2003). doi:10.1145/954339.954342
5. Database of Faces. AT&T Laboratories, Cambridge. http://www.cl.cam.ac.uk/research/dtg/attarchive/facedatabase.html (2002). Accessed 11 Jan 2016

Accurate and Versatile Automation of Industrial Kitting Operations with SkiROS

Athanasios S. Polydoros$^{(\boxtimes)}$, Bjarne Großmann, Francesco Rovida,
Lazaros Nalpantidis, and Volker Krüger

Robotics Vision and Machine Intelligence (RVMI) Lab.,
Department of Mechanical and Manufacturing Engineering, AC Meyers Vaenge 15,
2450 Copenhagen SV, Denmark
{athapoly,bjarne,francesco,lanalpa,vok}@m-tech.aau.dk
http://rvmi.aau.dk/

Abstract. The low automation level of industrial manufacturing processes, in conjunction with the limited research in the field of object manipulation for placing tasks, rise the need for adaptive and accurate industrial robotic systems that can manipulate a variety of objects in an uncertain environment. We deal with those issues by providing insights into two modules of a skilled-based architecture, the SkiROS. Those modules are applied on the placing part of an industrial kitting operation and can lead to versatile perception of the environment and the accurate placing of objects in confined areas. We evaluate both modules in terms of accuracy, execution time and success rate in various setups of the environment. Also we evaluate the whole kitting pipeline – including the proposed modules – as a unit in terms of repeatability and execution time. The results show that the proposed system is capable to both accurately localize the kitting box and place the object in its narrow compartments.

Keywords: Robotic systems · Industrial robotics · Kitting operations · Object localization · Skill-based architecture

1 Introduction

Despite the large scientific interest in object manipulation, real-life applications of such systems are rather limited. For-instance, the object manipulation task is automated by less than 30 % in the automotive industry[1]. The prominent deterrent factors are the high variety of manipulated objects and the unstructured environments of manufacturing facilities. Those issues raise the need for robotic systems that are able to adapt to environmental changes and also be accurate enough in order to cope with the demands of a manufacturing process.

In this paper, we focus on logistics operations and specifically consider the automation of a kitting task. There are two main reasons for this choice.

---

[1] http://stamina-robot.eu/about-stamina.

© Springer International Publishing Switzerland 2016
L. Alboul et al. (Eds.): TAROS 2016, LNAI 9716, pp. 255–268, 2016.
DOI: 10.1007/978-3-319-40379-3_26

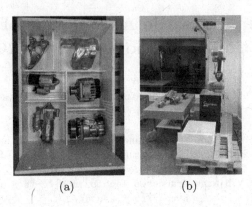

(a)   (b)

Fig. 1. (a) The kitting box containing automotive parts for the industrial kitting task and (b) the physical robot setup used for our evaluation. The setup consists of two long range RGB-D sensors for monitoring the picking and placing locations, one short-range RGB-D sensor mounted on the Universal Robots UR10 arm and a Robotiq 3-finger adaptive gripper.

First, it is a common task in a variety of manufacturing processes since it involves the placement of the appropriate objects in their corresponding box compartments (Fig. 1(b)). Secondly, it involves a challenging placing task – placing receiving considerably less research attention compared to picking. The limited research indicates that placing is considered as a trivial task, but real-life challenges prove the exact opposite. We argue that object placing can be a complex and challenging task depending on the application, which can vary from a simplistic table-top placing to the most complex assembly of parts. Thus, we consider kitting as a special case of a placing task where the manipulated objects have to be placed precisely enough in specific and narrow compartments without collisions.

The proposed robotic system addresses those challenges by utilizing its following three features: a skilled-based system architecture, an arm motion and a perception module. The skill-based system, namely the Skill system for ROS (SkiROS) [9], guarantees the information flow across the skills. Furthermore, its world model provides all needed information about the kitting operation. Such information includes the manipulated object, the destination compartment, the pose of the kitting box etc. The perception module provides the world model with the needed information about the pose of the kitting box. The planning pipeline ensures that the executed trajectories are collision-free and precise enough in order to place the manipulated object in the compartment without any damage. The system's software is build on Robot Operating System (ROS)[2] and exploits capabilities of the MoveIt software[3]. On the hardware side, a robot cell is used which physical setup is illustrated in Fig. 1(b).

---

[2] http://www.ros.org.
[3] http://moveit.ros.org.

This paper presents the supporting concepts and evaluates the operation of a complete robotic system for kitting operations. The presented system combines components that range from the high-level architecture – allowing for integration with manufacturing execution systems (MES) as a cyber-physical system [6], to the most low-level primitive of planning. Additional to previously proposed systems, the placing task is versatile and accurate ensuring collision-free placement of the objects in the kitting box.

2   Related Work

Previous research on robotic systems for kitting operations is mainly focused either on the architecture of the system or on depalletizing of parts. In [1] the authors propose a knowledge representation model for industrial robots using a kitting operation as case-study. Their approach integrates a Planning Domain Definition Language (PDDL) planner to automatically concatenate actions. In our solution, we also integrate a PDDL planner and a knowledge representation model to represent the objects of interest in the environment to simplify the kitting operation, but we go further by defining a general architecture to integrate any number of skills. Moreover, we automatically generate the PDDL domain depending on the skills available to the robot.

In [3] the authors use SkiROS as a system architecture for depalletizing parts and placing into a kit. For planning long motions, they use a set of precomputed paths between the initial pose of the manipulator and a dense grid of locations over the picking and placing areas. This ensures fast and stable operation of the planning module. However, it is not flexible enough to be transferred to other robotic manipulators, since the computed paths are robot-specific. Furthermore, they only provide a perception pipeline for depalletizing [4], while the kitting part of that system does not include perception. This can be problematic if the pose of the kitting box changes and will result in collisions with the walls of the narrow compartments.

In [8], the authors propose a robotic system that performs kitting operation in a toy-case where 3D-printed parts with various geometries have to be placed in narrow compartments of the kitting box. There is not any prior knowledge about either the pose of the kitting box or the geometry of the manipulated objects. A perception pipeline is employed in order to derive appropriate picking and placing poses. The 3D printed parts are manipulated easier compared to real industrial objects due to their size and weight. Also, real industrial objects need to be grasped only in certain ways; otherwise potential damages might occur.

3   Skill-Based System Architecture

The SkiROS system [9] is designed to control mobile manipulators in the industrial environment of the future. In this context, the robot should receive goals and high-level description of the environment directly from the Manufacturing Execution System (MES) of a factory and operate accordingly, with little or no support from a human operator [6]. SkiROS constitutes a 3-layered architecture implemented with a modular and scalable structure on top of the ROS middleware. Despite the initial focus on industrial applications, SkiROS remains a general framework that can be used for many different applications.

The skills framework is organized into several layers, each one represented by a manage. The task manager and the World model run in separate ROS nodes, meanwhile the Skill manager and the Primitive manager are grouped in one ROS node, defined as a *robot subsystem*. The *device manager* resides at the lowest layer, and loads proxies—drivers that conform to a standard interface—and presents standard interfaces for similar devices (e.g., gripper, arm, camera, etc.). The *primitive manager* loads and manages the list of available primitives. The primitives basically represent the building bricks for skills and usually implement one certain algorithm like arm movement, planning, segmentation or locating objects. Skills are loaded from the *skill manager*, the coordinator of a robotic subsystem. When started, the skill manager registers the robot subsystem on the world model, specifying the available hardware and skills. This information is then collected from the top layer of the architecture, i.e. the *task manager*. The task manager acts as the general robot coordinator. It presents an interface to receive goals from external systems and given the goal and the world state—as retrieved from the world scene—generates the proper sequence of skills for all robot subsystems using a PDDL planner [7].

3.1 SkiROS for Kitting

In our use-case, the SkiROS framework is used to model a kitting task. The kitting operation consists of two skills—pick and place—which are modelled by multiple primitives, as shown in Fig. 2. Hardware related primitives like *OpenGripper*, *CloseGripper* or *MoveArm* usually constitute implementations of drivers. More interesting, however, are the generative primitives like *Locate*, *Locate kit*, *Register* or *PlanMove* which may contain more complex algorithms.

Perception and motion planning primitives are hence the key aspects in the kitting pipeline. As much work has been done related to the picking skill in previous work Furthermore, due to the modular structure of SkiROS, all implementations of primitive are exchangeable and thus they can be used in multiple

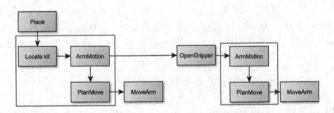

Fig. 2. The sequence of primitives composing our place skill. The red-line boxes include the new modules presented in this paper. (Color figure online)

skills. Our new motion planning primitive, has been without further development successfully integrated into the picking skill as well.

4 Perception and Arm Motion

As discussed in Sect. 3.1, the two crucial primitives of the kitting pipeline are the motion planning of the manipulator and the pose estimation of objects. In this section, we present the proposed approach that makes the robotic system more versatile on changes of its environment and guarantees accurate placing in the narrow compartments of the kitting box.

4.1 Perception

The perception of the environment plays a crucial role in the kitting operation. On the one hand, vision is needed to detect and localize the industrial parts to be picked and on the other hand, the location for placing skill has to be determined, namely a compartment in the kitting box. However, the nature of these two cases of object/kitting box pose estimation is fundamentally different. In the former case, the object can be easily separated from the plane it is lying on and we have an unoccluded close view of its surface. In the latter case, a container has to be extracted from a further distance where most of the box is occluded by itself (the sides of the box are vanishing when observed from a top view) and by objects already inside the container.

Locating the Object. The pose estimation of the object to pick from a pallet has been implemented in an earlier stage [3]. Using the workspace camera which captures the whole pallet to pick from, a plane detection using integral images is applied to identify the supporting plane of the industrial parts [2]. This plane is then used in a classic tabletop segmentation algorithm to extract the target object cluster. A wrist camera mounted on the robot arm is then moved close to the object such that the same procedure can be applied which results in a more accurate object cluster. After computing a rough pose estimate using PCA, a registration refinement step is applied using Multi-Resolution Surfel Maps (MRSMaps) [10]. The object model (a snapshot of the part) for the registration process and its corresponding grasping pose have been trained in an offline phase beforehand.

Locating the Kitting Box. The localization of the kitting box usually needs to be executed when a new box arrives in the placing area of the robot. Even though a rough pose of the kitting box is known, slight changes in its position or orientation might lead to collision between the robot and the box while placing, as the parts in most cases have a very tight fit. Additionally, the pose estimation of the kitting box can be rerun after placing parts in the compartments as slight displacements might occur during the placement.

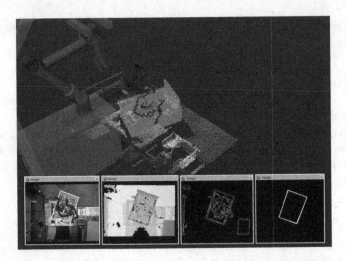

Fig. 3. The 3D scene shows a highly cluttered kitting box with the final registered kitting box (blue) overlaid. The images in the bottom (left to right) show different steps of the registration process: (a) raw RGB image (b) normalized depth image (c) after edge extraction with box candidates overlaid in green (d) final voting space for box pixels (Color figure online)

For the pose estimation of the box, an additional workspace camera with a top view on the kitting area is used. Due to the pose of the camera and objects which are already placed in the kitting box, most of it is not visible except for the top edges. Additionally, parts of the box edges are distorted or missing in the 3D data caused by the noise of the camera. Our approach is therefore based on a 2D edge detection on a cleaned and denoised depth image which applies a pixelwise temporal voting scheme to generate potential edge points. After mapping these back to 3D space, a standard ICP algorithm is utilized to find the kitting box pose.

In detail, the kitting box pose estimation can be split into following steps:

1. *Edge extraction:* The edge detection is base on the classical Canny edge detector. However, before applying the edge detector, we normalize depth image and then remove depth shadows and noise using morphologic filters and a bilateral filter respectively. Only then the canny edge detector is applied. Afterwards, another morphological closing operation is executed to close edge gaps due to missing depth data. Based on the edge image, all inner and outer contours are computed and used in the hypothesis generation. They are then approximated by a polygon to get rid of minor defects in the contour.
2. *Model generation:* Before computing probable kitting box candidates, a model of the kitting box has to be generated. As the kitting box might be adjusted during the project, we use a CAD model of it as the reference. The model is processed in a similar way as the scene: We create a pointcloud from the CAD model by densely sampling the surfaces and then project the upper

half of it to an image plane to retrieve (visible) edges representing the rim of the container. Afterwards, the outer contour is computed and saved as the reference model. Note that this process basically allows us to define a container of arbitrary shape and is not only restricted to boxes.

3. *Hypothesis generation:* For each contour found in the scene, we can now match the shape with the model and compute a confidence value for it. The matching of the polygonised contours is mainly based on the seven Hu invariants as proposed in [5]. However, since the Hu invariants are translation, rotation and scale invariant, but a container of a certain scale is used, we need to add features to enable the discrimination of differently scales. Therefore, we reintroduce the non normalized area of the contour and its perimeter to compute the final confidence.

4. *Temporal Pixelwise Voting Scheme:* The generated candidates are used to cast pixelwise weighted votes in a 2D histogram by rasterizing the contours and add a vote for each pixel with the weight of the contour's confidence. Additionally, the votes are propagated to neighboring cells by applying a Gaussian kernel. To achieve temporal consistency, we introduce an additional vote accumulator which is updated by weighting old and new votes according to a given "forget" and "learn" rate. Thereby, stable votes or pixels belonging to the shape model are kept and enhanced. After each processed frame, the accumulated vote space is thresholded and when it contains enough votes, the stable pixels are extracted and used to identify the corresponding 3D points. This allows us to extract only the rim of the container and neglect the clutter inside and outside of it.

5. *Candidate Registration:* Once back in 3D space, the potential points are cleaned by removing statistical outliers (too few neighbors) and clustered using the Euclidean distance. The principle axes of the largest cluster are then roughly aligned and refined with an ICP registration which results in the final pose estimate.

4.2   Arm Motion

Once the pose of the kitting box is provided by the perception module, its collision model is added to the planning scene (see Fig. 4(a)). Thus, the motion planning pipeline takes a precise model for collision checking into account instead of a noisy and inaccurate voxel representation. The high precision of the kitting box model makes the definition of placing locations within the narrow compartments feasible (see Fig. 4(b)). Hence, the planning pipeline should be able to find a path for poses that are not in collision and at the same time, located closely to the bottom layer of the kitting box. This is a desirable feature in industrial cases since it significantly decreases the possibility of damaging the manipulated object, the gripper, or the kitting box during placing.

The arm motion module of the placing skill is responsible for the execution of the motion from the initial pose of the robot to the placing pose and back to the initial. It first receives the object's placing pose for the specific manipulated object from the world model of SkiROS. At the next step, it creates a pre-placing

(a)

(b)

Fig. 4. (a) Visualization of the planning scene during a placing operation. A CAD-based model of the kitting box (blue) is added for ensuring precise representation of its structure. (b) Example of placing location in narrow compartment of the kitting box. (Color figure online)

pose, located slightly (30 cm) above the placing pose. Thus, the pre-placing pose is a waypoint of the trajectory after which the motion of the end-effector is linear and perpendicular towards the compartment. Such a Cartesian space motion is crucial for avoiding collisions with the inner walls of the compartments. When the final and pre-placing poses are available, the module propagates them to the planning pipeline which creates and concatenates the two trajectories (from initial to pre-placing and from pre-placing to placing). Once the motion is complete, the placing path is reverted and executed for moving back to the initial pose without the need of replanning. An overview of the arm motion module is illustrated in Fig. 5.

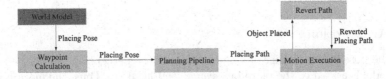

Fig. 5. Schematic representation of the arm motion module

The proposed motion planning pipeline augments the traditional stochastic planning algorithms by introducing two features which can be anticipated as planning reflexes. The need for that addition arises due to the instability of traditional planning algorithms. A preliminary benchmarking of those is presented in Table 1. The Probabilistic Road-maps (PRM), Expansive-Spaces Tree (EST) and Rapidly exploring Random Tree (RRT) motion planning algorithms are tested on finding a placing path between the initial pose and randomly selected kitting compartments. The results indicate that PRM performs better than the others on that task. However, its success rate is not desirable for industrial applications where system stability is a crucial factor. Moreover, the created motion paths are not optimal for placing operations in narrow compártments since they do not provide linear trajectories of the end-effector in the operational space. Thus the gripper follows a curved motion which results on collisions with the inner walls of the kitting box.

Table 1. Benchmarking of motion planning algorithms. The evaluation includes 20 runs per algorithm for randomly selected placing compartments. There are requested two plans, one from the initial to a pre-placing pose and from the pre-placing to the final placing pose.

Planner	Mean time ± Var(sec)	Success rate
PRM	6.73 ± 0.51	70 %
EST	6.97 ± 0.59	55 %
RRT	8.57 ± 1.17	60 %

One reason behind the low success rates of the planning algorithms is the inverse kinematics solutions. Such solutions are not guaranteed to be optimal, i.e. close to the initial joint configuration. Thus, this fact can create non-optimal motion plans that make the end-effector follow unnecessary long trajectories. An other issue of those planners is the stochastic approach of trajectory shaping. Both PRM and tree-based methods are randomly sampling the planning space for possible waypoints which can be used for shaping the trajectory. This procedure can create variations on trajectories planned for the same set of initial and final poses. Furthermore it does not ensure the linear motion which is needed for collision-free kitting.

The developed planning pipeline deals with those problems by employing two deterministic planning reflexes, the joint and operational space linear interpolations. The first ensures that the robot's joints will rotate as less as possible in order to reach the pre-placing configuration. This happens by linearly interpolating between the initial and the pre-placing configurations with a pre-defined interpolation step. So this planning reflex minimizes the energy consumption of the motion.

On the other hand the operational space interpolation provides a linear motion of the end-effector in the Cartesian space. This is achieved by interpolating linearly between the pre-placing and the placing position of the end-effector. The interpolated orientations are derived by performing spherical linear interpolation (slerp) between the quaternions of the pre-placing and placing orientation. Finally the inverse kinematics solution is calculated for each pose in the path. In order to ensure that the calculated solution is as close as possible to the pre-placing joint configuration, we sample multiple solutions with different random seeds from the kinematics solver and use the one that refrain the least from the pre-placing configuration. Once the path has been generated, it is checked for singularities, collisions with the environment and for constrains' violation. If any of those fault-criteria are met, the pipeline uses the PRM as a fallback for solving the planning problem.

5   Experimental Evaluation

The described arm motion and perception modules were separably evaluated on 50 runs and also as a unit within the kitting pipeline. The evaluation criteria for the kitting box detection are the relative pose error and the execution time while for the arm motion module was the successful path planning, collision-free object placements and execution time. Furthermore we evaluate the whole pick and place pipeline with various manipulated objects and measure its performance in terms of successful parts' kitting and execution time.

5.1   Locate Kitting Box

The kitting box pose estimation was tested independently of the kitting pipeline. For the evaluation, we used the a long-range Xtion RGBD Pro Live camera mounted 1.8 m above the empty kitting box. We chose three different randomly chosen angles between −45 to +45 degrees for the kitting box and performed 50 pose estimations of the kitting box in total. On the one hand, the precision and repeatability is measured in terms of the pose estimation, i.e. the relative error of the estimated position and orientation are captured. On the other hand, the time for the registration is measured, as it has an impact on the overall runtime of the kitting pipeline. Moreover, we conducted additional tests with a filled kittingbox and overly cluttered kitting box (see Fig. 3) which had no negative effect on the registration. Since the cluttered kittingbox was merely a test to see the limits of our approach and does not reflect the real use-case, the following evaluation neglects those results.

Figure 6 depicts the relative errors measured during the 50 trials of the kitting box pose estimation. The positional error (Fig. 6(b)) shows clearly that the precision is generally high with an average less than 1 cm (see Table 2). However, in unfavorable situations due to reflection, light or other interferences, we measured errors up to 3 cm as seen for the blue markers in Fig. 6(a) which might cause difficulties when placing the parts in very tight compartments albeit we did not observe any collisions during the evaluation of the whole pipeline.

The relative rotational error is shown in Fig. 6(c). The error is measured as the minimal angle between two orientations represented as quaternions. Except for two outliers which are still below 1 degree off, the angular error is consistently less than 0.3 degrees with an average of about 0.1 degree (Table 2) which results in almost no visible variance of the orientation.

The second part of the evaluation, the execution time, is listed in Table 2. The timing for the kitting box registration is quite high and takes in average 7.25 s and hence does not run in real-time. However, this time is measured without optimizations in code. The timing therefore includes the generation of the model

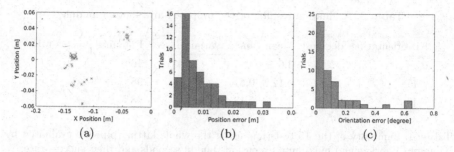

Fig. 6. Absolute positioning errors from camera perspective for the three different kittingbox poses (a). Precision of the kitting box registration showing (b) the relative translational error and (c) the relative angular error (Color figure online)

contour in each frame which is a major reason for the slow execution of the process. This could be easily avoided by implementing a caching function to save the precomputed model which will definitely be done in the future. Even though there is still room for increasing the performance of the kitting box registration primitive, in the use-case, the pose estimation can actually be done in parallel to the picking of the part and thereby does not increase the overall runtime.

Table 2. Summary of the position error, orientation error and execution time for the kitting box registration

Position error [cm]		Orientation error [degree]		Execution time [sec]	
avg	var	avg	var	avg	var
0.86	0.46	0.11	0.02	7.25	0.55

5.2 Arm Motion

The evaluation results for the proposed arm motion module are illustrated in Table 3. The proposed approach performs optimal when the kitting box is oriented perpendicular to the basis axis of the robot (Fig. 5). On the other hand, there are slight chances of collisions when the orientation of the kitting box has more extreme values ($-45°, +45°$). The main reason is the low accuracy of the perception module in those cases (Fig. 6(a)). Despite those cases, the proposed module performs better on kitting operations compared to the classic approaches (Table 1).

5.3 Kitting Pipeline

The final test evaluates the newly developed modules in interaction within the skill framework. For this purpose, we use the setup as depicted in Fig. 1(b) where

Table 3. Evaluation results for the proposed arm motion module

Kit orientation [degrees]	Mean time ± variance [sec]	Planning success rate
0°	3.6 ± 0.24	100 %
+45°	4.12 ± 0.51	98 %
−45°	4.15 ± 0.18	98 %

Table 4. Summary of the 15 test-runs using the whole kitting pipeline evaluated by minimum, average and maximum execution time in seconds and their success rate.

Evaluation	Pick	Place	Kitting pose	Planning	Overall
Min time	36	18	6.5	3.2	58
Mean time	40	19	7.5	3.7	59
Max time	43	21	9	4.2	62
Success rate	86 %	92 %	100 %	100 %	80 %

the picking pallet is on right-hand side of the robot base and the kitting box in front of it. For locating the object, the right workspace and the wrist cameras are used.

The picking and placing poses for the object were taught beforehand. The compartments we used for placing was two extreme cases, The bottom-left and the top-right. The other compartments are too small to contain the manipulated objects. The whole pipeline was executed 15 times. Table 4 summarizes the trials with the execution time and success rates. The overall time is given by the sum of the picking and placing skill, as the kitting box detection is done in parallel during the picking whereas the planning is part of the placing skill.

The qualitative evaluation of the kitting pipeline shows stable behavior in terms of execution time. The picking only needs about 40 s to detect, register and pick the object. The placing works much faster in about half the time as no object has to be registered, but also due to the fast planning of the proposed arm motion module which only needs 3.7 s in average to plan a valid trajectory to the kitting box compartment. The kitting box pose estimation behaves similar to the individual evaluation which is to be expected, as it runs independently from the picking or placing.

However, 3 out of the 15 trials are not executed successfully. Test-run 10 and 12 failed during the picking due to wrong object registration Test-run 13 failed during the execution of the placing due to a potential collision with the kitting box. This might have been caused by a slightly wrong pose estimation of the kitting box registration. Another issue which occurred during test-run 2 and 6, which we did not consider a failure of the integration of our developed modules, is a bad object recognition during the picking phase.

In summary, despite the difficulties which occur during some test-runs, the overall performance is very good, especially in respect to our newly developed modules. Lack of robustness in perception task and motion planning should be object of future improvement and development.

6 Conclusion

In this work we present two novel enhancements which are crucial for the successful execution of a kitting operation: (i) the implementation of a robust and accurate pose estimation algorithm for kitting boxes or other placing containers and (ii) an arm motion that moves a robot arm with high precision to the placing pose by avoiding collisions in narrow spaces. Moreover, these modules are embedded in the skill-based architecture SkiROS such that they can not only share information with other actors in the system, but also be easily reconfigured and reused in new environments. In combination with the picking skill, our developed placing pipeline leads to a fast and adaptable kitting operation which can be used in the high-level task planning context. This kitting operation is developed for a industrial scenario within the STAMINA project, hence is not only used in research labs, but is applied to real world problems.

Within this context, we claim that the problem of placing objects in a broader sense is underestimated in current research and that there is a need for fast high-precision placing skills. We prove our thesis by evaluating classical, but still state-of-the-art planning algorithms like PRM and show that they are incapable of finding valid trajectories in narrow and too constrained environments. It becomes clear that, in order to deal with confined placing areas, a dynamic placing operation has also to be supported by a precise perception system to estimate the pose of the placing location – in our case the kitting box.

In the evaluation, we show that both implemented modules independently satisfy these demands. Furthermore, we integrate those modules in the existing robotic architecture SkiROS and prove our concept by evaluating the kitting pipeline as a whole. The results underlie that the system is capable to perform versatile kitting operations in confined placing areas.

Acknowledgments. This work has been supported by the European Commission through the research project "Sustainable and Reliable Robotics for Part Handling in Manufacturing Automation (STAMINA)" (FP7-ICT-2013-10-610917).

References

1. Balakirsky, S., Kootbally, Z., Kramer, T., Pietromartire, A., Schlenoff, C., Gupta, S.: Knowledge driven robotics for kitting applications. Robot. Autonomous Syst. **61**(11), 1205–1214 (2013). ubiquitous Robotics
2. Holz, D., Holzer, S., Rusu, R.B., Behnke, S.: Real-time plane segmentation using RGB-D cameras. In: Röfer, T., Mayer, N.M., Savage, J., Saranlı, Ulu (eds.) RoboCup 2011. LNCS, vol. 7416, pp. 306–317. Springer, Heidelberg (2012)

3. Holz, D., Topalidou-Kyniazopoulou, A., Francesco Rovida, M.R.P., Krüger, V., Behnke, S.: A skill-based system for object perception and manipulation for automating kitting tasks. In: Proceedings of the 20th IEEE International Conference on Emerging Technologies and Factory Automation (ETFA). Luxemburg (2015)

4. Holz, D., Topalidou-Kyniazopoulou, A., Stueckler, J., Behnke, S.: Real-time object detection, localization and verification for fast robotic depalletizing. In: Proceedings of the IEEE/RSJ International Conference on Intelligent Robots and Systems (IROS), Hamburg, Germany (2015)

5. Hu, M.K.: Visual pattern recognition by moment invariants. IRE Trans. Inf. Theory 8(2), 179–187 (1962)

6. Krueger, V., Chazoule, A., Crosby, M., Lasnier, A., Pedersen, M.R., Rovida, F., Nalpantidis, L., Petrick, R., Toscano, C., Veiga, G.: A vertical and cyber–physical integration of cognitive robots in manufacturing. Proc. IEEE 104(5), 1114–1127 (2016)

7. McDermott, D., Ghallab, M., Howe, A., Knoblock, C., Ram, A., Veloso, M., Weld, D., Wilkins, D.: PDDL - The Planning Domain Definition Language (Version 1.2). Technical Report CVC TR-98-003/DCS TR-1165, Yale Center for Computational Vision and Control, October 1998

8. Rofalis, N., Nalpantidis, L., Andersen, N.A., Krüger, V.: Vision-based robotic system for object agnostic placing operations. In: International Conference on Computer Vision Theory and Applications (VISAPP) (2016)

9. Rovida, F., Krüger, V.: Design and development of a software architecture for autonomous mobile manipulators in industrial environments. In: 2015 IEEE International Conference on Industrial Technology (ICIT) (2015)

10. Stückler, J., Behnke, S.: Model learning and real-time tracking using multi-resolution surfel maps. In: Proceedings of the AAAI Conference on Artificial Intelligence (AAAI-12) (2012)

Introducing a 3D Physics Simulation Plugin for the ARGoS Robot Simulator

Richard Redpath[1,2](✉), Jon Timmis[1,2], and Martin A. Trefzer[1,2]

[1] Department of Electronics, University of York, York, UK
richard.redpath@york.ac.uk
[2] York Robot Laboratory, York, UK

Abstract. We present a plugin for the ARGoS robot simulator which enables the use of 3D physics simulation and definition of objects and entities without the need to develop new entities in C++. We provide the facility for entities to be loaded and added to simulations using a URDF inspired XML format. Loading of entities, physics simulation and rendering is all handled by the plugin, removing the need for knowledge of specific engines or OpenGL. This paves the way for ARGoS to be used not only for development and evaluation of controllers but also for evolution of robot morphologies and co-evolution of controllers and morphologies which was previously not possible without a considerable development overhead.

1 Introduction

The ARGoS simulator [3] has found traction in the robotics community recently as a highly configurable, light weight simulator for experimenting with swarm robotics. ARGoS is currently used by several groups around the world however is currently lacking support for including 3D physics simulations and for simulating entities which can be edited quickly and included without a working knowledge of physics engines or visualisation frameworks.

The current process to add a new robot or entity involves writing a series of new classes in C++ which represent the state of the object in ARGoS, a means of interaction between ARGoS and the underlying physics engine, and a visualisation of the entity. This means that adding new entities requires knowledge of ARGoS' architecture, the physics simulator that you wish to use (as new models need to be written for each), and how to render objects (in OpenGL if the default visualisation is used). While it could be argued that alternative simulators exist which do not require this knowledge they are often too slow to run experiments with more than a handful of robots and do not provide the same level of flexibility as ARGoS.

With this in mind we have developed a plugin which adds these features to the ARGoS simulator. Using it we are able to add entities to simulations from XML definition files using a format based on URDF, write controllers for loaded robots in Lua or C++, and have these loaded entities simulated using the 3D physics engine bullet [1]. The plugin also handles rendering these entities in the default Qt visualisation provided with ARGoS.

© Springer International Publishing Switzerland 2016
L. Alboul et al. (Eds.): TAROS 2016, LNAI 9716, pp. 269–274, 2016.
DOI: 10.1007/978-3-319-40379-3_27

2   Motivation

We required a simulator which was light weight enough to have multiple running instances, yet also flexible enough to be modified and adapted to a range of requirements and experiments if necessary. A number of the more well known simulators are closed source, limiting modification, or not designed for swarm robotics, causing poor performance when simulating a number of robots. This led us to ARGoS, a light weight simulator designed specifically for swarm robotics which is open source and could be modified or extended as required.

Given our intended use case of on-board simulations our first priority was to be able to perform 3D physics simulations, a feature which was not well supported by ARGoS. Due to the complexity of adding new types of entity to the simulator, and our need to be able to simulate potentially hundreds of different types of object, we added the ability to load solid objects from OBJ mesh files. We started with OBJ for ease of parsing and widespread use and support in various editors. We also needed to be able to interact with these loaded objects which would have required writing code for each type of robot we intended to simulate. As such we chose to implement loading of robots from an external definition file. While URDF [2], a format used by ROS [4], seemed quite well suited to this we felt that some modifications could be made to make it more usable.

3   Current Features

Here we present the features which have currently been implemented. We start by describing the definition of materials which will be used in both simulation and visualisation. We then move on to the addition of single rigid bodies before extending this to include multi-part, controllable entities. Full access to the bullet engine to create custom entities from compiled code is also possible although we do not present it here. We do not intend for URDF models to be fully compatible. There are a number of features in the URDF format which we do not feel would be used often in swarm robotics and would only serve to complicate our implementation. We would be open to exploring the addition of features as demand for them arises.

3.1   Entity Definition File

All entities, single and multi-body, should be described in an XML file which provides definitions for the required materials, along with any bodies and joints which describe the entity to be used. In keeping with URDF terminology we refer to these as materials, links and joints respectively. All definition files have a root 'entity' element which provides a name for the entity being described as shown below. Entity 'prototypes' are lazy-loaded and entities are added to simulations by cloning this prototype. This allows multiple instances of an entity to be added without re-parsing definitions.

```
<entity name="my_entity">
   <!-- Entity definition -->
</entity>
```

Materials. Typically, material definitions will appear as a direct child of the entity element and are separated into 'color' and 'contact_coefficients'. The 'color' element provides the colour the material will be rendered in. It has a single string attribute, 'rgba', which provides the red, green, blue and alpha values of the colour in the range 0-1. The 'contact_coefficients' element allows specification of 2 physical attributes, namely dampening (kd) and friction (mu). These values can also be provided with a standard deviation to add noise into the definition of the material. A sample material definition is shown below.

```
<material name="sample">
  <color rgba="0.8 0 0 1"/>
  <contact_coefficients kd="0.1" mu="0.2" kd_std="0" mu_std="0"/>
</material>
```

Materials can be defined in a hierarchical manner with default values being provided in the root of the document and different values being overridden in more specific nodes. For example, if we wanted to use the material above but render in blue then we can write the following inside of a link element.

```
<material name="sample">
  <color rgba="0 0 1 1"/>
</material>
```

This link would inherit its contact coefficients from the earlier definition of "sample" however would have its colour overridden. A similar technique can be used to override individual contact_coefficient values, allowing a single link to have a different frictional value or more noise in a parameter to all others using that material.

Single Rigid Bodies. To add a single rigid body to the arena we provide a definition file which contains a single link element. This link element specifies the material of the body as well as 'collision' and 'visual' elements which define how the link should be simulated and rendered. Each link may contain multiple visual elements and their union will be used for visualisation. Similarly multiple collision elements can allow more complex collision shapes to be built up. This provides us with the power to break down shapes into a collection of primitives which can be efficiently simulated yet still render a high definition, textured mesh. A sample link definition may be as follows.

```
<link name="sample_link">
  <material name="sample"/>
  <visual>
```

```
      <origin xyz="0 0 0" rpy="0 0 0" />
      <geometry>
        <box size="1 0.5 0.25"/>
      </geometry>
    </visual>
    <collision>
      <origin xyz="0 0 0" rpy="0 0 0" />
      <geometry>
        <cylinder radius="1" length="0.2"/>
      </geometry>
    </collision>
  </link>
```

This defines a link which will be rendered as a box with dimensions 1,0.5,0.25 but which will be simulated as a cylinder with radius 1 and length 0.2. Collision elements also support two other classes of children. These are 'sphere', which only has a radius parameter, and 'mesh' which has a single 'filename' parameter. At present only OBJ mesh files are supported however other formats may be added in the future. If either the 'visual' or 'collision' element is missing then the one which is provided is used for both. At least one must be provided.

Complex Entities. Each entity can, and typically will, contain multiple links which are connected by joints. A joint is defined by its type (only 'continuous' supported at time of writing), the links it connects (parent and child), the transform between the parent's origin and the child's, an axis in which it acts and a number of limits which specify operating parameters of the motor. A sample continuous joint definition is given below which would connect 'link_2' to 'link_1'.

```
<joint name="motor_front_left" type="continuous">
  <parent link="link_1"/>
  <child  link="link_2"/>
  <origin xyz="0.1 0.3 0" rpy="1.570795 0 0" />
  <axis xyz="0 1 0"/>
</joint>
```

We use the 'origin' element to specify where and in what orientation the child is connected to the parent, in this case the child is connected at an offset of 0.1,0.3,0.0 from the parent's origin and rotated by $\frac{\pi}{2}$ radians about the X axis. The axis element specifies the direction in which this joint operates in the child's coordinate frame, in this case the child would rotate about its Y axis when the motor is used.

Lua Integration. ARGoS includes support to develop controllers in the Lua scripting language as well as in C++ which provides a mechanism for rapid prototyping of algorithms. Lua scripting also allows controllers to be developed without the need to compile any code which allows for more flexible evolution of controllers. As such we provide access to all joints through the scripting interface with each joint being accessed using its defined name. Continuous joints currently have setTargetVelocity and getCurrentVelocity methods which can be accessed through Lua. New methods could be added if a need for them was found. If we consider a simple 4 wheeled robot with motors called 'motor_front_right', 'motor_front_left' and so on a simple controller step function could be defined as follows.

```
function step()
    robot.motor_front_left.setTargetVelocity(2);
    robot.motor_rear_left.setTargetVelocity(2);

    robot.motor_front_right.setTargetVelocity(-2);
    robot.motor_rear_right.setTargetVelocity(-2);
end
```

Clearly, more complex controllers could be developed which make use of the sensors each entity has access to but we provide a simple example here to show how individual joints are accessed.

3.2 Adding Entities to Scenes

Entities which have been defined in XML can be added to a scene using the experiment's .argos file much like any other entity. An example of adding an entity is given below with a screenshot of the simple entity which has been added shown in Fig. 1.

```
<xml_entity id="xml_entity" definition_file="test_entity.xml">
  <body position="0,0,0.3" orientation="0,0,0"/>
  <controller config="xml_lua_controller"/>
</xml_entity>
```

4   Future Work

Currently, we are improving joint support and adding linear actuators and positional servos. The next step towards further improving the plugin is likely to be allowing the incorporation of sensors into the entity definition file allowing more self contained, reusable robot definitions. From our experience ARGoS is a very powerful tool which allows flexibility without compromising execution speed and we hope that our contribution can help it to gain traction in the wider community. We are open to requests and suggestions for new features and will look to add those which appear to be popular or would provide sufficient benefit to users.

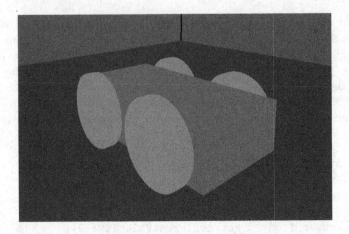

Fig. 1. A simple 4 wheeled entity added using our plugin. Each of the robot's wheels are attached and can be controlled as shown in the sample Lua controller.

5   Conclusion

We have introduced a new plugin to the ARGoS simulator which allows entities and controllers to be defined without knowledge of the ARGoS architecture, OpenGL or any physics engines. This plugin also brings 3D physics simulations to ARGoS using the bullet physics engine. The plugin is currently in active development and will continue to incorporate features as appropriate additions and modifications are found as a result of use in the community. Further details, including how to access source code, can be found at http://goo.gl/KnyH5v.

Acknowledgements. The authors would like to acknowledge support from the EPSRC and the Department of Electronics, University of York. JT is partially funded by The Royal Society.

References

1. Bullet Physics Engine. http://bulletphysics.org/. Accessed 11 Feb 2016
2. XML Robot Description Format. http://wiki.ros.org/urdf/XML/model. Accessed 11 Feb 2016
3. Pinciroli, C., Trianni, V., O'Grady, R., Pini, G., Brutschy, A., Brambilla, M., Mathews, N., Ferrante, E., Di Caro, G., Ducatelle, F., Birattari, M., Gambardella, L.M., Dorigo, M.: ARGoS: a modular, parallel, multi-engine simulator for multi-robot systems. Swarm Intell. **6**(4), 271–295 (2012)
4. Quigley, M., Conley, K., Gerkey, B.P., Faust, J., Foote, T., Leibs, J., Wheeler, R., Ng, A.Y.: Ros: an open-source robot operating system. In: ICRA Workshop on Open Source Software (2009)

Co-operative Use of Marine Autonomous Systems to Enhance Navigational Accuracy of Autonomous Underwater Vehicles

Georgios Salavasidis[1,2]($\boxtimes$), Catherine A. Harris[2], Eric Rogers[1],
and Alexander B. Phillips[2]

[1] University of Southampton, Southampton SO17 1BJ, UK
etar@ecs.soton.ac.uk
[2] National Oceanography Centre, Southampton SO14 3ZH, UK
{geosal,cathh,abp}@noc.ac.uk

Abstract. This paper gives the first results from a research programme that aims to develop efficient navigation techniques for long range Autonomous Underwater Vehicles (AUVs) conducting ocean basin transits using low power sensors. The results results relate to the development of and on-line and low power cooperative navigation scheme between an AUV and an Unmanned Surface Vehicle (USV), where the latter acts as an acoustic tether for localization error reduction. Most of the computations are executed on the USV in order to reduce the power requirements of the AUV. The methodology is assessed by numerical simulation, where the Dead Reckoning (DR) error is modelled using data obtained from field trials and the result show that considerable estimation error reduction is possible even in very deep water operations.

1 Introduction

Autonomous underwater vehicles (AUVs) are becoming increasingly popular for exploring and monitoring the oceans, offering unparalleled opportunities for data-collection. Applications range from performing oceanographic surveys to inspection of pipelines and the monitoring of sea floor assets. Conventionally, AUVs are deployed and operated from a support ship which significantly increases the cost of undertaking AUV missions. With the recent development of long-range platforms, such as Autosub Long Range (ALR) [1] (see Fig. 1.a), capable of operating for several months and travelling thousands of kilometres in a single deployment, the need for a dedicated support ship is much reduced.

Such long-range platforms open up a world of new AUV applications, including persistent monitoring of decommissioned oil and gas structures and data collection in some of the remotest areas on earth, such as under the Arctic ice. However, the increased endurance of ALR reduces the power available for onboard sensors, including those used for navigation.

© Springer International Publishing Switzerland 2016
L. Alboul et al. (Eds.): TAROS 2016, LNAI 9716, pp. 275–281, 2016.
DOI: 10.1007/978-3-319-40379-3_28

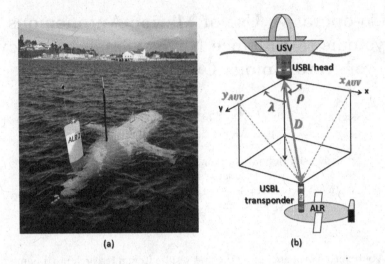

Fig. 1. (a) Autosub Long Range [1] - Field Trials, (b) USBL acoustic positioning.

2   AUV Localization

It is well known that without external aids, existing AUV localization techniques experience unbounded estimation error growth. Although inertial sensors are capable of providing excellent short term accuracy, the estimation degrades over time due to sensor drift. As a result, an external absolute positioning aid is required in order to reduce the estimation error. However, in the case of GPS this requires the vehicle to regularly surface, a considerable undertaking when the vehicle is operating at significant depth. Paull *et al.* [2] provide a comprehensive review of existing localization techniques and sensor technologies.

Conventional methodologies usually either: constrain the operating area to within a few kilometres [3,4], due to fixed sensor infrastructure, e.g. long baseline (LBL) aided localization; or perform well only in shallow waters. Range measurements from a single moving surface or seafloor mounted transponder may be used to provide position updates, but the system experiences observability issues, as reported in [5]. Ultra Short BaseLine (USBL) [6] systems have been widely utilized for localization purposes, but most of existing work has focused on shallow water and constrained operating areas due to, e.g., fixed transponders [7].

This paper considers an on-line and low power localization scheme for an AUV conducting long range and deep water operations, where simulations are used to assess the impact of water depth on the navigational accuracy of an acoustically coupled AUV-USV system. The aim is to minimize AUV power requirements and operational cost, whilst maximising the navigational accuracy.

3   Cooperative Localization Scheme - System Description

We consider a deep-rated, long-endurance ALR [1] AUV aided, for localization purposes, by a single manoeuvring USV. Utilising a USBL system (typical set up - head mounted on the USV), the USV is able to calculate the position of the AUV using the observed range and bearing (see Fig. 1.b). On-board the AUV, the Inertial Measurement Unit (IMU) and a pressure sensor allow the 3D localization problem to be transformed to an equivalent two dimensional problem. The AUV maintains an estimate of its own position using a simple dead reckoning algorithm. This integrates velocity and heading measurements, provided by a Acoustic Doppler Current Profiler (ADCP) and a magnetic compass, according to the following kinematic equation:

$$\mathbf{x}_k = \begin{bmatrix} x_{k-1} + (u_k cos\psi_k - v_k sin\psi_k)dt + w_k^x \\ y_{k-1} + (u_k sin\psi_k + v_k cos\psi_k)dt + w_k^y \end{bmatrix} \tag{1}$$

where $\mathbf{x}$ is the AUV position (x and y translational motion) expressed in north east down (NED) reference frame, dt is sampling time (assumed to be sufficiently small), and $w \sim \mathcal{N}(0, Q)$ is Additive Gaussian White Noise (AWGN). The input vector $\mathbf{u}_k = [u_k, v_k, \psi_k]^T$ is taken as the horizontal AUV speed (in body frame reference) in both the forward u and starboard v directions, and the heading of the AUV ψ at time-step k. However, as the actual input command is unknown, it must be estimated based on the sensor readings. The actual input vector $\mathbf{u}$, then, is taken to be equal to the sensor reading $\bar{\mathbf{u}}$ plus an AWGN with $\mathcal{N}(0, V)$.

When the estimation error becomes significantly large, the vehicle will require USBL position updates. The USBL head interrogates the USBL transponder/acoustic model mounted on the AUV, which replies with both a USBL response and its internal position estimate. On receiving this information, the USV calculates AUV position based on both the USBL measurement and the AUV estimate. The USBL measurement is modelled according by the following equation:

$$z_k^{usbl} = \mathbf{x}_k^{USV} + {}^n R_b(\psi_k^{USV})(p_k + e_k) \tag{2}$$

where ${}^n R_b$ is the body-to-NED transformation matrix, $\mathbf{x}^{USV}$ is the USV absolute position, ψ^{USV} is the USV heading and z_{usbl} is the AUV position in NED as calculated by the USBL system. Also p is the noise-free two dimensional USBL measurement[1] given by:

$$p = \begin{bmatrix} D\cos(\rho) \\ D\cos(\lambda) \end{bmatrix} \tag{3}$$

where D is the measured slant range and λ, ρ are two acoustic inclination angles (see Fig. 1.b). Finally, $e_k \sim \mathcal{N}(0, R)$ accounts for the USBL measurement noise.

---

[1] It is assumed that the USBL reference frame coincides with AUV body frame.

Given the sensor readings, the USV runs an Extended Kalman Filter (EKF) for data fusion. The EKF is an recursive estimation algorithm consisting of two phases, prediction and update. In this system, the EKF is run partially on the AUV and USV, reducing the AUV computational power requirements.

The predicted state and the associated error covariance matrix are calculated on-board the AUV using (4) and (5), respectively.

$$\hat{\mathbf{x}}_k^- = f(\mathbf{x}_{k-1}^+, \bar{\mathbf{u}}_k, 0) \tag{4}$$

$$P_k^- = F_k P_{k-1}^+ F_k^T + G_k V_k G_k^T + Q_k \tag{5}$$

where F_k, G_k are the Jacobian matrices of $f(\cdot)$ with respect to $\mathbf{x}$ and $\mathbf{u}$ respectively (evaluated at $\mathbf{x}_{k-1}^+$ and $\bar{\mathbf{u}}_k$).

On completion of the above computations, the information generated is transmitted to the USV, where measurements update the predicted values according to the update step of the EKF [8]. Manipulating (2), the observation model can be written as:

$$z_k^{usbl} = \mathbf{x}_k^- + {}^n R_b(\psi_k^{USV}) e_k \tag{6}$$

Once the USBL measurement z^{usbl} becomes available, the Kalman gain K_k is computed and used to update the predicted state and error covariance matrix. Estimating the AUV's position, the USV sends position updates to the AUV which re-initializes its internal estimate accordingly, before proceeding with its own dead reckoning estimation process.

4   Simulation Results

We assume the ALR is operating in the horizontal plane at a constant depth following a pre-defined, though noisy, trajectory (an inverted "8") with a constant forward speed and appropriately chosen heading (obeying the angular speed constraints of the ALR). Two different operating depths are considered - shallow (100 m) and deep water (5000 m). Throughout the simulations, the USV is perfectly localized and operates perpendicularly above the ALR (eliminating the acoustic ray bending effect). Table 1 summarizes the simulation parameters.

Initially, the ALR accurately knows its pose and the error covariance matrix is initialized to the value of the expected process uncertainty. The input covariance matrix, V, is defined based on data collected from ALR field trials at Portland Harbour, UK (October 2015) and the USBL update frequency, f, and the covariance matrix, R, are obtained using the sensor data sheet [9] (the deeper the AUV operates, the less frequent and accurate the measurements).

Figure 2 gives the simulation results (averaging multiple executions), giving the estimation error (Euclidean distance between estimated and the actual/ground-truth trajectory), Fig. 2.a, and the time evolution of the trace of error covariance matrix, Fig. 2.b, for both depth scenarios. Furthermore, to

Table 1. Simulation & EKF parameters

Simulation duration	$t_{max} = 5000\,\text{s}$
Measured forward speed	$\bar{u} = 0.77\,\text{m/s}$
DR update frequency	$f_{DR} = 1\,\text{Hz}$
Depth operation - Shallow	$z_{shallow} = 100\,\text{m}$
Depth operation - Deep	$z_{deep} = 5000\,\text{m}$
USBL frequency - Shallow	$f_{usbl}^{shallow} = 0.5\,\text{Hz}$
USBL frequency - Deep	$f_{usbl}^{deep} = 0.1\,\text{Hz}$
Measurement covariance - Shallow	$R_{shallow} = diag\{0.01\text{m}^2, 0.01\text{m}^2\}$
Measurement covariance - Deep	$R_{deep} = diag\{25\text{m}^2, 25\text{m}^2\}$
Process covariance	$Q = diag\{0.0009\text{m}^2, 0.0009\text{m}^2\}$
Input covariance	$V = diag\{0.000077\text{m}^2/\text{s}^2, 0.000077\text{m}^2/\text{s}^2, 0.000264\text{rad}^2\}$

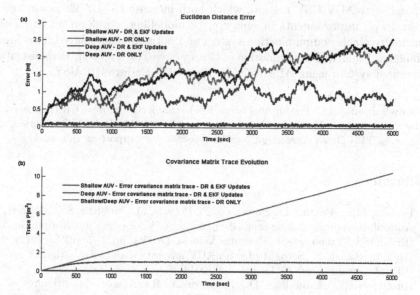

Fig. 2. Simulation results for both depth scenarios (100 m & 5000 m) with and without USBL/EKF updates. (a) Estimation error evolution, (b) Time evolution of the trace of the error covariance matrix. (Color figure online)

compare the error reduction by using the USBL system and the EKF update step to alternatives, the solo DR algorithm (4) was also computed and also shown in Fig. 2. It is clear that even for the complex case of 5000 m depth the estimation error remains within the (0-1 m) error region (red line) throughout the mission while the corresponding single DR experiences unbounded error growth (black line). Similar behaviour has been observed with the trace of the error covariance matrix. For both scenarios, EKF greatly reduces position uncertainty whereas the simple DR displays monotonically increasing position uncertainty.

5   Conclusion and Future Research

This paper has given preliminary simulation results from research on the development of an on-line low power cooperative localization scheme between an AUV and a USV, acoustically coupled using a USBL system. The USBL and DR sensor data fusion was treated as an optimal stochastic filtering problem, for which an EKF was used. Most of the computations were executed on-board the USV in order to minimise AUV power requirements and extend its endurance. The results show considerable error reduction even in short missions (≈ 1.4 h) demonstrating both the potential of the USV aiding system and navigation accuracy in extended missions.

Ongoing/future work will include the development of a more accurate simulator by removing unrealistic assumptions (such as perfectly localized USV and perpendicular AUV-USV motion, which both increase the USBL measurement uncertainty), improvements in sensor error modelling, development of an efficient inter-vehicle communication system and optimal time-delayed data fusion techniques. Finally, the culmination of this work will be a real-life demonstration of the final system using ALR AUV [1] and the C-Enduro [10] USV.

Acknowledgements. The authors would like to thank the Innovate UK (Autonomous Surface, Sub-surface Survey System project) and ROBOCADEMY (FP7 Marie Curie Programme ITN Grant Agreement Number 608096) for supporting this work.

References

1. Furlong, M.E., Paxton, D., Stevenson, P., Pebody, M., McPhail, S.D., Perrett, J.: Autosub long range: A long range deep diving AUV for ocean monitoring. In: 2012 IEEE/OES Autonomous Underwater Vehicles (AUV), pp. 1–7. IEEE (2012)
2. Paull, L., Saeedi, S., Seto, M., Li, H.: AUV navigation and localization: A review. IEEE J. Oceanic Eng. **39**(1), 131–149 (2014)
3. Kussat, N.H., Chadwell, C.D., Zimmerman, R.: Absolute positioning of an autonomous underwater vehicle using GPS and acoustic measurements. IEEE J. Oceanic Eng. **30**(1), 153–164 (2005)
4. Jakuba, M.V., Roman, C.N., Singh, H., Murphy, C., Kunz, C., Willis, C., Sato, T., Sohn, R.A.: Long-baseline acoustic navigation for under-ice autonomous underwater vehicle operations. J. Field Robot. **25**(11–12), 861–879 (2008)
5. Gadre, A.S.: Observability analysis in navigation systems with an underwater vehicle application. PhD thesis. Virginia Polytechnic Institute and State University (2007)
6. Milne, P.H.: Underwater acoustic positioning systems. Gulf Publishing Company (1983)
7. Caiti, A., Di Corato, F., Fenucci, D., Allotta, B., Bartolini, F., Costanzi, R., Gelli, J., Monni, N., Natalini, M., Pugi, L., et al.: Fusing acoustic ranges and inertial measurements in AUV navigation: The typhoon AUV at commsnet13 sea trial. In: OCEANS 2014-TAIPEI, pp. 1–5. IEEE (2014)
8. Simon, D.: Optimal state estimation: Kalman, H infinity, and nonlinear approaches. Wiley, New York (2006)

9. Sonardyne company: Ranger 2 USBL system data sheet. http://www.sonardyne. com/images/stories/system_sheets/sonardyne_ranger_2.pdf. Accessed 02 Mar 2016

10. Autonomous Surface Vehicles (ASV) company: C-Enduro unmanned surface craft. Available: http://asvglobal.com/product/c-enduro/. Accessed 02 Mar 2016

Human Management of a Robotic Swarm

Nicole Salomons[(✉)], Gabriel Kapellmann-Zafra, and Roderich Groß

Sheffield Robotics and Department of Automatic Control and Systems Engineering,
The University of Sheffield, Sheffield, UK
nicolejsalomons@gmail.com

Abstract. This paper proposes a management algorithm that allows a
human operator to organize a robotic swarm via a robot leader. When
the operator requests a robot to become a leader, nearby robots suspend
their activities. The operator can then request a count of the robots, and
assign them into subgroups, one for each task. Once the operator releases
the leader, the robots perform the tasks they were assigned to. We report
a series of experiments conducted with up to 30 e-puck mobile robots.
On average, the counting and allocation algorithm correctly assigns 95 %
of the robots in the swarm. The time to count the number of robots
increases, on average, linearly with the number of robots, provided they
are arranged in random formation.

1 Introduction

Swarm robotics is concerned with the study of a large number of robots which are
usually simple, and of limited capabilities. When deploying a swarm of robots in
a real-world scenario, the swarm is likely to get separated into different groups,
for example, due to obstructions in the environment. Moreover, some of the
robots might even get lost. A challenge of such real-world environments is that
the operator typically has no birds-eye perspective. Yet, the operator would like
to know how many robots are currently in an area and within communication
reach of one another.

Methods for counting robots have been proposed in [1,4]. Brambilla et al. [1]
use an algorithm similar to firefly signalling. The algorithm counts the number of
nearby robots that are signalling during a period of time. Ding and Hamann [2]
use an algorithm that counts the number of locally connected robots and causes
them to self-organize into two different areas based on their assigned colors. The
robots have unique IDs, and exchange these IDs via pair-wise communication
until all robots within communication range share the same set of IDs.

Different algorithms for task allocation are presented in [3]; our paper uses an
algorithm similar to Extreme-Comm. Extreme-Comm is considered reliable and
fast compared to the other algorithms, but requires a large number of exchanged
messages between the robots. Each robot spreads its ID and keeps a list of the
other robots' IDs. According to their position in the list, the robots choose their
task.

In both [1,2], the counting experiments are performed in simulation, with
no human operator involved. This paper proposes an algorithm that allows an

© Springer International Publishing Switzerland 2016
L. Alboul et al. (Eds.): TAROS 2016, LNAI 9716, pp. 282–287, 2016.
DOI: 10.1007/978-3-319-40379-3_29

operator to select any robot from a group of real robots, and receive a count of the robots which are in communication range. Subsequently the operator can manage the group of robots by assigning different tasks to subgroups of them, and then releasing them to perform the tasks.

2 Management Algorithm

The main purpose of the management algorithm is to provide the operator with advanced capabilities of organizing a robotic swarm. The algorithm was implemented on a swarm of e-puck robots [5]. The e-puck is a differential drive robot, around 75 mm in diameter. It is shown in Fig. 1a. It has eight infrared (IR) transmitters, located around its perimeter, which the robots use to communicate with surrounding robots. Their maximum communication range is limited to 15 cm, creating a fairly stable connection between the robots. The e-puck also has Bluetooth, which allows an operator to send and receive information wirelessly via a computer.

2.1 Human Interaction

The management algorithm allows the operator to reassign groups of robots to different tasks during run-time, and to verify the status of the robot swarm without requiring direct visual contact with the swarm. The commands are given to the e-puck robots through a Bluetooth connection with the leader robot. The following commands are available:

- **Request Robot:** The operator can request either a random robot or a specific robot, by providing its ID. The selected robot, hereafter called *leader*, stops its movement, and requests the neighboring robots to stop as well. If a robot receives a stop request, it stops and relays the message.[1] By suspending the

(a)   (b)

Fig. 1. (a) The e-puck robot used in the experiments; (b) a group of 30 robots, which are performing a count.

---

[1] The stop message is relayed periodically so if other robots enter communication range, they also stop and participate in the counting of the robots.

movement within the group of connected robots, their connectivity can be maintained for the duration of the operation.

– **Request Count**: The operator can request a count of robots belonging to the same group as the leader.
– **Assign Tasks**: For each task, the operator sends the task ID and how many robots should be assigned to it. However the operator has no control over which specific robots are selected. The leader may also assign a task to itself if all other robots have already been assigned tasks.
– **Release Robots**: The robots resume their suspended tasks, or perform their newly assigned task, if any.

2.2 Robot Swarm Counting

The counting algorithm is based on [2], but here implemented on physical rather than simulated robots. Each robot has a list of IDs. Upon initialization, a robot's list contains only its own ID, which is assumed to be unique. The robot iterates through its list and broadcasts each ID. Once reaching the end of the list, the robot starts over again. Whenever receiving an ID, the robot adds it to its list, if not already present.

Figure 2 illustrates the counting process for a group of four robots. At the beginning, each robot has only its own ID. The list is updated when receiving messages from neighboring robots. Notice how robots 1 and 2 are not in communication range of each other, but are both in range of robot 3. Robot 3 will pass the ID of robot 1 to robot 2 and vice versa, and thus robot 1 and robot 2 will, after two iterations, also have each other's IDs in their lists.

2.3 Task Assignment

After the operator connected to a leader and performed a count, a number of robots can be assigned to perform a certain task. Similar to counting, each robot

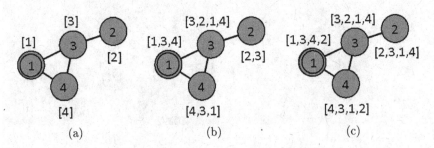

Fig. 2. Example of robots counting (lines denote pairs of robots within communication range). (a) Initially, only their own ID is in the list. (b) The robots include the IDs received by their neighbors. (c) After a further iteration, each robot contains everyones' IDs in their list. Note that perfect communication is assumed. In practice, the e-pucks may be unable to receive multiple messages simultaneously.

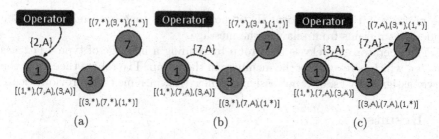

Fig. 3. Example of the leader robot assigning tasks (lines indicate pairs of robots within communication range). The leader robot receives a request by the operator to assign two robots to task A. The leader robot chooses two robots (a), and broadcasts the task assignment information through the swarm (b–c).

keeps a list of IDs which also includes a task ID associated with each robot ID. The list will be broadcast continuously until all the robots know which task all robot IDs are assigned to.

Figure 3 demonstrates how tasks are assigned to subgroups of robots. To start the process, the operator sends the leader task A and requests that 2 robots should be assigned to it. The leader assigns task A to robots 3 and 7 in the list. In our example, the leader and these two robots are not yet assigned any tasks. The leader starts broadcasting the IDs and associated tasks to its neighboring robots. Robot 3 upon receiving a task assignment for robot 7, broadcasts the message and robot 7 receives it. When robots 3 and 7 receive a message with their ID, they update their task but remain static. Once they are released by the leader robot (robot 1), they start task A. When the robots are released, they reset their ID list.

2.4   Message Structure

The robots exchange four types of messages through infrared communication. All messages consist of 5 digits: the first indicates the message type (1–4); the next 3 are the robot ID (001—999) if applicable, otherwise 000; the last digit can be a task ID, checksum or 0.

Once the operator requests a robot, the robot starts broadcasting message type 1 (*stop*) to all surrounding robots. When receiving this message, a robot stops moving and relays the message to surrounding robots periodically.

If the leader receives a request for a count of the swarm, it triggers the counting process described in Sect. 2.2. The robots use message type 2 (*count*), which contains an ID and checksum. This improves the reliability of the IR communication, and minimizes the risk of wrong IDs being added to the list. If the maximum message size permits, it is possible to send multiple IDs per message instead of just one.

Message type 3 (*task assignment*) contains a robot ID and a task ID. Whenever a robot receives this message, it saves the message to its list. If the ID in the message is the robot's own ID, it performs this task once released by the leader, but refrains from relaying the message.

The operator can release the group by sending a message of type 4 (*release robots*), which is relayed by the members in the group. The robots then start the newly assigned tasks, or if not assigned a new task, resume their previous tasks.

3   Results

Experiments were conducted to examine the counting and task assignment performance of swarms of physical e-puck robots. Unless otherwise stated, the robots were organized in random formations, that is, they were arbitrarily placed. For an example formation, see Fig. 1b.

3.1   Counting and Task Assignment Accuracy

We analyzed the accuracy of the counting and task assignment process. We assumed that 30 % of the swarm are to be assigned to tasks A and B each, and the remaining 40 % to no task. We define the overall accuracy as the percentage of robots that ended up with a correct assignment, averaged over multiple trials.

We performed 10 trials with 10 robots and 10 trials with 20 robots. For groups of 10 robots, the accuracy was 95 %. For groups of 20 robots, the accuracy was 96 %. Errors in the task assignment can happen, for example, when a low battery causes a robot to reset, and hence lose its list information. Errors can also be introduced during message transmission, causing a robot to be assigned the wrong algorithm, or include an ID in its list that does not exist.

3.2   Counting Time

We estimated the time it takes to obtain the count. Figure 4a shows the time (in s) it took to count 10, 15, 20, 25 and 30 robots organized in random formations.

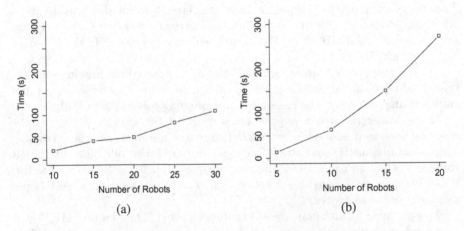

(a)   (b)

Fig. 4. Counting time for swarms in (a) random and (b) linear formations.

For each group size, 10 runs were conducted and the mean time is reported. The counting time seems to increase linearly with the number of robots.

We also considered linear formations. In other words, the robots were sequentially arranged, starting with the leader robot. Intermediate robots were in communication range of only their two neighbors. This formation can be considered as the worst case. Figure 4b shows the time it took to count 5, 10, 15 and 20 robots organized in linear formation. As can be seen, the counting algorithm scales less favorably when the robots are organized in this worst case formation.

4   Conclusions

We implemented a counting algorithm proposed by [2] on a swarm of physical e-puck robots and expanded it to also include a role assignment mechanism. The algorithm could be easily adapted for different platforms. The only requirements are that the robots have local communication, memory to store the lists and an interface to interact with the operator.

Our results show that the robots are on average 95 % accurate when counting and sequentially assigning tasks. The counting time seems to increase linearly with the number of robots, as long as they are sufficiently mixed. For linear formations, however, the algorithms took substantially longer to complete.

Future work will implement these algorithms on other robotic platforms as well as consider refined broadcasting protocols.

References

1. Brambilla, M., Pinciroli, C., Birattari, M., Dorigo, M.: A reliable distributed algorithm for group size estimation with minimal communication requirements. In: Proceedings of the International Conference on Advanced Robotics (ICAR 2009). pp. 1–6. IEEE (2009)
2. Ding, H., Hamann, H.: Sorting in swarm robots using communication-based cluster size estimation. In: Dorigo, M., Birattari, M., Garnier, S., Hamann, H., Montes de Oca, M., Solnon, C., Stützle, T. (eds.) ANTS 2014. LNCS, vol. 8667, pp. 262–269. Springer, Heidelberg (2014)
3. McLurkin, J., Yamins, D.: Dynamic task assignment in robot swarms. In: Proceedings of the Robotics: Science and Systems (RSS 2005). Cambridge, USA (2005)
4. Melhuish, C., Holland, O., Hoddell, S.: Convoying: using chorusing to form travelling groups of minimal agents. Robot. Auton. Syst. 28(2), 207–216 (1999)
5. Mondada, F., Bonani, M., Raemy, X., Pugh, J., Cianci, C., Klaptocz, A., Magnenat, S., Zufferey, J.C., Floreano, D., Martinoli, A.: The e-puck, a robot designed for education in engineering. In: Proceedings of the 9th Conference on Autonomous Robot Systems and Competitions. vol. 1, pp. 59–65. IPCB: Instituto Politécnico de Castelo Branco (2009)

Texture Recognition Using Force Sensitive Resistors

Muhammad Sayed[✉], Jose Carlos Diaz Garcia, and Lyuba Alboul

Sheffield Hallam University, Sheffield, UK
m.sayed@shu.ac.uk

Abstract. This paper presents the results of an experiment that investigates the presence of cues in the signal generated by a low-cost Force Sensitive Resistor (FSR) to recognise surface texture. The sensor is moved across the surface and the data is analysed to investigate the presence of any patterns. We show that the signal contains enough information to recognise at least one sample surface.

Keywords: Texture recognition · Force Sensitive Resistor · Active sensing

1 Introduction

Humans perform a repetitive lateral rubbing motion across a surface to feel its texture, an action known as *Lateral Motion Exploratory Procedure* [5]. The physiology of the human sense of touch suggests that the information the human brain receives during this motion is coming from force sensory elements embedded in the skin and encoded by frequency modulation [4].

Researchers were able to interface a Force Sensitive Resistor (FSR) sensor, installed on a fingertip of a prosthetic hand, with the user's nerves [8]. The user reported the ability to perceive "texture" of surfaces. This suggests that the single point force data acquired by the FSR hold enough information to perceive surface texture.

We propose that the same ability can be replicated in an artificial system using the same sensor. It would be particularly useful to achieve this ability using FSRs due to their low cost and low thickness that enables superficial installation on robotic hands.

2 Related Work

In [3], researchers constructed a low-profile fabric tactile sensor which was able to differentiate between three surface textures. The sensor was moved across the surface with constant velocity and contact pressure. The data was acquired through a Wheatstone bridge circuit and sampled at 100 Hz. The signal processing was performed in the time response domain.

L. Alboul et al. (Eds.): TAROS 2016, LNAI 9716, pp. 288–294, 2016.
DOI: 10.1007/978-3-319-40379-3_30

In [7], researchers used a metal probe instrumented with an accelerometer and two FSR sensors to classify 69 surface textures "during human freehand movement" with non-constant speed and contact pressure. The accelerometer signal was sampled at 10 kHz, the FSR data was only used to estimate surface friction by measuring lateral forces exerted by the operator's hand and was not in contact with the surface.

Results of two different experiments conducted using a tactile array force sensors attached to a robotic fingertip and moved across the test surfaces in a rubbing motion are presented in [1, 2]. The signals were processed using Neural-Networks and were able to recognise surface textures.

3 Theory

The human sense of touch is achieved through four nerve channels that connect the brain to four types of sensory elements in the skin known as *mechanorecep-tors* [4]. The nerve channels are categorised according to receptor receptive field diameter into type I (receptors with a small receptive field) and type II (receptors with a large receptive field) and according to temporal response into Slowly Adapting (SA) and Fast Adapting (FA). These channels encode information in terms of nerve "firing" frequency.

Our hypothesis, based on the working principle of the channels, is that the frequency of the signals mediated by one or more of the channels contains enough information to recognise surface texture and that this ability can be replicated in an artificial system using only force sensors. We propose to use Force Sensitive Resistors (FSR) to investigate this hypothesis due to their low cost and low thickness which enables superficial installation on robotic hands.

4 Experimental Setup

An FSR acts as a resistor whose resistance changes when *pressure* is applied on the sensor's active area. It is made of two flexible layers joined by an adhesive spacer (Fig. 1 Left). One layer contains a printed open circuit while the other contains a printed semiconductor layer whose conductivity increases with pressure. When pressure is applied, the two layers come in contact thus closing the circuit. The circuit's resistance is inversely proportional to the applied pressure with a non-linear hyperbolic relation (Fig. 1 Right).

The sensor readings were recorded using an oscilloscope and an operational amplifier circuit. The signals were sampled at 10 kHz. Four sample surfaces were selected to have textures that vary from smooth to rough. The selected surfaces are shown in Fig. 2: (a) Velcro loops, (b) Velcro hooks, (c) rough sandpaper, and (d) smooth plastic-coated cork.

In the human case, the motion is performed with **non-constant speed and contact pressure**. Therefore, our experiment investigates the presence of patterns in signals acquired during motion with non-constant speed and contact pressure. For this reason, a set of signals was recorded while performing

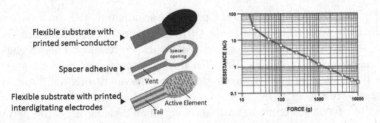

Fig. 1. (Left) exploded view of the sensor, and (Right) resistance-force relation [6]

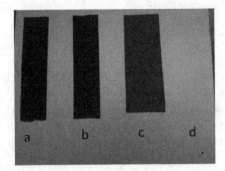

Fig. 2. Surface samples

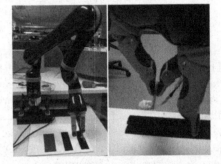

Fig. 3. FSR held by a robotic arm

the motion by holding the sensor and manually moving it across the surfaces. However, for verification purposes, a second set was recorded while the motion is performed using a robotic arm to minimise variation in speed and contact pressure (Fig. 3).

5   Results

The recorded samples were initially processed using fast Fourier transform (FFT) to look for patterns in the signal frequency response. The data varied between samples of the same surfaces; however, most samples showed a relatively consistent pattern of very high magnitude at low frequencies followed by low magnitude at the higher frequency with a region of relatively high magnitude at frequencies between 2250 Hz and 3250 Hz (Fig. 4). Surprisingly, this pattern only appeared in samples recorded with non-constant motion speed and contact pressure but not in samples recorded with constant speed and pressure.

The high magnitude at low frequencies is probably due to variation in contact pressure. The regions of high magnitude at high frequencies appears to be related to surface texture and are centred around 2500 Hz for smoother surfaces (Velcro loops and cork) and 3000 Hz for rougher surfaces (Velcro hooks and sandpaper).

The FFT analysis results prompted a second analysis; therefore, the Power Spectral Density was estimated by using the covariance method. The obtained

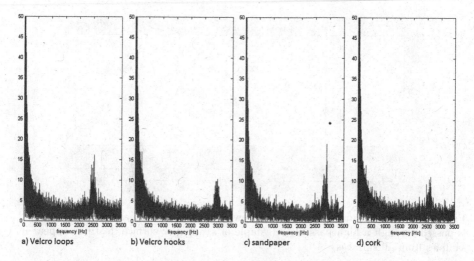

a) Velcro loops   b) Velcro hooks   c) sandpaper   d) cork

Fig. 4. FFT analysis results

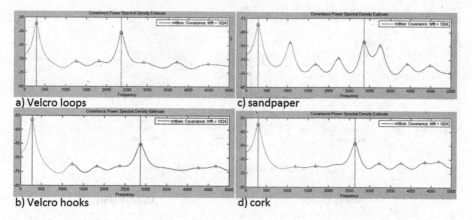

a) Velcro loops   c) sandpaper

b) Velcro hooks   d) cork

Fig. 5. Covariance Power Spectral Density analysis results

results (Fig. 5) show the same peaks of activity close to 2500 Hz for smooth surfaces and close to 2800–3000 Hz for rough surfaces.

The results also suggest another region with potential identifying features within the range of 900–1600 Hz (Fig. 6). The peaks and valley points between the two frequencies for 120 samples (30 samples per surface) are plotted in Figs. 7 and 8. The plots show that the points are apparently randomly distributed; however, there is a large area (about 50 %) that is exclusively occupied by sandpaper points. In the case of peaks points, this area contains 18 of the 30 points from sandpaper samples. Also, more than 60 points of the other surfaces' samples lie outside the total sandpaper area. In the case of valley points, the exclusive sandpaper area contains only 11 of the 30 sandpaper points, and only 40 of the other surfaces data points lie outside the total sandpaper area.

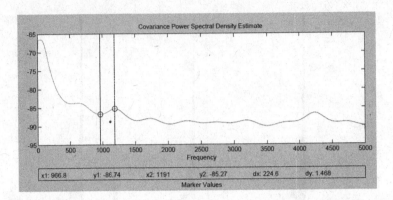

Fig. 6. Potential features in Covariance Power Spectral Density estimates after signal is passed through a high pass filter (sample of sandpaper surface, features appear in samples from all surfaces)

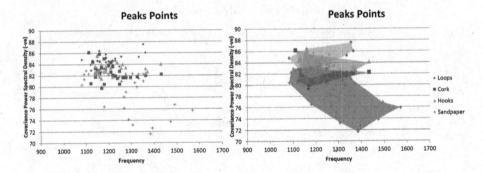

Fig. 7. Distribution map of peaks points

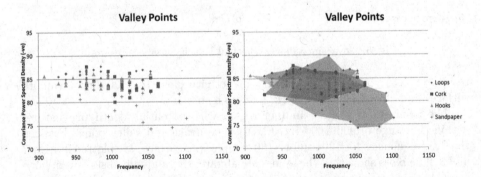

Fig. 8. Distribution map of valley points

6   Discussion

The results show that some information does exist in the signal from a single point FSR to differentiate between surfaces with different textures. One particularly interesting observation is that the observed patterns mainly exist in samples recorded with non-constant motion speed and contact pressure.

While the experiment would not completely isolate these features, the obtained results clearly show a potential to differentiate between textures, particularly between rough and smooth textures.

Sandpaper, in particular, showed promising results. The distribution of the peaks points in Fig. 7 suggest that sandpaper can be positively detected 60 % of the time while it can be correctly ruled out 67 % of the time.

7   Conclusion and Future Work

In conclusion, a single point FSR sensor provides enough information to detect differences in the roughness of surface texture when the sensor is moved in a lateral motion across the surface with non-constant speed and contact pressure. This information is not sufficient to quantify the roughness of the surface texture.

In future experiments, we plan to investigate the effect of adding soft textured surfaces on either side of sensor representing the skin and its surface irregularities (fingerprint ridges). We also plan to investigate the effect of using two layers of FSRs, one made of a single large sensor while the other is made of four small sensors, to approximate the differences in receptive field diameters. These experiments will also include low-pass and high-pass filters to imitate the slow and fast adapting behaviour of human receptors and the signal will be processed using other techniques, such as neural networks, to eliminate the possibility that the features may be results of other factors and to investigate the possibility of real-time surface recognition.

References

1. Cretu, A.M., De Oliveira, T.E.A., Prado Da Fonseca, V., Tawbe, B., Petriu, E.M., Groza, V.Z.: Computational intelligence and mechatronics solutions for robotic tactile object recognition. In: Proceedings of the IEEE International Symposium on Intelligent Signal Processing, WISP 2015, vol. 2 (2015)
2. De Oliveira, T.E.A., Cretu, A.M., Da Fonseca, V.P., Petriu, E.M.: Touch sensing for humanoid robots. IEEE Instrum. Measur. Mag. 18(5), 13–19 (2015)
3. Ho, V.A., Araki, T., Makikawa, M., Hirai, S.: Experimental investigation of surface identification ability of a low-profile fabric tactile sensor. In: IEEE International Conference on Intelligent Robots and Systems, pp. 4497–4504 (2012)
4. Jones, L.A., Lederman, S.J.: Human Hand Function. Oxford University Press, USA (2006)
5. Lederman, S.J., Klatzky, R.L.: Hand movements: a window into haptic object recognition. Cogn. Psychol. 19(3), 342–368 (1987)

6. Fried, L.: Force Sensitive Resistor Overview (2012). https://learn.adafruit.com/force-sensitive-resistor-fsr

7. Strese, M., Schuwerk, C., Steinbach, E.: Surface classification using acceleration signals recorded during human freehand movement. In: IEEE World Haptics Conference, WHC 2015, pp. 214–219 (2015)

8. Tan, D.W., Schiefer, M.A., Keith, M.W., Anderson, J.R., Tyler, J., Tyler, D.J.: A neural interface provides long-term stable natural touch perception. Sci. Transl. Med. 6(257), 257ra138 (2014)

First Experiences Towards Potential Impact of an Outdoor Shopping Assistant

Johannes Schmölz[✉], Barbara Kühnlenz, and Kolja Kühnlenz

Department of Electrical Engineering and Computer Sciences,
Coburg University of Applied Sciences and Arts,
Friedrich-Streib-Str. 2, 96450 Coburg, Germany
{johannes.schmoelz,barbara.kuehnlenz,kolja.kuehnlenz}@hs-coburg.de

Abstract. In this paper we present the results of a preliminary study conducted to investigate the acceptance of a robotic shopping assistant among elderly people. The results show that most participants acclaim the development of assistive robots.

1 Introduction

A self-determined life is important to all of us. Through the ongoing demographic change and the increased life expectancy through improvements in the health care sector the average age in our society increases and self-determination of the individual may become very limited because of physical restrictions caused by age or disease. To maintain self-determination, the need for assistive technologies increases, especially for elderly people. Our project aims at supporting people with physical restrictions by an autonomous shopping robot to retain their self-determined lives.

The robot we are developing within the project shall act as autonomous companion supporting the user with her daily shopping. The aim of the project is to develop an affordable robot acting autonomously in an unobtrusive and reliable way that works in actual urban environments.

In this paper we present the results of our preliminary study about the acceptance of a robotic shopping assistant among elderly people. For the study we simulated a typical task, taking empty bottles to the recycling bin and buying new ones, as part of the daily shopping. The robot carried the bottles to release the subjects of this tedious task and was remote-controlled by us within a Wizard-of-Oz setting. The participants were acquired with help of the AWO Mehrgenerationenhaus in Coburg, a social meeting point, especially for elderly people. Our focus is on the acceptance of a shopping robot and the interaction with it. Furthermore, we want to investigate if the robot is seen as a status symbol or a stigma by the participants.

The remainder of this paper is structured as follows: In Sect. 2 a brief overview of related work is given. In Sect. 3 the experimental design of the study will be explained. Selected results of the study will be presented in Sect. 4 and will be discussed in Sect. 5. Conclusions are given in Sect. 6.

© Springer International Publishing Switzerland 2016
L. Alboul et al. (Eds.): TAROS 2016, LNAI 9716, pp. 295–300, 2016.
DOI: 10.1007/978-3-319-40379-3_31

2   Related Work

In previous works we focused on the perception and the appropriate representation of dynamic urban environments, the identification of knowledge gaps and the retrieval of missing information by pro-actively approaching humans and interacting with them in a natural way to fill these gaps in the context of the EU-funded FP7 project IURO [10,12].

Garcia-Arroyo et al. describe a robotic assistant located at supermarkets which helps the customers to carry products and to keep track of their shopping list [6]. In [14] and [15] Nishimura et al. outline a shopping robot based on a robotized shopping cart, while Tomizawa et al. describe a tele-operated shopping robot [19,20].

The Robot-Era project [2,16] developed different robotic services for the elderly, including shopping, garbage disposal and walking assistance.

There are also some works dealing with indoor shopping robots, e.g. [4,11,13], which provide specific services to the customers like guiding.

3   Experimental Design

Our experiment was aimed at elderly people without a mobility impairment and who are still able to go shopping on their own. For the scenario we selected a task which is a typical part of the daily shopping: Exchanging empty bottles with new ones. This task is divided into two sub-tasks: First, the subject has to take empty bottles to the recycling bin and return with new ones using our robot. After completing the first run, the subjects have to start over using a wheeled walker instead of the robot. We use a within-subjects design for our study. The route for the scenario lies in the historic city of Coburg, Germany. It forms a circular course exposing the subjects to different surfaces that a very common within urban environments, e.g. asphalt or cobblestones. There are also some changes in height and a curbside which has to be overcome. An overview of the defined course indicated by the red lines is depicted in Fig. 1.

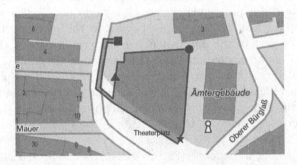

Fig. 1. Map of the route we defined for the scenario (Source: Bayerische Vermessungsverwaltung, http://www.bayernatlas.de). (Color figure online)

The starting point of the course lies in front of the AWO Mehrgenerationen-haus and is marked with a bullet (•) as shown in Fig. 1. The subjects have to complete the course counterclockwise. The downward gradient of the footpath is marked with a triangle (▲), the upward gradient with a star (★). The square (■) identifies the recycling bin, where the empty bottles have to be exchanged with new ones. To reach the recycling bin, a curbside has to be overcome.

The robot we used in the study is a Jaguar 4×4 Wheel by Dr Robot Inc. [5] with a plastic box mounted on top of it acting as a shopping basket.

The questionnaires are divided in pre- and post-questionnaires: As pre-questionnaires, dispositional factors of personality are prompted by items related to the "Big Five personality traits" that are used to describe human personality [9], according to the five-factor-model (FFM) [3]: openness, conscientiousness, extraversion, agreeableness, and neuroticism. The IPIP40 questionnaire is used [8]. In order to check for dispositional and personality-related interrelations in perceiving a shopping robot as a status symbol or a stigma, a combination of relevant item-constructs is chosen from the validated FAHR-questionnaire [17] and ISE-questionnaire [18] to capture the subject's individual affinity to status symbols as well as subjective stigma experiences.

As post-questionnaires, the same item-constructs for the dispositional status- and stigma experiences of the pre-questionnaires are formulated in direct rela-tion to the real-world experiences of the subjects using the shopping robot and the wheeled walker during the experiment in order to see if and how the direct utilization changed their minds. Additionally, validated item-constructs for User Experience (UX) are added in order to assess different dimensions of the sub-jective task-performance showed by the technical devices, as perceived by the users [1, 7].

4 Results

For the study, we recruited 14 participants (3 male and 11 female) with an average age of 68.4 ($SD = 11.66, Min = 41, Max = 83$). Half of the participants lived alone, while the other half lived with a partner and/or children.

Eleven participants are bothered by the amount of time shopping consumes. Long ways to shopping malls, heavy shopping bags and stress caused by crowds of people are further annoyances. Nearly all participants, 12 out of 14, buy groceries on their own. Two participants would never let another person do their shopping, while 6 of them would accept help of other people if absolutely necessary. For the remaining 6 participants help is not an issue. Only 2 of the participants consider weekly shopping as an annoying obligation.

Eleven participants approve the development of assistive robots. Two par-ticipants are uncertain about the development and another one refuses it at all. Ten out of 14 participants are interested in new and evolving technologies, while most of them are not willing to deal with complex technical devices.

Ten out of 14 participants desired a possibility to interact with the robot by any means. The preferred method of interaction with 12 out of 14 was voice control.

Fig. 2. A participant interacting with the robot at the recycling bin. Image courtesy of Madelaine Ruska (Press Office, Coburg University of Applied Sciences and Arts).

Two participants would prefer a classical user interface or remote control. For all participants it was import that the robot acts autonomously (Fig. 2).

For most of the participants the robot has to be convenient in its usage. There is no significant change in the opinion of the participants before and after using the robot. The performance of the robot was also an important criterion. Eleven out of 14 participants were interested in customization with respect to the design and the features of the robot. Before the experiment the participants were excited about the upcoming interaction with the robot. The excitement decreased after the experiment. Most of the participants did not think that they will receive negative feedback from their environment using an assistive robot.

5   Discussion

The annoying aspects of daily shopping like the amount of time consumed, covering long distances, heavy shopping bags and stress can be addressed, at least to some extent, by a shopping robot acting in a supporting or autonomous mode.

More than half of the participants go shopping on their own and would not accept help from other people or only if absolutely necessary. A shopping robot is perfectly suited to help these people to keep their status quo.

The participants' opinions about assistive robots are mostly positive. Many of them are interested in new technologies but at the same time do not want to deal with complex devices. Thus, a shopping robot must be reliable, unobtrusive and very easy to operate.

The majority of the participants desire a good performance of the robot. Additionally, the need for customization in terms of design and features came up during the study. The importance of these two requirements is an indicator that the robot could be seen as a status symbol analogously to a premium car with custom interior and/or exterior. Apart from that, most of the participants consider that the robot does not reflect the social status of its owner.

Especially after interacting with the robot most of the participants answered that they do not expect negative feedback from their environment while using an assistive robot. So the participants do not expect a stigmatization.

The excitement about the robot decreased during the study. This could be caused by the fact that the robot was remote-controlled and did not act autonomously. Furthermore the design and the appearance of the current robotic platform we used for the study is quite military.

6   Conclusion and Outlook

In the paper we presented selected results from our preliminary study about the acceptance of a robotic shopping assistant among elderly people. The majority of the participants acclaim the development of assistive robots. They are interested in new technologies, but at the time do not want to deal with complex devices. Therefore, a shopping robot has to be reliable and easy to use. The participants tend to think about robots as status symbols, while they do not expect to get stigmatized using one.

The positive feedback received from the participants has to be confirmed in further studies with a larger number of subjects. For our future research we will focus on HRI and acceptable social behavior of autonomous robots.

Acknowledgment. We like to thank Johanna Thomack and her team of the AWO Mehrgenerationenhaus in Coburg for supporting our study and all the volunteers participating. This work is supported in part by the Bavarian State Ministry for Education, Science and the Arts (Bayerisches Staatsministerium für Bildung und Kultus, Wissenschaft und Kunst).

References

1. Bartneck, C., Croft, E., Kulic, D.: Measurement instruments for the anthropomorphism, animacy, likeability, perceived intelligence, and perceived safety of robots. Int. J. Soc. Robot. 1(1), 71–81 (2009)
2. Cavallo, F., Limosani, R., Manzi, A., Bonaccorsi, M., Esposito, R., Rocco, M., Pecora, F., Teti, G., Saffiotti, A., Dario, P.: Development of a socially believable multi-robot solution from town to home. Cogn. Comput. 6(4), 954–967 (2014). http://dx.org/10.1007/s12559-014-9290-z
3. Costa, P., McCrae, R.: Revised NEO Personality Inventory (NEO PI-R) and NEO Five-Factor Inventory (NEO-FFI). Psychological Assessment Resources, Odessa (1992)
4. Datta, C., Kapuria, A., Vijay, R.: A pilot study to understand requirements of a shopping mall robot. In: 2011 6th ACM/IEEE International Conference on Human-Robot Interaction (HRI), pp. 127–128 (March 2011)

5. Dr Robot Inc: Jaguar 4x4 Wheel. http://jaguar.drrobot.com/specification_4x4w. asp

6. Garcia-Arroyo, M., Marin-Urias, L., Marin-Hernandez, A., de J. Hoyos-Rivera, G.: Design, integration and test of a shopping assistance robot system. In: 2012 7th ACM/IEEE International Conference on Human-Robot Interaction (HRI), pp. 135–136 (March 2012)

7. Gonsior, B., Sosnowski, S., Mayer, C., Blume, J., Radig, B., Wollherr, D., Kühnlenz, K.: Improving aspects of empathy and subjective performance for hri through mirroring facial expressions. In: Christensen, H.I. (ed.) IEEE International Symposium on Robot and Human Interactive Communication (RO-MAN), pp. 350–356. IEEE, Atlanta (2011)

8. Hartig, J., Jude, N., Rauch, W.: Entwicklung und Erprobung eines deutschen Big-Five-Fragebogens auf Basis des International Personality Item Pools (IPIP40). Arbeiten aus dem Institut für Psychologie der Johann-Wolfgang-Goethe-Universität Frankfurt am Main, Institut für Psychologie der Johann-Wolfgang-Goethe-Universität (2003). in German

9. International Personality Item Pool. http://ipip.ori.org

10. IURO Consortium: IURO (Interactive Urban Robot). http://www.iuro-project.eu

11. Kanda, T., Shiomi, M., Miyashita, Z., Ishiguro, H., Hagita, N.: An affective guide robot in a shopping mall. In: 2009 4th ACM/IEEE International Conference on Human-Robot Interaction (HRI), pp. 173–180 (March 2009)

12. Kühnlenz, B., Sosnowski, S., Buß, M., Wollherr, D., Kühnlenz, K., Buss, M.: Increasing helpfulness towards a robot by emotional adaption to the user. Int. J. Soc. Robot. 5(4), 457–476 (2013). http://dx.org/10.1007/s12369-013-0182-2

13. Ludewig, Y., Doring, N., Exner, N.: Design and evaluation of the personality trait extraversion of a shopping robot. In: 2012 IEEE RO-MAN, pp. 372–379 (September 2012)

14. Nishimura, S., Itou, K., Kikuchi, T., Takemura, H., Mizoguchi, H.: A study of robotizing daily items for an autonomous carrying system-development of person following shopping cart robot. In: 9th International Conference on Control, Automation, Robotics and Vision, ICARCV 2006, pp. 1–6 (December 2006)

15. Nishimura, S., Takemura, H., Mizoguchi, H.: Development of attachable modules for robotizing daily items -person following shopping cart robot. In: IEEE International Conference on Robotics and Biomimetics, ROBIO 2007, pp. 1506–1511 (December 2007)

16. Robot-Era Consortium: Robot-Era. http://www.robot-era.eu

17. Ruhr-Universität Bochum Projektteam Testentwicklung: FAHR. http://www.testentwicklung.de/testverfahren/Fahr/fahr-revii.html.de

18. Schulze, B., Stuart, H., Riedel-Heller, S.G.: Das inventar subjektiver stigmaerfahrungen (ISE): ein neues Instrument zur quantitativen Erfassung subjektiven Stigmas. Psychiatr. Prax. 36(8), 379–386 (2009). in German

19. Tomizawa, T., Ohba, K., Ohya, A., Yuta, S.: Remote food shopping robot system in a supermarket -realization of the shopping task from remote places. In: International Conference on Mechatronics and Automation, ICMA 2007, pp. 1771–1776 (August 2007)

20. Tomizawa, T., Ohya, A., Yuta, S.: Remote shopping robot system. Development of a hand mechanism for grasping fresh foods in a supermarket. In: 2006 IEEE/RSJ International Conference on Intelligent Robots and Systems, pp. 4953–4958 (October 2006)

Evaluating Multi-Robot Teamwork
in Parameterised Environments

Eric Schneider[1]([⊠]), Elizabeth I. Sklar[2], and Simon Parsons[2]

[1] Department of Computer Science, University of Liverpool, Liverpool, UK
eric.schneider@liverpool.ac.uk
[2] Department of Informatics, King's College London, London, UK
{elizabeth.sklar,simon.parsons}@kcl.ac.uk

Abstract. The work presented here investigates the impact of certain environmental parameters on the performance of a multi-robot team conducting exploration tasks. Experiments were conducted with physical robots and simulated robots, and a diverse set of metrics was computed. The experiments were structured to highlight several factors: (a) single-robot versus multi-robot tasks; (b) independent versus dependent (or "constrained") tasks; and (c) static versus dynamic task allocation modes. Four different task allocation mechanisms were compared, in two different exploration scenarios, with two different starting configurations for the robot team. The results highlight the distinctions between parameterised environments (characterised by the factors above, the robots' starting positions and the exploration scenario) and the effectiveness of each task allocation mechanism, illustrating that some mechanisms perform statistically better in particular environment parameterisations.

Keywords: Multi-robot team · Auction mechanism · Task allocation

1 Introduction

A future is envisioned in which autonomous mobile robots work together across a wide range of scenarios, from disaster response to humanitarian de-mining to factory maintenance. As the set of possible environments expands, so do the demands for multi-robot teamwork, requiring robots to operate in increasingly complex and challenging settings. A key requirement for deployment of such teams is a comprehensive understanding of the factors that contribute to these challenges. For example, is it important that the team knows what tasks it will be asked to complete beforehand? Do dependencies between tasks contribute to the efficacy of a task allocation? If some tasks require two robots to work together, how does that impact the organisation of the team? These questions form an open area of research, which falls under the heading of *multi-robot task allocation*: given a mission composed of a number of tasks, what is the best way to assign tasks to robots so that the mission is executed in an efficient way according to some performance metric(s) (e.g., minimising distance travelled)?

© Springer International Publishing Switzerland 2016
L. Alboul et al. (Eds.): TAROS 2016, LNAI 9716, pp. 301–313, 2016.
DOI: 10.1007/978-3-319-40379-3_32

There are various approaches to multi-robot task allocation, ranging from centralised and fully connected (where a single controller talks to and coordinates all robots) to distributed and partially connected (where communication is limited [2]). *Market-based approaches*, which use estimates of cost or utility to distribute tasks, as goods are priced and distributed in an economic market, fall somewhere in the middle of this range. *Auctions* are a popular form of market for task allocation. Tasks are advertised to team members, who then bid on them using private valuation functions. Many different auction mechanisms (i.e., the rules and procedures that describe how, when and to whom tasks are advertised, bid upon and awarded) have been studied [3].

Most of the existing work studies environments in which tasks are known ahead of time, are independent, can be completed in any order, and require only one robot. Existing taxonomies [6,16] suggest three task dimensions, labelling well-studied environments as *single-robot* (*SR*), *independent* (*IT*) and *static* (*SA*). In our work, we are investigating task allocation in more complex environments using several auction-based mechanisms found in the literature. Our earlier work evaluated *static* versus *dynamic* task allocation factors, comparing situations where tasks were all known ahead of time and were allocated before execution of any task commenced (*SR-IT-SA* [21,24]) and situations where tasks appeared *during* execution, meaning that allocation occurred dynamically, after some tasks had commenced (*SR-IT-DA* [25]). Here, we consider two additional confounding factors: *multi-robot* (*MR*) tasks, where more than one robot is required (e.g., moving a heavy object); and *constrained* (*CT*) tasks, where a task may be dependent on others to be completed before it can be executed (e.g., clearing debris from a doorway before being able to enter a room).

Our long term goal is to develop a comprehensive understanding of the factors that contribute to multi-robot team performance in varied environments. Here, we test two hypotheses. The first hypothesis is that *within* a single environment, the different mechanisms evaluated here produce statistically significantly different results, according to particular performance metrics. Thus, *for any one point* in the environment landscape, we can identify one task allocation mechanism that reliably performs the best for a given metric. The second hypothesis is that *across* multiple environments, there is no definitive or consistent ranking of mechanisms across the metrics. Thus, *across all points* in the environment landscape, none of the task allocation mechanisms evaluated performs the best for a given metric. As shown here, we have proven both these hypotheses through empirical results obtained on physical robots, backed up with results obtained in simulation experiments.

2   Related Work

The use of market mechanisms in distributed computing can be considered to have begun with Smith's *contract net* protocol [27], and this was followed by Wellman and Wurman's *market-aware agents* [29]. A primary strength of market-based approaches is their reliance only on local information and/or the self-interest of agents to arrive at efficient solutions to large-scale, complex problems that are otherwise intractable [3]. The most common instantiations of market mechanisms in multi-robot systems are *auctions*. Auctions are commonly used for distribution tasks, where resources or roles are treated as commodities and auctioned to agents. Existing work analyzes the effects of different auction mechanisms [1,3,13,30], bidding strategies [23], dynamic task re-allocation [7] and levels of commitment to the contracts [18] on the overall solution quality.

In domains where there is a strong synergy between items, single-item auctions can result in sub-optimal allocations [1]. In multi-robot exploration, studied here, strong synergies may exist between tasks. Combinatorial auctions remedy this limitation by allowing agents to bid on bundles of items, and minimise the total travel distance because they take the synergies between tasks into account [10]. Combinatorial auctions suffer, however, from the computational costs of bid generation and bundle valuation by the agents, and winner determination by the auction mechanism itself, all of which are NP-hard [11]. As a result, a body of work has grown up around the *sequential single-item auction* (SSI) [11], which has been proven to create close to optimal allocations, handling synergies while not suffering from the complexity issues of combinatorial auctions.

This paper is a further contribution to the body of work around SSI, extending the use of SSI and related mechanisms to task environments that are, according to the taxonomies developed by Gerkey and Matarić [6] and Landén et al. [16]: *multi-robot (MR)*, *constrained (CT)* and *dynamic (DA)*. Auction-based approaches to task allocation have been proposed for tasks with precedence [17], with temporal [8,20] constraints, and for dynamic environments [25] with single robot tasks. Environments that contain multi-robot tasks, with and without constraints, are less well investigated than their single-robot counterparts [12].

Of all the literature on auctions in multi-robot teams, [19] and [26] are the most closely related to our work. Both evaluate SSI in simulation and so one could argue that their work evaluates the practical cost of solutions generated using SSI. However, the focus of both [19] and [26] is on finding optimal mechanisms for dynamic task reallocation during execution rather than being concerned with the quality of team performance across a full set of tasks, which is our focus. In addition, neither [19] nor [26] consider the range of metrics that we do and so are unable, for example, to comment on the load balancing that our measurement of "idle time" exposes or the amount of time one robot waits for another to arrive at a joint location so they can execute a multi-robot task together.

3   Methodology

Formal Description. We extend the notation of [11], where a set of n **robots**, $R = \{r_0, \ldots, r_{n-1}\}$, forms a *team*; a set of m **tasks**, $T = \{t_0, \ldots, t_{m-1}\}$, comprises a *mission*; and $T(\rho)$ is the set of tasks assigned to robots $\rho \in R$. We make three extensions.

First, we specify dependencies between tasks in T. Following [16], *Independent Tasks (IT)* can be executed in any order, whereas *Constrained Tasks (CT)* have an implicit ordering. For example, suppose that t_p is a task to clear debris blocking the entrance to a passageway and t_q is a task to collect sensor data inside that passageway: t_p must be completed *before* t_q can proceed. Formally: the set of *constrained tasks CT* is a set of pairs of tasks (t_p, t_q), $t_p, t_q \in T$, such that t_p must be completed before t_q can proceed. The set of relations $(t_p, t_q) \in CT$ defines a partial order over T.

Second, we specify the number of robots required to execute each task. Following [6], *Single-Robot (SR)* tasks need only be assigned to one robot, whereas *Multi-Robot (MR)* tasks need more than one robot. Formally, each task, t, has an associated value $t.req$ defining the number of robots required to complete that task; thus if $t.req = 1$, then t is an SR task. If $t.req > 1$, then t is an MR task and $t \in T(\rho)$ where $\rho \subseteq R$ such that $|\rho| = t.req$.

Third, we specify the *arrival time* of each task. Each task t has an associated value $t.arr$ which defines the time, τ, at which any of the robots $\rho \subseteq R$ become aware of t. Following [16], we distinguish between *Static Allocation (SA)* environments, where every $t.arr$ time is before the execution of any task begins, and *Dynamic Allocation (DA)* environments, where $t.arr$ values may occur after the execution of at least one task begins. Each mission consists of an initial step where tasks are announced; next, tasks are assigned to robots; and then, tasks are executed. In an SA environment, the allocation of all tasks in the mission occurs before any task is executed, whereas, in DA, the steps may overlap.

Previous work has experimentally evaluated auction-based mechanisms for task allocation in SR-IT-SA [11] and SR-IT-DA [25] settings. Korsah et al. [12] provide a comprehensive discussion of prior work in this domain, across a range of task allocation methodologies (not just auction-based). In the work presented here, we focus on dynamic allocation environments and present experimental evaluations of MR-CT-DA, MR-IT-DA, SR-CT-DA and SR-IT-DA.

Metrics. To evaluate the performance of a team, we consider metrics that measure the performance of both individual robots and the team as a whole. In any work with robots, power consumption is the fundamental scarce resource that a robot possesses. Robot batteries only last for a limited time, and so, all other things being equal, we prefer task allocations and subsequent executions that minimise battery usage. As in [10,11,14,15,28], therefore, we measure the **distance travelled** by the robots in executing a set of tasks—both individually and as a group—since this is a suitable proxy for power consumption.[1]

Time is also important, which we measure in several ways: **run time** is the time between the start of an experiment and the point at which the last robot on the team completes the tasks allocated to it; **deliberation time** is a component of run time, the time that it takes for tasks to be allocated amongst the robots; **execution time** is another component of run time, the time it takes robots to complete tasks once they have been allocated; **movement time** is the

---

[1] Note that we compute distance not by looking at the shortest distances between the task locations, but is (as closely as we can establish) the actual distance travelled by the robots during task execution. We collect frequent position updates, compute the Euclidean distance between successive positions, and sum these.

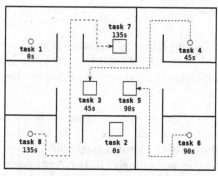

(a) MR-CT-DA Scenario

(b) The physical arena

Fig. 1. A MR-CT task scenario and lab in which physical experiments were conducted. Bold lines indicate the walls of the arena. Task locations are shown as circles for single-robot tasks or squares for two-robot tasks. A dashed line from task t_p to task t_q indicates a precedence constraint (t_p, t_q). Arrival times for each task $(t.arr)$ are also shown.

time robots spend actually moving, without interruption, toward task locations; **delay time** is the amount of time robots spend avoiding collisions with each other; **waiting time** is the amount of time robots wait for others to arrive at MR task locations; and **idle time** is the amount of time that robots wait for others to complete tasks, having completed their own.

Mechanisms. Our experiments employ four task allocation mechanisms:

(1) In round-robin (RR), tasks (T) and robots (R) are placed in two ordered lists. The first task, t_0, is allocated to the first robot, r_0. If $t_i.req > 1$ (an MR task), then it is also allocated to the next robot and so on until $t_i.req$ is reached. If $t_i.req = 1$ (an SR task), then the next task, t_{i+1}, is allocated to the next robot.

(2) In ordered single-item auction (OSI) [21], the tasks are placed in an ordered list. Each task t_i in turn is advertised to all the robots. Each robot makes a bid for the task, where the bid value is the increase in the total path cost that the robot estimates (using A*[9]) it would incur if it were to win that task. The task is allocated to the $t_i.req$-lowest bidding robots and the next task is auctioned.

(3) In sequential single-item auction (SSI) [11], unallocated tasks are presented to all robots simultaneously. Each robot bids on the task with the lowest cost (computed as in OSI), and the task with the lowest bid is allocated to the robot that made the bid. If the winning task, t_w, has been completely allocated (to $t_w.req$ robots), it is removed from the set of tasks to be advertised in the next round and the process is repeated until all tasks have been allocated.

(4) In parallel single-item auction (PSI) [11], allocation starts like SSI: all robots bid on all tasks from their current locations. All the tasks are allocated in one round, with each task t_i going to the $t_i.req$-lowest bidding robots that bid on it.

4   Experiments

We conducted a series of experiments comparing the task allocation mechanisms described earlier in a structured set of $\langle SR|MR\rangle\langle IT|CT\rangle\langle SA|DA\rangle$ environments. Here we describe the system used to conduct these experiments, the specific scenarios evaluated, and our results.

System Description. Task allocation is conducted by a central *auctioneer* agent, which communicates the start of an auction and awards tasks to *robot controller* agents.[2] Each robot controller submits bids, the auctioneer determines the winner(s) of the auction and allocates tasks accordingly. Robot controllers then execute tasks autonomously. Our software architecture is agnostic about whether the team executes its tasks on real robot hardware or in simulation. Our physical platform is the Turtlebot2,[3] which has a differential drive base and a colour/depth-sensing Microsoft Kinect camera. The ROS [22] navigation stack provides communication, localisation and path planning capabilities. Our simulated robot (using Stage [5]) has the same properties as its physical counterpart.

The operating environment for our robots is an office-like setting with rooms opening off a central hallway. The layout of this environment is shown in Fig. 1(a), and a photograph is given in Fig. 1(b). While this is a smaller environment than that studied by some others (e.g., [11]), our setup allows us to run parallel experiments on physical robots and—on a larger scale—in simulation, which produces more statistically significant results.

Experimental Setup. An experimental condition is defined by the starting locations of the robots and the task scenario (defined by task locations, task arrival times, constraints and robot requirements). This work investigates routing tasks—a robot executes a task simply by driving to the task's location. All of the experiments reported here involve a team of $n = 3$ robots. We used two sets of *starting configurations* for the robot team: one *clustered* the robots in the "room" in the lower left corner of the arena, while the other *distributed* the robots at three corners of the map. We examined four different *parameterised environments*, all with dynamic allocation (DA), combining single-robot (SR) vs. multi-robot (MR) and independent (IT) vs. constrained (CT) tasks: *SR-IT-DA, SR-CT-DA, MR-IT-DA* and *MR-CT-DA*. We employed two different *task scenarios*. Fig. 1(a) shows a diagram of the first task scenario.

The aim in choosing this combination of parameterised environments was to see how performance of the four task allocation mechanisms varied along the MR/SR and CT/IT dimensions. In total, 192 physical and 960 simulation trials were performed: 2 *starting configurations* × 4 *parameterised environments* × 2 *task scenarios* × 4 *allocation mechanisms* × {3 *physical* | 15 *simulation*} *trials*. For each experiment, we recorded the metrics described in Sect. 3.

---

[2] Though bidding and winner determination are managed centrally, there is no centralised control in the usual sense. The auction could also be distributed among the robots as in [4].

[3] www.turtlebot.com.

Results. Figures 2, 3 and Table 1 contain representative results from the experiments. Figure 2 shows the average *distance travelled* by the team in eight variations of the scenario shown in Fig. 1(a). In each plot, average travel distances resulting from allocations produced by RR, OSI, SSI, and PSI are shown from left to right. Error bars indicate 95 % confidence intervals. Figure 2(a) and (b) show how, in the SR-clustered conditions of this scenario, PSI allocations result in distances that are significantly shorter than those produced by the other mechanisms. As we move to distributed-start conditions of the scenario (Fig. 2(c) and (d)), differences among three of the mechanisms diminish but remain statistically significantly different, while RR continues to lead to dramatically longer distances. This result is similar to those reported in [25], where it was shown that a starting configuration that distributes team members more evenly amongst tasks tends to lessen the advantages of mechanisms such as SSI that exploit clustering properties of task locations. In MR conditions of the same scenario, the results are somewhat different. For example, RR doesn't always result in the longest distances nor does PSI always result in the shortest. The relative rankings of the mechanisms are much less predictable than in the SR conditions. Our second experimental task scenario produced similar results.

We can choose other of our performance metrics to examine individually. But with nine metrics and a large number of combinations of environments and experimental configurations, we want to make sense of the results as a whole. Do any of the mechanisms produce the best performance across environments or experimental configurations? Do clear patterns emerge? We address these questions in the following section by examining the data in aggregate.

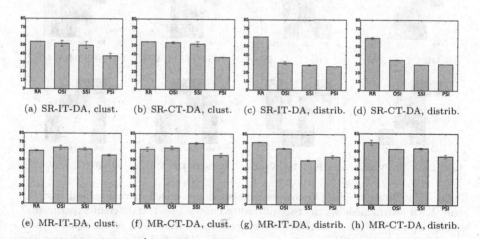

(a) SR-IT-DA, clust.   (b) SR-CT-DA, clust.   (c) SR-IT-DA, distrib.   (d) SR-CT-DA, distrib.

(e) MR-IT-DA, clust.   (f) MR-CT-DA, clust.   (g) MR-IT-DA, distrib.   (h) MR-CT-DA, distrib.

Fig. 2. Average distance (meters) travelled in physical experiments for variations of the scenario shown in Fig. 1(a). Mechanisms are ordered RR, OSI, SSI and PSI.

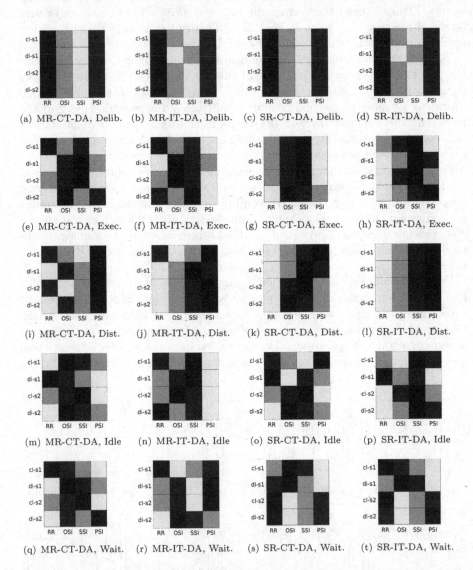

Fig. 3. Heat maps for the physical experiment data on each parameterised environment. Each heatmap shows the two different scenarios and two different experimental conditions. For a given scenario/condition pair (row) the colour of the squares indicates the rank order of the mechanism (column). The darkest square indicates the lowest value of the metric (best mechanism), the lightest square indicates the highest value (worst mechanism). (a)–(d) show deliberation time, (e)–(h) show execution time, (i)–(l) show distance, (m)–(p) show idle time, and (q)–(t) show waiting time.

Table 1. F-ratios for 5 different metrics

	Physical					Simulation				
	(a) Deliberation time										
	$F(3,8)$	p		$F(3,8)$	p		$F(3,56)$	p		$F(3,56)$	p
MR-CT-DA-			SR-CT-DA-			MR-CT-DA-			SR-CT-DA-		
cl-s1	83.96	0.010	cl-s1	71.77	0.010	cl-s1	28709.89	0.010	cl-s1	241.51	0.010
di-s1	158.13	0.010	di-s1	43.87	0.010	di-s1	54561.93	0.010	di-s1	213.79	0.010
cl-s2	3901.58	0.010	cl-s2	1766.23	0.010	cl-s2	18630.69	0.010	cl-s2	30977.14	0.010
di-s2	3080.90	0.010	di-s2	3708.91	0.010	di-s2	22404.35	0.010	di-s2	21734.58	0.010
MR-IT-DA-			SR-IT-DA-			MR-IT-DA-			SR-IT-DA-		
cl-s1	93.79	0.010	cl-s1	5038.94	0.010	cl-s1	24591.32	0.010	cl-s1	174.05	0.010
di-s1	1150.34	0.010	di-s1	45.53	0.010	di-s1	44307.68	0.010	di-s1	2089.02	0.010
cl-s2	5124.26	0.010	cl-s2	37639.65	0.010	cl-s2	15842.79	0.010	cl-s2	23317.28	0.010
di-s2	5364.80	0.010	di-s2	146.10	0.010	di-s2	44591.27	0.010	di-s2	27112.49	0.010
	(b) Execution time										
	$F(3,8)$	p		$F(3,8)$	p		$F(3,56)$	p		$F(3,56)$	p
MR-CT-DA-			SR-CT-DA-			MR-CT-DA-			SR-CT-DA-		
cl-s1	1.39	**0.950**	cl-s1	19.70	0.010	cl-s1	30.43	0.010	cl-s1	60.39	0.010
di-s1	9.58	0.010	di-s1	3.27	**0.950**	di-s1	5.94	0.010	di-s1	22.79	0.010
cl-s2	5.49	0.050	cl-s2	19.72	0.010	cl-s2	24.02	0.010	cl-s2	19.82	0.010
di-s2	3.19	**0.950**	di-s2	24.63	0.010	di-s2	9.88	0.010	di-s2	39.51	0.010
MR-IT-DA-			SR-IT-DA-			MR-IT-DA-			SR-IT-DA-		
cl-s1	2.82	**0.950**	cl-s1	18.58	0.010	cl-s1	36.39	0.010	cl-s1	33.09	0.010
di-s1	1.54	**0.950**	di-s1	11.17	0.010	di-s1	9.53	0.010	di-s1	14.14	0.010
cl-s2	3.92	**0.950**	cl-s2	9.93	0.010	cl-s2	6.62	0.010	cl-s2	28.28	0.010
di-s2	0.77	**0.950**	di-s2	79.93	0.010	di-s2	5.10	0.010	di-s2	15.93	0.010
	(c) Distance travelled										
	$F(3,8)$	p		$F(3,8)$	p		$F(3,56)$	p		$F(3,56)$	p
MR-CT-DA-			SR-CT-DA-			MR-CT-DA-			SR-CT-DA-		
cl-s1	7.76	0.010	cl-s1	30.83	0.010	cl-s1	35.88	0.010	cl-s1	312.84	0.010
di-s1	13.04	0.010	di-s1	784.63	0.010	di-s1	4817.66	0.010	di-s1	75593.00	0.010
cl-s2	12.90	0.010	cl-s2	7.70	0.010	cl-s2	33.75	0.010	cl-s2	60.12	0.010
di-s2	9.39	0.010	di-s2	996.79	0.010	di-s2	132.54	0.010	di-s2	1395.83	0.010
MR-IT-DA-			SR-IT-DA-			MR-IT-DA-			SR-IT-DA-		
cl-s1	10.38	0.010	cl-s1	6.01	0.050	cl-s1	390.48	0.010	cl-s1	436.66	0.010
di-s1	68.46	0.010	di-s1	173.25	0.010	di-s1	3121.61	0.010	di-s1	98521.39	0.010
cl-s2	13.30	0.010	cl-s2	29.16	0.010	cl-s2	122.99	0.010	cl-s2	231.39	0.010
di-s2	10.21	0.010	di-s2	2823.98	0.010	di-s2	527.39	0.010	di-s2	3676.25	0.010
	(d) Total idle time										
	$F(3,8)$	p		$F(3,8)$	p		$F(3,56)$	p		$F(3,56)$	p
MR-CT-DA-			SR-CT-DA-			MR-CT-DA-			SR-CT-DA-		
cl-s1	0.72	**0.950**	cl-s1	8.44	0.010	cl-s1	40.63	0.010	cl-s1	40.08	0.010
di-s1	2.17	**0.950**	di-s1	4.33	0.050	di-s1	36.85	0.010	di-s1	34.25	0.010
cl-s2	8.28	0.010	cl-s2	6.23	0.050	cl-s2	112.31	0.010	cl-s2	7.00	0.010
di-s2	4.30	0.050	di-s2	29.89	0.010	di-s2	70.23	0.010	di-s2	29.02	0.010
MR-IT-DA-			SR-IT-DA-			MR-IT-DA-			SR-IT-DA-		
cl-s1	111.22	0.010	cl-s1	2.19	**0.950**	cl-s1	117.09	0.010	cl-s1	14.87	0.010
di-s1	7.90	0.010	di-s1	6.69	0.050	di-s1	40.33	0.010	di-s1	52.47	0.010
cl-s2	20.62	0.010	cl-s2	4.12	0.050	cl-s2	99.37	0.010	cl-s2	12.82	0.010
di-s2	16.53	0.010	di-s2	90.31	0.010	di-s2	16.58	0.010	di-s2	8.40	0.010
	(e) Total waiting time										
	$F(3,8)$	p		$F(3,8)$	p		$F(3,56)$	p		$F(3,56)$	p
MR-CT-DA-			SR-CT-DA-			MR-CT-DA-			SR-CT-DA-		
cl-s1	26.38	0.010	cl-s1	100.07	0.010	cl-s1	10.02	0.010	cl-s1	1260.61	0.010
di-s1	1.28	**0.950**	di-s1	9.19	0.010	di-s1	30.23	0.010	di-s1	100.39	0.010
cl-s2	0.15	**0.950**	cl-s2	6.01	0.050	cl-s2	16.90	0.010	cl-s2	38.08	0.010
di-s2	8.92	0.010	di-s2	22.55	0.010	di-s2	20.93	0.010	di-s2	6.94	0.010
MR-IT-DA-			SR-IT-DA-			MR-IT-DA-			SR-IT-DA-		
cl-s1	4.21	0.050	cl-s1	0.25	**0.950**	cl-s1	28.44	0.010	cl-s1	0.63	**0.950**
di-s1	0.26	**0.950**	di-s1	0.42	**0.950**	di-s1	23.39	0.010	di-s1	0.64	**0.950**
cl-s2	0.30	**0.950**	cl-s2	2.49	**0.950**	cl-s2	14.05	0.010	cl-s2	0.00	**0.950**
di-s2	1.00	**0.950**	di-s2	1.69	**0.950**	di-s2	8.03	0.010	di-s2	1.32	**0.950**

5   Analysis

Here, we focus on five different performance metrics. *Deliberation time* is a component of overall run time and a good measure of how well an allocation mechanism scales with the number of tasks and the size of the team. *Execution time* is another component of run time and one of the main measures we would like to minimise, the other being *distance*. We also look at *idle time* as a measure of how well balanced the task load is among the team. Finally, we look at *waiting time*. This is a feature specific to the MR and CT environments. A key contribution of our work is extending experimental results, particularly with physical robots, into MR and CT environments.

One of our long term goals is to develop task allocation mechanisms, or methods of choosing mechanisms, that perform well in different environments. Underlying this is the assumption that some mechanisms lead to better performance outcomes in some environments than others, and that there may not be a single mechanism that is best suited for all environments. We suggest two research hypotheses to evaluate this assumption and use the results of experiments discussed here to provide evidence for them.

The first hypothesis is that *within* a single $\langle sc, pe, ts \rangle$ tuple (where sc =starting configuration, pe =parameterised environment, and ts =task scenario), the four mechanisms examined here produce statistically significantly different results, according to our performance metrics. It is important to show that performance differences between mechanisms exist in the first place before examining the effects of varying environments. To evaluate this first hypothesis, we apply *analysis of variance (ANOVA)* to determine if there are significant differences between the different mechanisms. We ran ANOVA on the four samples—one for each mechanism in each $\langle sc, pe, ts \rangle$ tuple. If the null hypothesis were true and the differences among the four samples were due to chance, then the likelihood of producing the F-ratio would be less than $p\%$. The F-ratios of samples from both physical and simulation experiments are shown in Table 1. These F-ratios (p-value = 0.01) indicate a significant performance difference between the populations (mechanisms). For example, in the case of deliberation time (Table 1(a)), very large F-ratio values are the result of comparing RR, a simple mechanism that runs very quickly, with the others. In contrast, F-ratios for distance travelled (Table 1(c)) are lower but still above the critical value for the significance level and degrees of freedom tested. This supports our first hypothesis.

The second hypothesis is that *across* multiple $\langle sc, pe, ts \rangle$ tuples, there is no definitive ranking amongst the metrics for each mechanism. Figure 3 shows performance rankings obtained from physical experiments in the form of heat maps. Each row of heat maps in the figure corresponds to one of the five metrics discussed above. Within each heat map, the four columns correspond to the four task allocation mechanisms (RR, OSI, SSI, PSI, from left to right). The rows of each heat map are labelled with a variation of a particular scenario. For example, *cl-s1* indicates *clustered, scenario 1*. The shading of a cell indicates its rank: darker shades indicate lower values for that metric. While the ANOVA results mentioned in support of the first hypothesis don't directly measure the degree to which any pair of mechanisms differed in performance, they do provide evidence that the rankings shown in the heat maps are based on statistically significant differences. The heat maps for *deliberation time* (Fig. 3(a)–(d)) reveal some consistency when comparing environments and experimental conditions (rows within a single heat map, and across heat maps in the same row of the figure). RR is always the quickest to run, followed by PSI, while OSI and SSI trade ranks depending on the experimental condition. Apart from deliberation time, this type of performance ranking does not hold in a consistent way for the other metrics when comparing across environments and experimental conditions. This supports our second hypothesis. Our next steps involve looking at more environments and experimental conditions that vary in systematic ways, to help discover correspondences between parameters of environments and the performance characteristics of different task allocation mechanisms. The type of analysis presented here is likely to be a useful tool for this future endeavour.

6 Summary

The work presented here tests two hypotheses: (1) within a single parameterised environment, a given task allocation mechanism can be proven to consistently outperform others for certain metrics; and (2) across a varied set of parameterised environments, no single task allocation mechanism will consistently outperform others for any metrics. We conducted experiments with physical robots, as well as simulated robots in an environment that parallels our physical setup. Our empirical results support both of these hypotheses.

Future work will involve exploration of more complex scenarios in simulation, with larger teams and more tasks, as well as additional experiments with physical robots. In addition, we will be assessing how *task duration* affects the metrics presented here, by varying the time it takes to complete different tasks. Our long term goal is to identify the features of parameterised environments and/or task scenarios that influence the performance of the different mechanisms, so that the differences in rankings highlighted here can be attributed to particular features of a given experimental environment.

References

1. Berhault, M., Huang, H., Keskinocak, P., Koenig, S., Elmaghraby, W., Griffin, P.M., Kleywegt, A.: Robot exploration with combinatorial auctions. In: IROS (2003)
2. Bonabeau, E., Dorigo, M., Théraulaz, G.: Swarm Intelligence: From Natural to Artificial Systems. Oxford University Press, New York (1999)
3. Dias, M.B., Zlot, R., Kalra, N., Stentz, A.: Market-based multirobot coordination: a survey and analysis. Proc. IEEE **94**(7), 1257–1270 (2006)
4. Ezhilchelvan, P., Morgan, G.: A dependable distributed auction system: architecture and an implementation framework. In: International Symposium on Autonomous Decentralized Systems (2001)

5. Gerkey, B., Vaughan, R.T., Howard, A.: The player/stage project: tools for multi-robot and distributed sensor systems. In: International Conference on Advanced Robotics (2003)

6. Gerkey, B.P., Mataríc, M.J.: A formal analysis and taxonomy of task allocation in multi-robot systems. Int. J. Robot. Res. **23**(9), 939–954 (2004)

7. Golfarelli, M., Maio, D., Rizzi, S.: A task-swap negotiation protocol based on the contract net paradigm. TR-005-97, DEIS/CSITE/Università di Bologna (1997)

8. Gombolay, M., Wilcox, R., Shah, J.A.: Fast scheduling of multi-robot teams with temporospatial constraints. In: RSS (2013)

9. Hart, P., Nilsson, N., Raphael, B.: A formal basis for the heuristic determination of minimal cost paths. IEEE Trans. Sys. Sci. Cybern. **4**(2), 100–107 (1968)

10. Koenig, S., Keskinocak, P., Tovey, C.: Progress on agent coordination with cooperative auctions. In: AAAI (2010)

11. Koenig, S., Tovey, C., Lagoudakis, M., Kempe, D., Keskinocak, P., Kleywegt, A., Meyerson, A., Jain, S.: The power of sequential single-item auctions for agent coordination. In: AAAI (2006)

12. Korsah, G.A., Stentz, A., Dias, M.B.: A comprehensive taxonomy for multi-robot task allocation. Int J. Robot. Res. **32**(12), 1495–1512 (2013)

13. Kraus, S.: Automated negotiation and decision making in multiagent environments. In: Luck, M., Mařík, V., Štěpánková, O., Trappl, R. (eds.) ACAI 2001 and EASSS 2001. LNCS (LNAI), vol. 2086, pp. 150–172. Springer, Heidelberg (2001)

14. Lagoudakis, M., Berhault, M., Koenig, S., Keskinocak, P., Kelywegt, A.: Simple auctions with performance guarantees for MRTA. In: IROS (2004)

15. Lagoudakis, M., Markakis, V., Kempe, D., Keskinocak, P., Koenig, S., Kleywegt, A., Tovey, C., Meyerson, A., Jain, S.: Auction-based multi-robot routing. In: RSS (2005)

16. Landén, D., Heintz, F., Doherty, P.: Complex task allocation in mixed-initiative delegation: a UAV case study. In: Desai, N., Liu, A., Winikoff, M. (eds.) PRIMA 2010. LNCS, vol. 7057, pp. 288–303. Springer, Heidelberg (2012)

17. Luo, L., Chakraborty, N., Sycara, K.: Multi-robot assignment algorithm for tasks with set precedence constraints. In: ICRA (2011)

18. Mataric, M., Sukhatme, G., Ostergaard, E.: Multi-robot task allocation in uncertain environments. Auton. Robots **14**(2–3), 255–263 (2003)

19. Nanjanath, M., Gini, M.: Repeated auctions for robust task execution by a robot team. Robot. Auton. Syst. **58**(7), 900–909 (2010)

20. Nunes, E., Gini, M.: Multi-robot auctions for allocation of tasks with temporal constraints. In: AAAI (2015)

21. Özgelen, A.T., Schneider, E., Sklar, E.I., Costantino, M., Epstein, S.L., Parsons, S.: A first step toward testing multiagent coordination mechanisms on multirobot-teams. In: AAMAS Workshop: ARMS (2013)

22. Quigley, M., Conley, K., Gerkey, B.P., Faust, J., Foote, T., Leibs, J., Wheeler, R., Ng, A.Y.: Ros: an open-source robot operating system. In: ICRA Workshop: Open Source Software (2009)

23. Sariel, S., Balch, T.: Efficient bids on task allocation for multi-robot exploration. In: FLAIRS (2006)

24. Schneider, E., Balas, O., Özgelen, A.T., Sklar, E.I., Parsons, S.: Evaluating auction-based task allocation in multi-robot teams. In: AAMAS Workshop: ARMS (2014)

25. Schneider, E., Sklar, E.I., Parsons, S., Özgelen, A.T.: Auction-based task allocation for multi-robot teams in dynamic environments. In: Dixon, C., Tuyls, K. (eds.) TAROS 2015. LNCS (LNAI), vol. 9287, pp. 246–257. Springer, Switzerland (2015)

26. Schoenig, A., Pagnucco, M.: Evaluating sequential single-item auctions for dynamic task allocation. In: Li, J. (ed.) AI 2010. LNCS, vol. 6464, pp. 506–515. Springer, Heidelberg (2010)

27. Smith, R.G.: The contract net protocol: high-level communication and control in a distributed problem solver. In: Distributed Artificial Intelligence (1988)

28. Tovey, C., Lagoudakis, M.G., Jain, S., Koenig, S.: Generation of bidding rules for auction-based robot coordination. In: International Workshop on Multi-Robot Systems (2005)

29. Wellman, M.P., Wurman, P.R.: Market-aware agents for a multiagent world. Robot. Auton. Syst. **24**, 115–125 (1998)

30. Zlot, R., Stentz, A., Dias, M.B., Thayer, S.: Multi-robot exploration controlled by a market economy. In: ICRA (2002)

How to Build and Customize
a High-Resolution 3D Laserscanner Using
Off-the-shelf Components

Stefan Schubert[(⊠)], Peer Neubert, and Peter Protzel

TU Chemnitz, 09126 Chemnitz, Germany
{stefan.schubert,peer.neubert,peter.protzel}@etit.tu-chemnitz.de

Abstract. 3D laserscanners are well suited sensors for different perception tasks like navigation and object recognition. However, ready-to-use 3D laserscanners are expensive and offer a low resolution as well as a small field of view. Therefore, many groups design their own 3D laserscanner by rotating a 2D laserscanner. Since this whole process is done frequently, this paper aims at fostering other groups' future research by offering a list of necessary hardware including an online-accessible mechanical drawing, and available software. As it is possible to align the rotation axis and the 2D laserscanner in many different ways, we present an approach to optimize these orientations. A corresponding Matlab toolbox can be found at our website. The performance of the 3D laserscanner is shown by multiple matched point clouds acquired in outdoor environments.

Keywords: 3D laserscanner · 3D reconstruction · Sensors · Environment perception

1 Introduction

Perceiving the three dimensional world is an essential capability for reconstructing unstructured environments and for autonomous mobile robots operating within them. The complexity of tasks like mapping or traversability estimation is strongly influenced by the properties of the used sensors. The availability of a suitable sensor system may considerable simplify a certain task. For example, the advent of cheap and ready-to-use RGB-D cameras (most notably the Kinect) fostered the progress in areas like object recognition, indoor 3D reconstruction and mobile manipulation.

For large scale outdoor areas, the available RGB-D sensors are less useful due to their illumination sensitivity and limited range. In contrast, laserscanners (or LiDAR) work independently of the lighting conditions and provide depth information for large distances. 2D laserscanners are established sensor systems for indoor robots and several off-the-shelf systems are available. In contrast, only few ready-to-use systems are available for obtaining *three dimensional* laser-based measurements, most notably the *Velodyne*[1] laserscanners and multi-layer scanners developed for advanced driver assistance systems.

---

[1] http://velodynelidar.com.

© Springer International Publishing Switzerland 2016
L. Alboul et al. (Eds.): TAROS 2016, LNAI 9716, pp. 314–326, 2016.
DOI: 10.1007/978-3-319-40379-3_33

However, these systems are expensive and/or provide low vertical resolutions, e.g. 4–64 layers. To obtain high-resolution scans at lower budget, several research groups built custom solutions by moving an off-the-shelf 1D [1,2] or 2D laser sensor [3–8]. Although, there are repeatedly used configurations, for none of these custom solutions all the necessary information for replication are available. Therefore, in this paper, we

1. Discuss properties of possible configurations including a comparison to available commercial solutions.
2. Provide all the necessary information required for building a customized 3D laserscanners with off-the-shelf components (i.e. a 2D laserscanner) and ROS support. Only a few parts, for which technical drawings are provided, require manufacturing.
3. Propose a simple optimization framework to find the best alignment of the laser's rotation axis for the task at hand.

Finally, we will show example results of 3D reconstructions using the presented sensor system on a mobile robot in different indoor and outdoor environments. The technical details and an implementation of the optimization framework as well as further supplementary material like videos will be available from our website[2].

2   Available Solutions and Related Work

Perceiving the environment can be accomplished with different hardware. 3D laserscanners have been established due to their positive properties like high ranges and illuminance independence. There are only few ready-to-use 3D laserscanners available. By offering scanning rates of 5–20 Hz they are suited for highly dynamic environments like urban streets [9]. The most popular 3D laserscanners are the *VLP-16*, *HDL-32* and *HDL-64* by *Velodyne*. Due to their valuable properties for advanced driver assistance systems (ADAS) there are further more specific 3D laserscanners like the *ibeo LUX* series[3]. However, since these systems are expensive (*Velodyne*: 8,000\$ to 75,000\$) and offer only a small number of layers (*Velodyne*: 16-64, *Ibeo*: 4-8) there are many custom-made 3D laserscanners.

Custom sensors are typically built by rotating either a 1D [1,2] or a 2D sensor [3–8]. Since building custom solutions for fast rotating a 1D sensor around two axes is more complicated and error-prone than slowly rotating a 2D sensor, the latter solution is usually preferred.

Dependent on the application, an oscillating motion of the 2D scanner (e.g. tilting in the range [15°, −15°]) or a continuous rotation may be preferable. Although the continuous rotation requires additional hardware (i.e. a slip ring, cf. Sect. 3.3), it enables e.g. a 360° surround view. In contrast to at most 64 layers with ready-to-use 3D laserscanners, dependent on the arrangement, a continuously rotating 2D

---

[2] https://www.tu-chemnitz.de/etit/proaut/3dls.html.
[3] http://www.ibeo-as.com.

scanner can provide more than 1000 vertical layers (e.g., vertically rotating *Hokuyo UTM-30LX*: 1080). Such high-resolution sensors are particularly suitable for tasks like object recognition [10] and traversability assessment [11].

Rotating a 2D laserscanner can be done in many different ways, however, usually the sensor is simply rotated by rolling or pitching [12]. We present an optimization method similar to [13]. They improve the 3D laserscanner by pitching the 2D sensor. In this paper, we extend their approach by additionally optimizing the rolling angle of the 2D laserscanner and the complete orientation of the rotation axis, this is the servomotor. Supplementary, we offer a concrete mechanical setup including a slip ring allowing continuous spinning. This is advantageous since a constant turn rate reduces mechanical strain by avoiding accelerations and improves the measurements by keeping the turn rate always constant. Nevertheless, the spinning rate can be changed if necessary.

3   Hardware Selection

This section presents the required hardware components to build a 3D laserscanner using a continuously rotating 2D scanner. In particular these are a 2D laserscanner, a servomotor, a slip ring, and the components for the mechanical linkage. In the following, we will present selected solutions for each component based on the experiences from our experiments. Additionally, available ROS drivers and a ROS package for 2D scan accumulation are referred which we use in our system.

3.1   2D Laserscanner

A 2D laserscanner consists of a spinning laser beam transceiver returning the distance to an obstacle by measuring the time of flight. Building an own 2D laserscanner is a challenging task due to the high rotation speed (e.g. 40rps), consequently, we highly recommend to use a purchasable sensor. There are different well suited commercial scanners. We have gained good experience with two different laserscanners from the manufacturer *Hokuyo*[4]. Depending on the tasks, we can recommend both *Hokuyo URG-04LX* and *Hokuyo UTM-30LX* (see Table 1). Each sensor can be simply attached to a computer with USB2.0 and an external power supply. The *Hokuyo URG-04LX* is a low budget scanner ($\sim 2,000\$$), however, it has a range of ≤ 4 m which constrains its usage to small range tasks like indoor navigation. In contrast, the *Hokuyo UTM-30LX* ($\sim 5,000\$$) promises a range of around 30 m, though, in our experience we noticed practically a smaller range of around 15 m to 20 m depending on the environment. Larger distances result in NaN-values (Not a Number). Nevertheless, this range can be considered sufficient for the most indoor applications and for many outdoor tasks, especially in semi-structured scenarios. Other often used 2D laserscanners are offered by *SICK*[5]. A comprehensive characterization

---

[4] http://www.hokuyo-aut.jp.
[5] http://www.sick.com.

Table 1. Technical details of two *Hokuyo* 2D laserscanners

	Hokuyo	
Type	**URG-04LX**	**UTM-30LX**
Range	4 m	30 m
Accuracy	0.02 m – 1 m: 10 mm	0.1 – 10 m: 30 mm
	1 m – 4 m: 1 % of measurement	10 – 30 m: 50 mm
Scan angle	240°	270°
Angular resolution	0.36°	0.25°
Frequency	100 ms/scan	40 Hz
Wavelength	785 nm	905 nm
Weight	160 g	370 g

of the mentioned *Hokuyo* scanners can be found in [14,15], and for the *SICK LMS 200* in [16]. Of course, there might be other suited 2D laserscanners from different manufacturers. For running the 2D laserscanners in a system, there are drivers as ROS-packages for all newer ROS-versions, i.e. *ROS hokuyo_node*[6], *ROS sicktoolbox*[7].

3.2 Servomotor

Basically, a self-made 3D laserscanner can apply every rotating motor with a position sensor. However, a good solution is a servomotor which enables the laserscanner to target a direction and to receive the corresponding exact angle or rotation speed, respectively. For the application of rotating a laserscanner, we recommend the *Robotis*[8] *Dynamixel* series. Especially, we have gained good experience with the *Dynamixel AX-12A* and the *Dynamixel MX-64R* (see Table 2). Both servomotors can rotate in a *Joint Mode* which enables the user to control the angle, and in a *Wheel Mode* enabling the user to control the rotation speed. The *Dynamixel AX-12A* is low priced ($\sim$ 50$) and possibly sufficient, but it tends to oscillate in Joint Mode. In contrast, the *Dynamixel MX-64R* ($\sim$ 300$) showed to work stable and has a higher angular resolution, which is better for high distances. A corresponding ROS driver, providing a convenient interface, is *ROS dynamixel_controllers*[9].

3.3 Slip Ring

A slip ring enables the 3D laserscanner to rotate continuously. There are many different products at different prices and with different connections. However,

---

[6] http://wiki.ros.org/hokuyo_node.
[7] http://wiki.ros.org/sicktoolbox.
[8] http://www.robotis.com.
[9] http://wiki.ros.org/dynamixel_controllers.

Table 2. Technical details of two *Robotis Dynamixel* servomotors

	Robotis Dynamixel	
Type	**AX-12A**	**MX-64**
Weight	54.6 g	126 g
Dimension	$32 \times 50 \times 40$ mm	$40.2 \times 61.1 \times 41$ mm
Angular range	300°	360°
Angular resolution	0.29°	0.09°
Position sensor	Potentiometer	Contactless absolute encoder
Number of steps	1024	4096

we have chosen the *SenRing*[10] *SNU11-P0210* ($\sim 130\$$) which is a slip ring for USB2.0 and power supply.

3.4 Mechanical Linkage

For the mechanical linkage of 2D laserscanner, servomotor, and slip ring, we require at least three parts (Fig. 1): linkages between servomotor and slip ring (Fig. 1 number 3), servomotor and 2D laserscanner (Fig. 1 number 2), and a component to align the 2D laserscanner in a preferred angle in relation to the rotation axis (Fig. 1 number 5). You can find the mechanical drawings at our website.

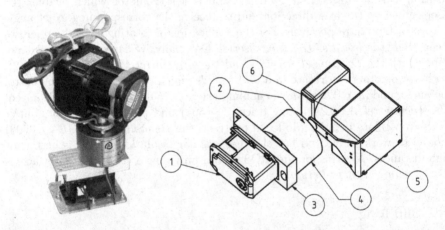

Fig. 1. Left: The proposed 3D laserscanner setting. Right: The mechanical drawing of the proposed 3D laserscanner. The complete drawing can be found at our website. 1: servomotor *Robotis Dynamixel MX-64R*, 2: shaft servo - 2D laserscanner, 3: linkage servo - slip ring, 4: slip ring *SNU11-P0210*, 5: linkage shaft - 2D laserscanner, 6: *Hokuyo UTM-30LX*

---

[10] http://www.senring.com.

3.5   A Complete 3D Laserscanner Setting

Figure 1 shows the complete 3D laserscanner using a *Hokuyo UTM-30LX* 2D laserscanner, a *Robotis Dynamixel MX-64R* servomotor, a *SenRing SNU11-P0210* slip ring and the mechanical linkages depicted in Fig. 1 leading to an overall cost of around 5, 500$ (*Velodyne HDL-32* ∼ 30, 000$). The hardware requires two USB connections as well as a 12V power supply. You can find the details for construction at our website. Feel free to download, rebuild, and modify it depending on your needs. In order to accumulate the 2D scans of the running system to get a full 3D scan, the ROS *laser_assembler* package[11] can be used. The required bandwidth and processing power are similar to 2D scan acquisition, however, the resulting 3D point clouds are much larger.

We want to emphasize that the design of the linkage which aligns the 2D laserscanner in a specific angle (part 5 in Fig. 1) has important influence on the resulting distribution and density of distance measurements. Finding an optimal design is the content of the following section. Finally, Sect. 5 will show example results that can be obtained with the presented hardware setup.

4   Optimization of the Alignment

Basically, it is sufficient to rotate a 2D laserscanner in any possible alignment to acquire a 3D scan of the environment (except if it rotates around the scan plane normal). The two most applied alignments are the so called Rolling Scanning and Pitching Scanning [12] which can be considered as standard alignments using simple 90°-angles. As you can see in Fig. 2, the result is a 3D scan with one or two focus points, respectively. In these focus points, there is a higher measurement density (a locally higher angular laser beam resolution). For other directions, a lower measurement density is obtained.

While the total number of laser beams per rotation is constant, the particular setup of the angles in the mechanical linkages from the previous section strongly influence the distribution of the measurements. Dependent on the task at hand, a different setup may be preferable. E.g., think of a mobile outdoor robot that may not be interested in many distance measures towards the sky but could benefit from a particularly high density on the ground ahead. This is particularly interesting regarding the fact that a better distribution of measurements allows for faster rotation of the laser and thus higher scan rates.

In this section, we present a simple approach to find an optimal configuration for a particular task: Given a task specific target distribution of distance measurements (the user has to tell the system in which directions measurements are particularly important or not), we incorporate simplifying assumptions to allow an exhaustive search over the resulting configuration space. For each configuration, the laser scan density is computed and evaluated based on the target distribution for this task.

---

[11] http://wiki.ros.org/laser_assembler.

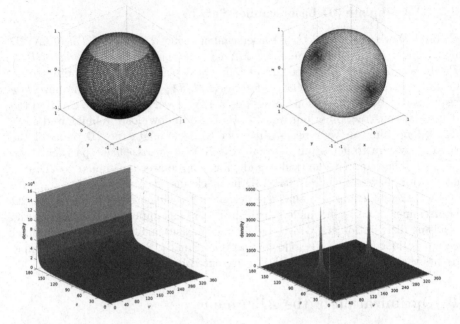

Fig. 2. Two simulated 3D scans with point cloud (top) and corresponding density map (bottom). Left: *Rolling Scanning*. Right: *Pitching Scanning*

However, it should be mentioned that, on the one hand, the Rolling Scanning is the only setting which enables the 3D laserscanner to return a complete 3D scan already after 180° rotation. On the other hand, every non standard alignment has the property that the scan lines cross each other avoiding large connected blind spot areas. The advantage is illustrated in Fig. 3: With crossing scan lines we can nearly not miss thin objects even for fast rotation speed of the servomotor, since we exploit the fast frequency of a 2D laserscanner in combination with its high angular resolution.

First of all, transformations (TF's) inside the 3D laserscanner have to be defined. As one can imagine, low shifts in any direction inside the alignment has only a small influence on the resulting 3D scan in relation to rotations. Accordingly for simplification, we assume that each part is only rotated and not translated, since this accelerates the optimization by reducing the number of parameters. Consequently, all rotation axes have to cross in one point. However, the transformations can be easily modified by extending the rotation matrices R with translation vectors t. Accordingly, we define two rotation matrices:

1. $R^{\mathrm{world,rot}}$ - Orientation of the spinning axis from the servomotor in relation to a world frame, called *rotation frame*. By definition, the servomotor rotates around the z-axis.
2. $R^{\mathrm{rot,laser}}$ - Orientation of the scan plane from the 2D laserscanner in relation to the rotation frame, called *laser frame*. The x-axis points to the front of the scan plane, and the z-axis is parallel to the scan plane normal.

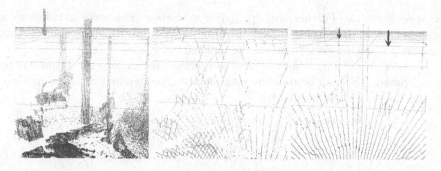

Fig. 3. Comparison of crossing and non-crossing scan lines. The complete 360° scan was acquired with a *Hokuyo UTM-30LX* 2D laserscanner (40 Hz frequency). Left: high resolution scan. Middle: 3D scan with crossing scan lines, acquired in approx. 2 s. Right: 3D scan with parallel scan lines, acquired in approx. 2 s - the red arrows mark an almost missing (left) and a missing object (right). (best viewed at high resolution) (Color figure online)

In the following subsections we describe the necessary steps in order to optimize the 3D laserscanner alignment

1. Simulating a 3D Laser Scan - Given a set of all possible parameters, a 3D scan has to be simulated for each combination.
2. Assessing a 3D Laser Scan - After simulating a 3D scan, it has to be assessed by a density map which represents the measurement point distribution.
3. Choosing the Optimal Alignment - In a last step, the density maps have to be transferred into assessment values $J_k \in \mathbb{R}$. The minimal value represents the best alignment.

4.1 Simulating a 3D Laser Scan

The matrices $R^{\text{world,rot}}$ and $R^{\text{rot,laser}}$ are concatenations of the basic rotations $R_x(\alpha)$, $R_y(\alpha)$ and $R_z(\alpha)$ around the axes x, y and z. With these basic rotations, we can define

$$R^{\text{world,rot}} = R_x(\alpha_{\text{rot}}) \cdot R_y(\beta_{\text{rot}}) \cdot R_z(\gamma_{\text{rot}})$$
$$R^{\text{rot,laser}} = R_z(\gamma_{\text{laser}}) \cdot R_y(\beta_{\text{laser}}) \cdot R_x(\alpha_{\text{laser}})$$

Accordingly, we receive six parameters α_{rot}, β_{rot}, γ_{rot}, α_{laser}, β_{laser}, and γ_{laser}. However, γ_{rot} describes the angle of the rotation of the servomotor, respectively, this angle is set for the simulation from 0° to 360° with $T \cdot f$ steps, whereby T is the time for a full 3D scan and f is the frequency of the 2D laserscanner - e.g. *Hokuyo UTM-30LX*: $f = 40$ Hz. We found a good choice for T to be 18 s in our specific setting. Furthermore, γ_{laser} is a redundant parameter as it follows a rotation around the z-axis although itself rotates around the z-axis, too. So, this parameter is not necessary and can be omitted.

In order to get single laser beams, a direction vector for each beam has to be computed. This can be done by rotating the concatenated rotation matrices around the scan plane normal by

$$R^{\text{world,beam}} = R^{\text{world,rot}} \cdot R^{\text{rot,laser}} \cdot R_z^{\text{laser,beam}}(\gamma_{\text{beam}})$$

$$= \begin{bmatrix} n_x & n_y & n_z \end{bmatrix}$$

The resulting rotation matrix $R^{\text{world,beam}}$ contains in its first column the direction vector n_x which describes the laser beams. With these preliminaries a spheric 3D scan can be simulated as depicted in Fig. 2 by projecting the single laser beams onto a spheric surface by varying the parameters γ_{rot} and γ_{beam}.

4.2   Assessing a 3D Laser Scan

Above, it is shown how a 3D laser scan can be simulated. The result is a point cloud projected onto a spheric surface. In the next step, this cloud can be transformed into a density map $d = f(\theta, \varphi)$ which contains the number of points per surface. This density map $d(\theta, \varphi)$ is calculated as follows. In a first step, the 3D scan is overlaid with a measurement grid defined by

$$x_{\text{grid}} = \begin{bmatrix} r \cdot \sin\theta \cdot \cos\varphi \\ r \cdot \sin\theta \cdot \sin\varphi \\ r \cdot \cos\theta \end{bmatrix}$$

with

$$r \dots \text{radius sphere}$$
$$\theta \in [0° \dots 180°] \dots \text{vertical angle}$$
$$\varphi \in [0° \dots 360°] \dots \text{horizontal angle}$$

For each grid point, we search all points within the 3D scan which are close to this point x by applying a ball search. Thereby, we receive a density map $d(\theta, \varphi)$, an example can be seen in Fig. 2.

4.3   Choosing the Optimal Alignment

In order to optimize the alignment of the 3D laserscanner, for all K existing parameter settings a density map $d_k(\theta, \varphi)$ can be computed. Subsequently, the K density maps are transferred into an assessment value

$$J_k = \sum_{\forall\theta, \forall\varphi} g(d_k(\theta, \varphi), w(\theta, \varphi))$$

The function $g(\cdot)$ can be any reasonable function for summarizing the density map. We chose a logarithm in order to avoid sparse and extremely dense regions. By weighting the density map d with a weight function $w(\theta, \varphi)$, a desired field of view of the 3D laserscanner can be preferred. Finally, we chose the optimal assessment value and its parameters by

$$J_{opt} = \min_k J_k$$

5 Results and Discussion

We use the presented 3D laserscanner on two of our autonomous mobile robots (Fig. 4). The sensors are applied for different navigation tasks like localization, mapping, and drivability assessment. The matched point clouds in Fig. 1 are solely acquired by the presented 3D laserscanner in combination with an ICP matching (iterative closest point).

Building a custom-made 3D laserscanner comes always along with small distortions. However, in order to get accurate 3D scans the system needs to know its exact alignment. Therefore, a calibration is necessary. Discussing this would be beyond the scope of this paper, however, you can refer to calibration publications for both rotating 2D laserscanners and 3D laserscanners [18, 19]. Furthermore, it could be interesting for some applications to add color information to the point cloud [20–22].

Another important aspect one should be aware is the distortion of the 2D laser scans during a continuous rotation of the 2D laserscanner. A possible simple mathematical solution can be found in [23]: By combining alignment, rotation speed of the servomotor, and measurement frequency of the 2D laserscanner, the 2D scan can be rectified easily.

The advantages of a rotating 2D laserscanner compared to the available ready-to-use 3D laserscanners are the lower price as well as a wider vertical field of view (fov) and a higher vertical resolution achieved by using much more layers (*Velodyne HDL-64*/*Hokuyo UTM-30LX*: 26.8°/180° vertical fov, 0.4°/0.25° vertical resolution, 64/1080 layers). However, 2D laserscanners generate this number of layers by rotating a 1D laser beam transceiver whereas purchasable 3D laserscanners use multiple laser beam transceivers concurrently. Applying multiple distance measurements at the same time enables the ready-to-use systems to spin much faster with multiple scans per second. Depending on the horizontal resolution, a rotating 2D scanner may require several seconds – however, if a lower resolution is sufficient, e.g., for relocalization in a given map or observing a particular part of the world, higher update rates are possible. Hence, a system with rotating 2D laserscanner should not move during the scanning time, otherwise the resulting 3D scans are distorted if there is no accurate robot motion estimation. Accordingly, purchasable 3D scanners are suited for highly dynamic scenarios like urban street scenes, whereas spinning 2D scanners are well suited for perception tasks with the requirement of high accuracy and measurement density like traversability assessment or object recognition.

Fig. 4. 3D point clouds acquired with the presented 3D laserscanner setup. The first row shows our mobile robots each equipped with the 3D sensor during a robot challenge called *SpaceBot Camp 2016* [17], and a single 3D scan showing an indoor laboratory environment. From the second to the last row three semi-structured and unstructured environments are shown with their corresponding 3D point cloud. Each 3D scan was acquired with the presented 3D laserscanner and matched to the map with an ICP matching. Corresponding videos can be found at our website.

6 Conclusion

In this paper, we aim to accelerate the process of designing a custom-made 3D laserscanner, as this is done usually by different groups. Therefore, we offered a list of well suited hardware in combination with a mechanical drawing for the linkage which can be found at our website. Since it is important to target measurement points to an area of interest, we presented an approach to optimize the rotation axis orientation as well as the 2D laserscanner orientation. The explained method is offered as Matlab toolbox at our website, too. In the final section, some results achieved with our 3D laserscanner were shown. Again, supplementary videos are accessible at our website.

References

1. Kimoto, K., Asada, N., Mori, T., Hara, Y., Ohya, A., Yuta, S.: Development of small size 3d lidar. In: IEEE International Conference on Robotics and Automation (ICRA) (2014)
2. Reyes, A.L., Cervantes, J.M., Gutiérrez, N.C.: Low cost 3d scanner by means of a 1d optical distance sensor. Procedia Technol. **7**, 223–230 (2013)
3. Bosse, M., Zlot, R.: Continuous 3d scan-matching with a spinning 2d laser. In: Proceedings of the IEEE International Conference on Robotics and Automation (2009)
4. Lingemann, K., Surmann, H., Nüchter, A., Hertzberg, J.: Indoor and outdoor localization for fast mobile robots. In: Proceedings of the IEEE/RSJ International Conference on Intelligent Robots and Systems (2004)
5. Pfotzer, L., Oberlaender, J., Roennau, A., Dillmann, R.: Development and calibration of karola, a compact, high-resolution 3d laser scanner. In: IEEE International Symposium on Safety, Security, and Rescue Robotics (SSRR) (2014)
6. Schadler, M., Stuckler, J., Behnke, S.: Multi-resolution surfel mapping and real-time pose tracking using a continuously rotating 2d laser scanner. In: IEEE International Symposium on Safety, Security, and Rescue Robotics (SSRR) (2013)
7. Schadler, M., Stückler, J., Behnke, S.: Rough terrain 3d mapping and navigation using a continuously rotating 2d laser scanner. KI **28**, 93–99 (2014)
8. Surmann, H., Nüchter, A., Hertzberg, J.: An autonomous mobile robot with a 3d laser range finder for 3d exploration and digitalization of indoor environments. In: Robotics and Autonomous Systems (2003)
9. Zhang, L., Li, Q., Li, M., Mao, Q., Nüchter, A.: Multiple vehicle-like target tracking based on the velodyne lidar. In: Proceedings of the IFAC Intelligent Autonomous Vehicles (2013)
10. Steder, B., Rusu, R.B., Konolige, K., Burgard, W.: Point feature extraction on 3d range scans taking into account object boundaries. In: IEEE International Conference on Robotics and Automation (ICRA) (2011)
11. Schwarz, M., Behnke, S.: Local navigation in rough terrain using omnidirectional height. In: Proceedings of the Joint 45th International Symposium on Robotics (ISR) and 8th German Conference on Robotics (ROBOTIK), München (2014)
12. Wulf, O., Wagner, B.: Fast 3d scanning methods for laser measurement systems. In: International Conference on Control Systems and Computer Science (2003)

13. Ohno, K., Kawahara, T., Tadokoro, S.: Development of 3d laser scanner for measuring uniform and dense 3d shapes of static objects in dynamic environment. In: IEEE International Conference on Robotics and Biomimetics (ROBIO) (2009)

14. Kneip, L., Tache, F., Caprari, G., Siegwart, R.: Characterization of the compact hokuyo urg-04lx 2d laser range scanner. In: IEEE International Conference on Robotics and Automation (ICRA) (2009)

15. Demski, P., Mikulski, M., Koteras, R.: Characterization of Hokuyo UTM-30LX laser range finder for an autonomous mobile robot. In: Nawrat, A., Simek, K., Świerniak, A. (eds.) Advanced Technologies for Intelligent Systems of National Border Security. SCI, vol. 440, pp. 143–154. Springer, Heidelberg (2013)

16. Ye, C., Borenstein, J.: Characterization of a 2d laser scanner for mobile robot obstacle negotiation. In: Proceedings of the IEEE International Conference on Robotics and Automation (ICRA) (2002)

17. Lange, S., Wunschel, D., Schubert, S., Pfeifer, T., Weissig, P., Uhlig, A., Truschzinski, M., Protzel, P.: Two autonomous robots for the dlr spacebot cup - lessons learned from 60 min on the moon. In: International Symposium on Robotics (ISR) (2016)

18. Muhammad, N., Lacroix, S.: Calibration of a rotating multi-beam lidar. In: IEEE/RSJ International Conference on Intelligent Robots and Systems (IROS) (2010)

19. Sheehan, M., Harrison, A., Newman, P.: Self-calibration for a 3d laser. Int. J. Robot. Res. **31**(5), 675–687 (2012)

20. Scaramuzza, D., Harati, A., Siegwart, R.: Extrinsic self calibration of a camera and a 3d laser range finder from natural scenes. In: IEEE/RSJ International Conference on Intelligent Robots and Systems (IROS) (2007)

21. Pandey, G., McBride, J., Savarese, S., Eustice, R.: Extrinsic calibration of a 3d laser scanner and an omnidirectional camera. In: 7th IFAC Symposium on Intelligent Autonomous Vehicles (2010)

22. Andreasson, H., Lilienthal, A.: Vision aided 3d laser scanner based registration. In: Proceedings of the European Conference on Mobile Robots (ECMR) (2007)

23. Pascoal, R., Santos, V.: Compensation of azimuthal distortions on a free spinning 2d laser range finder for 3d data set generation. In: Proceedings of the Robtica 2010 - Encontro Nacional de Robotica, Marco, Leiria (2010)

Effects of Residual Charge on the Performance of Electro-Adhesive Grippers

Jatinder Singh[1,2(✉)], Paul A. Bingham[1], Jacques Penders[1,2],
and David Manby[3]

[1] Materials and Engineering Research Institute,
Sheffield Hallam University, Sheffield, UK
jatinder.singh@student.shu.ac.uk
[2] Sheffield Robotics, Sheffield, UK
[3] Aylesbury Automation Ltd, Aylesbury, UK

Abstract. Electro-adhesion is the new technology for constructing gripping solutions that can be used for automation of pick and place of a variety of materials. Since the technology works on the principle of parallel plate capacitors, there is an inherent ability to store charge when high voltage is applied. This causes an increased release time of the substrate when the voltage is switched off. This paper addresses the issue of residual charge and suggests ways to overcome the same, so that the performance of the gripper can be improved in a cycle of pick and release. Also a new universal equation has been devised, that can be used to calculate the performance of any gripping solution. This equation has been used to define a desired outcome (K) that has been evaluated for different configurations of the suggested electro-adhesive gripper.

Keywords: Advanced technology · Robot grippers · Dielectric materials · Automation

1 Introduction

Automation has rapidly grown in the last century from small to large scale industries, impacting the global economy [1]. An important aspect of automation is the end effector or the gripper, which is the end part of the system that directly interacts with the environment. This has been previously compared to a human hand [2]. Traditional gripping techniques include vacuum suction, chemical adhesion and micro-spines that have been used in the pick and place mechanism. These are well established, but in the case of flexible material handling these techniques do not provide distortion free and efficient handling [3]. For handling flexible material, electro-adhesive grippers (EAG) have been previously explored [6]. This forms the basis for the current research and here we explore the possibility of a pick and place mechanism using an Electro-adhesive gripper.

Although force calculations have been the primary basis for analysis of performance of an EAG, yet there are many other key parameters that need to be considered for the same. Since EAG works as a parallel plate capacitor, the amount of storage of

© Springer International Publishing Switzerland 2016
L. Alboul et al. (Eds.): TAROS 2016, LNAI 9716, pp. 327–338, 2016.
DOI: 10.1007/978-3-319-40379-3_34

charge becomes one such key component, which needs to be taken into account while working with an EAG.

This paper presents a simplistic equation that can be used as a tool for analysis of desired outcome, thereby reflecting the performance of any gripping solution. Using this equation, experimental results can be evaluated and analyzed for a complete cycle of pickup and release of various substrates. Experiments have been performed to support the theory of EAG working as a parallel plate capacitor, thereby storing charge and recommendations have been made to optimize this solution of using EAG as an improved gripping solution for insulating material. The key focus is on the use of dielectric in the configuration of EAG that causes the storage of charge, thereby impacting the performance of gripping solution. These recommendations can be useful for further development and deployment of EAG in an industrial environment to achieve desired results.

2 Background

Grippers are divided into mainly four types depending upon the application they are deployed. Ranging from microspines to vacuum suction techniques, each gripper technology has its own advantages and disadvantages. For application of automating material pick up usually astrictive grippers are preferred. As the name suggests astrictive grippers use some kind of binding force for adhesion. Key techniques in this category of grippers include vacuum suction, electro-adhesion and Van Der Waals forces based gripper. With capability of electrostatic attraction known from ancient Greeks times and the ability to attract small particles after charging an ebony rod is known to every school child having an interest in science [4], research on electro-adhesion has led the technique to be used in different industries for applications including grippers [5, 6], electrostatic chucks [7–9], cloth manufacturing robotics [10] and, recently, wall climbing robots [11, 12, 14].

With substantial potential and efficient results (<50 ms clamping and unclamping, [11]) obtained in some applications, electro-adhesion may be an appropriate technique for automating the pick and place of different materials specifically insulating materials. It has the advantage of giving uniform and controlled grip on the materials. Also there is no contamination or deformation of any kind. Being light weight, this gripper can be used efficiently with repeatable and reliable results. The use of this technique in automating pick and place of different materials was inspired by research on the effects of electro-adhesive forces in fabric handling in 1986 by Monkman [12]. Different configurations were studied and positive conclusions were made with electro-adhesion shown to be a successful technique for handling materials.

Even though electro-adhesion is proven to be at a relatively advanced stage of practical development and researchers have explored some of its capabilities, yet analysis on key component of EAG on a production level scale has been missing from the literature till date.

3 Theory

3.1 Principle of an Electro-Adhesive Gripper

There are five key components of electroadhesion are listed below

1. Electrodes: Conductive metal that acts as the plates of the parallel plate capacitor. These are provided with high voltage so as to charge the EAG.
2. Dielectric: Insulating layer attached to the electrodes that allows for storage of charge between the plates.
3. Substrate: Object to be picked up by the EAG.
4. Base material: Material on which the substrate is placed.
5. Power Supply: High voltage supply used to charge the EAG (Fig. 1).

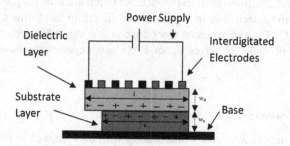

Fig. 1. Principles of electroadhesion

We follow the theory of parallel plate capacitor [18] for explaining the science behind EAG. The above setup represents a parallel plate capacitor formed due to charge being stored on the electrodes when power supply is switched on. Charge is allowed to collect due to the presence of the dielectric layer as it holds the charge helping in polarizing the substrate to be picked up. Thus the circuit acts as a parallel plate capacitor circuit with a finite time to charge and discharge. Attraction of the substrate takes place due to polarization of the substrate in the presence of the electric field generated between the electrodes as a result of storage of charge.

The principle of EAG has been best described as a parallel place capacitor in many research papers [2, 3, 18]. In [3], the concept is explained in terms of a single pole gripper, in which electrode plate forms one plate and the base material forms the second plate of the parallel plate capacitor. The dielectric material and the substrate together form the insulating material between the parallel plates of the capacitor. Thus the whole system forms a method of charge storage. In [3], it has been demonstrated that the net charge accumulation is directly proportional to the dielectric strength of the dielectric material used as well as the substrate. Mathematical modelling of dynamic properties (pick up and release time) has been shown for actual calculation which proves that the pickup and release times are directly proportional to the total number of charges developed.

The performance of an EAG is measured by doing force calculations based on parallel plate capacitance. A standard equation to calculate force is [15, 18]

$$f = \frac{\varepsilon_r \varepsilon_0 V^2}{2d^2} \tag{1}$$

Where f is the force between the two plates (N), ε_r is the relative permittivity of the dielectric, ε_0 is the relative permittivity of the vacuum, V is the voltage applied to the plate's measure in Volts (V) and d is the distance between the plates (mm).

As f is directly proportional to the dielectric constant therefore higher the dielectric constant more is the force generated by the EAG. This would suggest that an EAG with high dielectric constant is more efficient for picking heavier loads. Using a high dielectric constant causes more charge to build up in the electrodes as capacitance is increased [18] and that may result in the increase in release time [17], when the voltage is shut off (as will be seen later in experiments). Therefore this forms a trade off in the design of an EAG. The experimental work done in this report has been concentrated on the importance of dielectric layer in an EAG and experiments have been devised accordingly.

3.2   Safety Measures

As safety is an important aspect while dealing with high voltage, a Faraday's cage is necessary to conduct the trials in an isolated and safe environment. The voltage supplied to the electrodes is quite high (kV), but the current is expected to be very low (10–20 nA/N of force). For example, if a force of 0.5 N is required for the substrate to be lifted off, the current passing through the gripper will range from 5–10 nA [11], but as a safety measure a Faraday's cage is necessary to ensure any leakage of current does not lead to any accidents (electrocution). Earth plugs are typically used to ground the Faraday's cage at multiple places.

3.3   Parameter (K) that Defines an Effective Gripping Solution

For successful pick up and release of polymeric material such as gloves on a production line running 24 h a day, repeatable and reliable (R2) gripping results are required.

R2 can be defined as achieving consistent results (that is desired outcome is achieved) for every cycle of pick up and release. Thus the R2 needs to be analyzed in terms of efficiency of the gripping solution. A measure to calculate efficiency is needed to judge the effectiveness of the gripper on a production line scale. Therefore we define efficiency as the ratio of number of successful pick up and release cycles/trials to the total number of cycles/trials performed. Higher the efficiency (η), higher is the ability to achieve R2 results.

$$\eta = \frac{no.\ of\ trials\ with\ desired\ outcome}{total\ number\ of\ trials} * 100 \tag{2}$$

This is a universal equation that can be used to analyze not only electro-adhesive griping solution but any gripping solution. Thus to analyze η, first we need to define what is a desired outcome (successful pick and release). Since we are focusing on analyzing an effective gripping solution for pick and release cycle, in order to achieve R2, a successful outcome depends on the following relationship:

$$Desired\ outcome \propto \frac{assured\ pickup * assured\ release}{(Time\ to\ pick) * (Time\ to\ release)} \tag{3}$$

Thus we define parameter for desired outcome

$$K = \frac{P \times R}{T_P \times T_R} \tag{4}$$

Where,

$P = \{0,1\}$ 0 = Substrate not picked up
                 1 = Substrate picked up
$R = \{0,1\}$ 0 = Substrate not released
                 1 = Substrate released
T_P = Time to pickup
T_R = Time to release

(Units for desired outcome: s^{-2})

By setting desired values of above variables, the desired outcome (K) can be defined and thereby analysis on efficiency can be performed. Thus this equation can be used to analyze and compare experiments involved in this application for not only electro-adhesive grippers but for any gripping solution.

From the equation, it can be seen that in order to achieve a desired outcome, P and R must have a value 1. This means that in a cycle, if the substrate is not picked up, $P = 0$ and therefore desired outcome $K = 0$. Same case exists for release cycle (R is 0 if P is 0). Also, for maximum K, Tp and T_R value must be as small as possible (instantaneous pick and release). Tp and Tr are parameters which needs to be defined by the user and can vary depending upon the automation cycle or the application where automation is used. For analysis of the electro-adhesive gripper's used in this study, it has been agreed for Tp and T_R to have a minimum value of 1 (anything below 1 is rounded off to 1 due to measurement constrains and inefficiencies in the setup, (≥ 1 s = rapid release)). In this paper, Eq. (4) has been used to analyze the output of the EAG in two configurations; with dielectric and without dielectric layer. Further work will be done to calculate the overall efficiency by performing repeated experiments. This will therefore help in achieving a gripping solution with optimum R2 results.

4 Experimental Setup

The aim of the experiments is to show the impact of the dielectric layer on the performance of an EAG in completing one cycle of pick up and release of a substrate. For this purpose, we have chosen four different substrates. (i) High density polyethylene (HDPE) (ii) Polycarbonate (iii) Mobile phone screen glass (iv) Nitrile Glove. The selection of these substrates was based on the fact that electro-adhesion has different impact on different materials due to materials having different molecular structure, and different electrical properties.

Two different configurations of gripper were developed and these are shown in Fig. 2a, b. The difference between the two configurations is one of the configuration has a dielectric layer (Fig. 2b) and the other one is without the dielectric layer (Fig. 2a). Comparison of such configurations not, to our knowledge, been carried out for pick and release of substrates. The electrode structure in both was inter-digitated structure, as inter-digitated structure has been the most effective of all the structure [13]. Two configurations prepared are as follows

1. Bare Electrodes: B Electrodes as further defined in the study
2. Bare Electrodes and Polyimide (as dielectric): D electrodes as further defined in the study

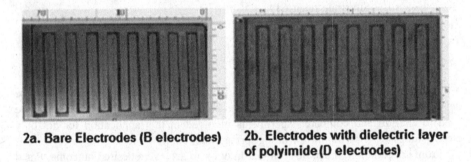

2a. Bare Electrodes (B electrodes)   **2b. Electrodes with dielectric layer of polyimide (D electrodes)**

Fig. 2. Two configurations of grippers

The bare electrode (B electrode) was prepared using copper and a printed circuit board (PCB) and milling process was used to obtain the required electrode dimensions, as shown in Fig. 2. A polyimide sheet (thickness 0.075 mm) with a dielectric constant 3.5 sourced from RS Components ltd [16], was used as the dielectric material to create the D electrode. The sheet was cut in size of 26 cm × 16 cm which covers the area of the electrodes fully (as shown in Fig. 2). It was glued to the plate using non-conductive glue. Polyimide was used as a dielectric since due to ease of availability and has been previously proven as a good dielectric medium for EAG experimentation [12].

The experiments were performed using a VP series Denso robot Fig. 3 to replicate the environment of a production line. The constructed electrode was placed as an End effector Fig. 3 on the Denso arm.

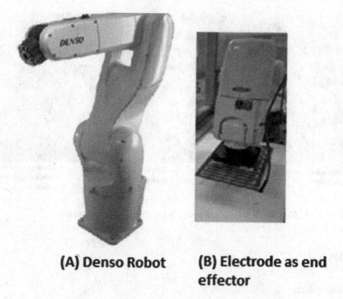

(A) Denso Robot   **(B) Electrode as end effector**

Fig. 3. VP Series DENSO robotic arm

A cycle for the experiment is defined as the pick-up and release of the substrate by the robot arm. The substrate is placed in a known location on the base. These coordinates are fed into the Denso arm. As shown in Fig. 4, the arm begins the cycle from a known position A to the position B of the substrate on the base. It then attempts to pick up the substrate by pressing the end effector on the substrate. The next step is for the arm to travel to a known position C. After this, the voltage is cut off and the time to release the substrate is noted down.

The experiments were performed for each substrate and electrode combination and results for the desired outcome were noted down (as per Eq. (4). The parameters for the desired outcome for our experiments are defined as:

P = 1 (substrate should be picked up)
R = 1 (substrate should be released)
T_P = 1 s (Pick up must be rapid)
T_R = 1 s (Release must be rapid when voltage is shut off)

It is the key requirement of universal equation method to define Tp and Tr based on the automation equipment used. It is not important for Tp and Tr to have same times. For the experiments presented in the paper the selection of 1 s for pick and 1 s for release have been based on the methodology of the experiments conducted. The experiments were performed with incremental voltage input (steps of 500 V) starting from 500 V.

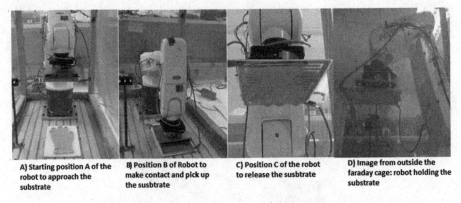

A) Starting position A of the robot to approach the substrate

B) Position B of Robot to make contact and pick up the susbtrate

C) Position C of the robot to release the susbtrate

D) Image from outside the faraday cage: robot holding the substrate

Fig. 4. Robotic arm positions during the experiment

5   Results

5.1   D Electrodes

The results for D electrodes show that the EAG is unable to pick up HDPE. For glass and polycarbonate, adhesion was possible only above 2500 V and 3000 V respectively, whereas the glove has been picked up at 1000 V. It was also observed that the release time for the glove increased considerably at high voltage. The D electrode started to arc at 3 kV restricting any further testing on this particular pad as arcing can lead to substantial damage to the properties of the substrate and the EAG, thereby providing misleading results to the experiment (Fig. 5).

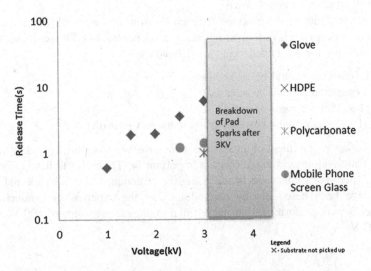

Fig. 5. Release time results of D electrodes

5.2   B Electrodes

When compared to the D electrodes, the results for B electrodes show that the pick-up voltage for gloves was increased to 2000 V, for glass it was 1000 V and for polycarbonate it was 1500 V. The B electrodes were also able to pick up HDPE at 1000 V. Also all the substrates demonstrated instantaneous releases. Also with increase in voltage, the release time for glove was not increased. Due to time constrains, only B electrodes experiments were repeated with glove and mobile phone screen glass as substrates and reproducibility was demonstrated for gloves and mobile phone screen glass by two measurements of each, which gave identical release times (Fig. 6).

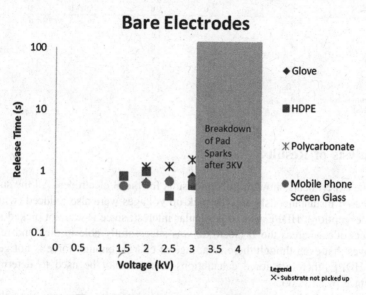

Fig. 6. Release time results for B electrodes

The performance of each gripper was done using the universal equation method presented in Eqs. 2–4. Desire outcome in the experiments presented is 1 in our case as Tp and Tr are both one second

$$K = \frac{P \times R}{T_P \times T_R} = \frac{1 \times 1}{1 \times 1} = 1$$

Tp and Tr are user defined times based on the automation pick and release application. It is not important for Tp and Tr to be of same value as well. e.g. if Tp is 0.5 s and Tr is 0.25 s the desired outcome following Eq. 4 will be

$$K = \frac{P \times R}{T_P \times T_R} = \frac{1 \times 1}{0.5 \times .25} = 8$$

So the total number of trials with desired outcome will depend on the total number of trials with 8 as outcome which further will be used to calculate the efficiency of the gripper using Eq. 2.

Following the above a desired outcome table has been made for the experimental work conducted which is shown below in Table 1.

Table 1. Desired outcome measurements for both gripper configurations (D and B electrodes) (Units for desired outcome: s^{-2})

Voltage (V)	Mobile phone screen glass		Nitrile		HDPE		Polycarbonate	
	D	B	D	B	D	B	D	B
500	0	0	0	0	0	0	0	0
1000	0	1	1	0	0	0	0	0
1500	0	1	0.5	0	0	1	0	0
2000	0	1	0.5	1	0	1	0	1
2500	.83	1	0.2	1	0	1	0	1
3000	.71	1	0.125	1	0	1	1	1
Number of trials desired outcome achieved	0	5	1	3	0	4	1	3

6 Analysis of Results

The above experiments show promising results for the B electrodes. All the substrates were released instantaneously and the pick-up voltages were also reduced (with nitrile glove as exception). HDPE was of particular interest, since it was not picked at all by the D electrode whereas the B electrode was successfully able to pick and release it. This proves that even though there is storage of charge, enough force is not generated to pick HDPE, therefore force calculations alone cannot be used to determine the performance of an EAG.

Since the desired outcomes were better for B electrodes, it supports our theory of storage of charge causing an increase in the release time of an EAG. Therefore for applications where rapid T_P and T_R are required, B electrodes are more suitable whereas for applications where it is necessary to hold the substrate for longer duration, D electrodes are to be used.

7 Conclusion

In this paper, we have determined how the performance of an EAG can be improved in an application involving pick up and release of various substrates (insulators). We have defined a desired outcome to be one cycle of successful pick up and release within the shortest time possible to complete the cycle.

We conclude that traditional method of mere force calculations are not enough for evaluating the performance of an EAG. There are various other parameters that must be taken into account. One such parameter is the storage of charge in the substrate and dielectric, which plays a vital role in achieving the desired outcome. To evaluate the effect of storage of charge on the EAG, two configurations of electrodes were built; electrodes with polyimide as dielectric (D electrodes) and bare electrodes, with no dielectric attached (B electrodes).

This paper also presents a unique method to evaluate the results in an efficient way and the new universal equation presented in this paper can not only analyze electroadhesive grippers but can be used on any robotic pick and place application. Through experimentation and evaluation of results using the new universal equation devised to check desired outcome, we concluded that the B electrodes provided more desired results, as they had a smaller pick up and release time, when compared to the D electrodes. This supports the theory of EAG working as a parallel plate capacitor. Since D electrodes contain the dielectric layer, they are able to store charge even after the voltage supply is shut off. Since the charge is not able to dissipate through the dielectric, it leads to a longer release time. Also this residual charge increases with increase in voltage that is applied to the EAG. The B electrodes on the other hand, do not have any charge storage mechanism and therefore release time for them is significantly less (Rapid release time = 1 s). In theory, there should be no release time delays with B electrodes since they only contain conductors, but practically some release time is seen as air between the substrate and the electrode also acts as a dielectric.

We also conclude that B electrodes can only be used to pick up substrates that are insulators since conductive substrates will create a short circuit on the electrodes, thereby causing large current to flow through them and destroying them. Therefore for applications that require quick pick up and release cycle for insulating materials, bare electrodes must be used in the construction of an EAG.

References

1. Hasegawa, Y.: Advances in robotics and automation: historical perspectives. In: Shimon, Y. (ed.) Springer Handbook of Automation. Springer, Berlin (2009)
2. Monkman, G.J., Hesse, S., Steinmann, R., Schunk, H.: Introduction to prehension technology. In: Robot Gripper. Wiley, Hoboken (2007)
3. Zhang, Z.: Modeling and analysis of electrostatic force for robot handling of fabric materials. IEEE Trans. Mechatron. 4(1), 39–49 (1999)
4. Trevena. D.H.: Static Fields in Electricity and Magnetism. Butterworths, London (1961)
5. Monkman, G.: Electroadhesive microgrippers. Ind. Robot Int. J. 30(4), 326–330 (2003)
6. Monkman, G.J.: Robot grippers for use with fibrous materials. Int. J. Robot. Res. 14(2), 144–151 (1995)
7. Shim, G.I., Sugai, H.: Dechuck operation of Coulomb type and Johnsen-Rahbek type of electrostatic chuck used in plasma processing. Plasma Fusion Res. 3, 051 (2008)
8. Asano, K., Hatakeyama, F., Yatsuzuka, K.: Fundamental study of an electrostatic chuck for silicon wafer handling. IEEE Trans. Ind. Appl. 38(3), 840–845 (2002)
9. Yoo, J., Choi, J.-S., Hong, S.-J., Kim, T.-H., Lee, S.J.: Finite element analysis of the attractive force on a Coulomb type electrostatic chuck. In: International Conference on Electrical Machines and Systems, pp. 1371–1375 (2007)

10. Monkman, G.J., Taylor, P.M., Farnworth, G.J.: Principles of electroadhesion in clothing robotics. Int. J. Clothing Sci. Technol. **1**(3), 14–20 (1989)

11. Prahlad, H., Pelrine, R., Stanford, S., Marlow, J., Kornbluh, R.: Electroadhesive robots - wall climbing robots enabled by a novel, robust, and electrically controllable adhesion technology. In: IEEE International Conference on Robotics and Automation, pp. 3028–3033 (2008)

12. Yamamoto, A., Nakashima, T., Higuchi, T.: Wall climbing mechanisms using electrostatic attraction generated by flexible electrodes. In: International Symposium on Micro-Nano Mechatronics and Human Science, pp. 389–394 (2007)

13. Tellez, J.P.D, Krahn, J., Menon, C.: Characterization of electro-adhesives for robotic applications. In: IEEE International Conference on Robotics and Biomimetics (ROBIO) (2011)

14. Liu, R., Chen, R., Shen, H., Zhang, R.: Wall climbing robot using electrostatic force generated by flexible interdigital electrodes. In: International Conference on Robotics and Biomimetics (ROBIO), pp. 2031–2036 (2011)

15. Jeon, J.U., Higuchi, T.: Electrostatic suspension of dielectrics. In: IEEE Transactions on Industrial Electronics, pp. 938–946 (1998)

16. DuPoint. Kapton Polyimide Film. H-38479 datasheet (2014)

17. Electrogrip. Principles of Electrostatic Chucks. Electrogrip, Pittsburgh, USA, March 2013. http://electrogrip.com/Egrip2013Support/Principles1no3.pdf

18. Koh, K.H., Sreekumar, M., Ponnambalam, S.G.: Experimental investigation of the effect of the driving voltage of an electroadhesion actuator. Materials **7**(7), 4963–4981 (2014)

A Data Set for Fault Detection Research on Component-Based Robotic Systems

Johannes Wienke[1]([⊠]), Sebastian Meyer zu Borgsen[2], and Sebastian Wrede[1]

[1] Research Institute for Cognition and Robotics (CoR-Lab),
Bielefeld University, Bielefeld, Germany
{jwienke,swrede}@techfak.uni-bielefeld.de
[2] Center of Excellence Cognitive Interaction Technology (CITEC),
Bielefeld University, Bielefeld, Germany
semeyerz@techfak.uni-bielefeld.de

Abstract. Fault detection and identification methods (FDI) are an important aspect for ensuring consistent behavior of technical systems. In robotics FDI promises to improve the autonomy and robustness. Existing FDI research in robotics mostly focused on faults in specific areas, like sensor faults. While there is FDI research also on the overarching software system, common data sets to benchmark such solutions do not exist. In this paper we present a data set for FDI research on robot software systems to bridge this gap. We have recorded an HRI scenario with our RoboCup@Home platform and induced diverse empirically grounded faults using a novel, structured method. The recordings include the complete event-based communication of the system as well as detailed performance counters for all system components and exact ground-truth information on the induced faults. The resulting data set is a challenging benchmark for FDI research in robotics which is publicly available.

1 Introduction

Like most other technical systems, robots are not free from faults that occur at runtime and affect the mission success as well as the safety for physically interacting systems. In contrast to pure software systems, classical testing and modelling methods which prevent faults in the first place are often only applicable to parts of robot systems and with a high overhead, i.e. due to the interaction of robots with their environment, in particular humans. Therefore, failures can frequently be observed in these systems and further measures need to be taken to improve the situation. For this purpose we use the definitions from Steinbauer [13], where "a failure is an event that occurs when the delivered service deviates from correct service. An error is that part of the system state that can cause a subsequent failure. A fault is the adjudged or hypothesized cause of an error."

This work was funded as part of the Cluster of Excellence Cognitive Interaction Technology 'CITEC' (EXC 277), Bielefeld University and by the German Federal Ministry of Education and Research (BMBF) within the Leading-Edge Cluster Competition "it's OWL" (intelligent technical systems OstWestfalenLippe) and managed by the Project Management Agency Karlsruhe (PTKA).

L. Alboul et al. (Eds.): TAROS 2016, LNAI 9716, pp. 339–350, 2016.
DOI: 10.1007/978-3-319-40379-3_35

One common method to address such failures is *autonomous fault detection*, which is an ongoing research topic since many years with several different directions. A significant portion of fault detection research for robotics so far has focused on sensor and actuator faults. However, on the level of the complete software system controlling complex robotic applications (e.g. mobile platforms like in the RoboCup@Home league) research is much more fragmented and results are often hardly comparable. Nevertheless, software faults occur and handling these would increase system stability. One reason for this is that common data sets and established benchmarking methods, as e.g. the Tennessee Eastman Process [2] for industrial FDI, do not exist (cf. Pettersson [11]). While erroneous behaviors are frequently observed when operating robots, gathering usable data from these executions is often impossible. Because of the much more dynamic nature of current robotics systems, such data sets usually cannot be acquired from existing systems, because either the appropriate execution traces were not recorded in a sufficient quantity at all, or in case recordings exist, ground-truth information about the exact faults that occurred and their timing is missing. Therefore, reference data sets need to be explicitly created.

This paper introduces a novel data set for developing and benchmarking fault detection approaches for autonomous robotics with a focus on the overarching system. The data set contains recorded executions of our RoboCup@Home system in a typical task for this RoboCup league to provide a realistic and challenging scenario. During execution, selected faults have been induced in the system. These faults are based on empirical findings from an online survey about faults in robotics to ensure realistic conditions and the scheduling is based on an algorithm which maximizes the amount of usable fault data while maintaining clear properties on the timing of faults. The data set comprises the complete system communication of the robot, detailed performance counters for all system components, ground-truth information about the induced faults and videos from a camera observing the scene for manual inspection and further annotation. All data is synchronized and available for download[1].

To provide a challenging and profitable benchmark, the data set specifically focuses on performance-related faults. Such faults do not immediately render the system unusable, e.g. through crashing important components, but instead slowly degrade its perceived or computational performance (following the ideas of Application Performance Management [15]). Hence, they are much harder to detect and easily missed during short testing cycles in active development work. Also, these non-catastrophic issues have a higher potential for being recovered at runtime and therefore are the most valuable ones to detect with FDI methods.

2   Related Work

Up to our knowledge, publicly available data sets for FDI research regarding the overarching software system of robots do not exist yet. Nevertheless, several

---

[1] At   https://doi.org/10.4119/unibi/2900912   and   https://doi.org/10.4119/unibi/2900911. Detailed technical usage instructions are given there.

authors have performed research in this direction and reported on evaluation methods and internal data sets that have been used. Steinbauer and Wotawa [14] propose an FDI approach based on observers which check the communication and process spawning behaviors of system components against modelled frequency or threshold rules. For the evaluation two distinct faults have been induced in a test system. Neither their origin nor the amount of recorded trials is mentioned. A closely-related approach is described in Peischl et al. [10]. Here, the evaluation consists of 20 experiments in which a process crash is simulated and 6 additional case studies which test faults designed to analyze properties of the detection method. Golombek et al. [3] learn the statistics for communication patterns of the robot system. For the evaluation, a data set with 4 different fault types was recorded within an existing scenario of a service robot. For each fault type 10 runs of the scenario were recorded, each approximately lasting 5 min. The induced faults were based on the experience of an expert user with the specific system. Similarly, Jiang et al. [6] describe an approach which learns communication invariants from successful trials. For evaluation, a UAV had to land on a moving target area. In contrast to other approaches, no explicit software faults were induced but instead the task conditions were made more challenging, e.g. by producing wind, and success was measured in the task space instead of measuring the detected faults.

3 Robot System and Recording Scenario

Our new data set has been recorded using the mobile robot platform *ToBi* [9] (cf. Fig. 1), which successfully participates at the RoboCup@Home challenge since 2009. The robot consists of a mobile base with differential drive and two laser range finders with 360° coverage for distance data. Mounted on top of the robot are two RGBD cameras for object-recognition, obstacle avoidance, gesture-recognition and scene interpretation as well as an RGB camera for face recognition. For manipulation, the robot is equipped with a 5 DOF manipulator. The robot carries two Linux-based laptops which are connected via Gigabit Ethernet. They run the distributed software system controlling the platform.

The software architecture of *ToBi* (cf. Fig. 1) consists of multiple sensor components that provide the system with information extracted from the scene like object recognition, person tracking and speech recognition. Actuator components that allow manipulation comprise navigation, text to speech and grasping. All components are distributed via an event-based middleware called RSB [17], which has tool support for transparently introspecting and recording the execution of the system at runtime. For the coordination, the BonSAI framework [12] abstracts the different components as software sensors and actuators and allows to model the system behavior as a finite state machine.

We chose a modified version of the restaurant task from the RoboCup@Home competition 2015 [1] as the recording scenario for the data set. RoboCup is one of the leading robot competitions and therefore provides a challenging and realistic environment for current capabilities of mobile robots. In the scenario the robot

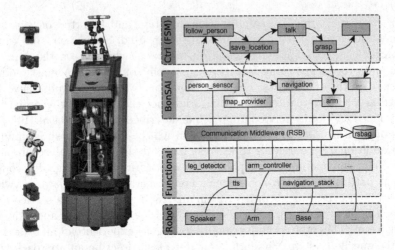

Fig. 1. Overview of *ToBi* and the system architecture. All inter-process communication is performed via the RSB middleware and can be recorded transparently using the `rsbag` utility.

acts as a waiter in a restaurant. The plot consists of three phases. At first, an operator shows the robot around and trains multiple locations i.e. where it can pick up drinks as well as the locations of the tables where drinks have to be delivered to. In the second phase the robot waits for somebody waving to take the order of that person. The robot asks for the person's name and the desired drink. After taking all orders *ToBi* can be told to enter the third phase in which the orders get executed one after another. During the different phases SLAM is used to build up the map of the environment, face recognition and object recognition are used to identify persons and objects, and the RGBD sensors are used for planing grasping tasks. The scenario thereby integrates a diverse range of behaviors and skills necessary for a mobile robot with a high complexity and variability, especially due to the involved human-robot interaction. One iteration of the scenario with 2 to 3 guests takes approx. 11 min.

4   Induced Faults

To create a data set for fault detection research in such a complex system, we did not only rely on our own experience to ensure that the resulting data set is representative for the domain. Instead, we created an online survey to acquire empirical and quantitative data on the types and distribution of different kinds of faults in comparable robotics and intelligent systems. For related domains like cloud computing or classical desktop software, a considerable amount of research on the types of faults and their occurrence frequencies exists. However, for robotics the situation is different. Up to our knowledge, there is only Steinbauer [13] which contains a systematic survey on faults in platforms participating at the

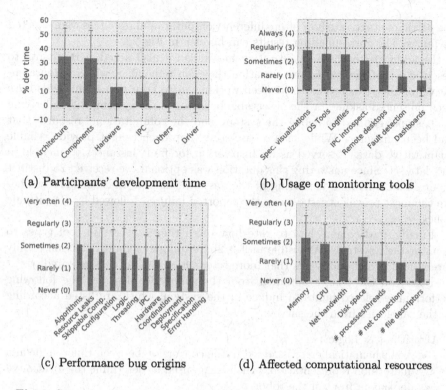

(a) Participants' development time

(b) Usage of monitoring tools

(c) Performance bug origins

(d) Affected computational resources

Fig. 2. Survey results as averages over all participants with standard deviation.

RoboCup challenges. Despite giving valuable insights, it is not detailed enough to support the construction of a data set with actual fault instances.

We implemented our survey as an online questionnaire (following methodology advices from Gonzalez-Bañales and Adam [4]) which was distributed around robotics researchers using well-known mailing lists. In total we received 61 completed submissions and 141 incomplete ones[2]. 86 % of the participants were researchers or PhD candidates at universities, 7 % regular students and 7 % from an industrial context. On average, participants had 5.8 years of experience in robotics (sd: 3.3). When asked with which activities participants spend their active development time, software architecture and integration as well as component development are the most frequent activities (cf. Fig. 2a).

To determine the types of faults to induce we asked how often certain faults were the origin for performance bugs in the systems participants have been working with. Different categories could be rated with 5 choices ranging from *Never* to *Very often*. The categories have been chosen based on existing surveys from related domains [5,7,8,13]. The results in Fig. 2c indicate that algorithmic faults are the most frequent cause for failures, followed by resource leaks (not limited to memory) and skippable computations. Along these lines we also asked

---

[2] Incomplete submissions include visitors which only opened the welcome page once.

the participants to rate how often different computational resources were affected by faults. The corresponding results can be seen in Fig. 2d.

In addition to these quantitative answers, we also asked participants to describe a prototypical failure situation they can remember, which is representative for the systems they have been working with. Four text input fields have been presented asking for the observable behavior of the system, the underlying fault, steps required to debug the system, and the computational resource that had been impacted. While such a case-study approach cannot provide reliable quantitative data, it served as an inspiration for fault instances to include in the data set. Since answering this question was optional, we received 25 distinct reports and a manual clustering of the answers suggests that software crashes are the most prevalent issue along the reported failures, followed by algorithmic issues.

Finally, participants had to rate how often they use certain tool types to monitor their systems. As visible in Fig. 2b, autonomous fault detection is only rarely used, which motivates that more research on this topic is required.

Based on the results obtained from the survey we designed the following performance-related faults to induce in the system, which are listed according to the categories from the survey:

- Algorithms & Logic[3]:
    - A mathematical error is added to the conversion between Euler angle and quaternion representation, which is e.g. used to determine the location of persons in front of the robot.
    - The grasping controller for the arm performs unnecessary movements due to a bug in generating a trajectory in a graph of valid postures.
- Resource Leak:
    - The central state machine did not deallocate unused IPC connections, which results in a leak in TCP connections.
    - The speech recognizer did not deallocate memory for the sound buffers, which results in a memory leak.
- Skippable Computation:
    - The component which tracks persons transforms egocentric coordinates for detected person into global SLAM coordinates multiple times instead of only once per person.
    - Removes the throttling of the main loop in the face detection component, which increases the CPU load.
    - The detector for legs in the laser scans performed operations multiple times.
- Configuration:
    - The configuration of the state machine used a wrong middleware address to communicate with the text to speech engine and had to wait for a timeout before resorting to the correct address.
    - To emulate a configuration issue with the clock synchronization via NTP, the clock of one computer was shifted at system runtime.

---

[3] Categories were combined as it turned out to be hard to distinguish between them.

- Threading:
  - An unnecessary sleep instruction was added to the object recognition component, which delayed the classification results for 5 s to simulate the effect of inefficient threading strategies in this component.
- IPC:
  - The central communication daemon of the middleware is affected by constantly adding and removing a participant to the daemon network, which is a costly operation. As a consequence, IPC messages have much higher latencies and jitter.

These instances represent the 7 most frequent kinds of performance-related faults according to the survey.

5 Data Acquisition Method

5.1 Recording Setup

For realizing the data set acquisition process, we have used a method along the lines of Wienke et al. [16]. The core idea is to use the transparent recording capabilities of the robot middleware whenever possible. This automatically solves the synchronization for most parts of the data set, assuming that the middleware records accurate timing information. Consequently, we used the rsbag tool, which is part of the RSB middleware, to record the communication of the system, as depicted in Fig. 1.

In order to include ground-truth information about induced faults and their exact timing within this recording method, the induced faults have been made triggerable via middleware events. A scheduling component was added to the system, which issued the required trigger events. This way fault ground-truth is included in the middleware recordings.

In the case of performance bugs, performance counters like CPU usage or network bandwidth are important information sources which had to be included in the data set. For this purpose we have added monitor components to the system, which uncover performance counters for all functional system components and the hosts and expose them via RSB for the inclusion in the data set. These monitors operate as external processes to the core system, which prevents changes in the system behavior due to the added monitoring. They are implemented as C++ programs with minimal processing overhead and acquire the required information via Linux interfaces like the /proc filesystem.

In addition to these data sources we have recorded each trial of the data set with an external HD camera to provide the possibility for human inspection and annotation. Since processing video inside the middleware infrastructure, in contrast to the aforementioned information, generates a significant load on the system, which differs from the usual usage of the system, we recorded the video out of band with the system communication. In order to synchronize this data, a special middleware event at the start of each trial was generated in parallel to a recognizable beep sound using a loudspeaker on the robot. As this sound was recorded in the external video camera as well, the videos could be synchronized to the system time using an automated cross-correlation analysis, which was manually checked for all videos.

To ensure the validity of the recorded data, each trial was automatically checked against completeness of the recorded middleware communication (RSB scopes and event rates). Additionally, the software system was restarted for each trial to prevent undesired effect from previous runs. Finally, in case unexpected errors or behaviors of the robot occurred during executing, these were annotated, so that trials without unexpected faults[4] are clearly recognizable.

5.2 Scheduling of Faults

Existing work like Golombek et al. [3] and Jiang et al. [6] used a simple strategy to induce faults into the system: at a certain point in time of each trial a single fault is induced and maintained until the end of the trial or system crash. While this approach provides an easy to analyze data set, it would be very time-consuming given the more complex scenario and number of faults in our data set, to provide a statistically feasible amount of fault occurrences. Therefore we opted for a strategy where multiple reversible faults were triggered during each recording trial, to maximize the amount of fault occurrences being recorded. This means that each fault that was triggered via the middleware communication could also be reverted back to a healthy system state as if the fault had never occurred. Such an approach forbids faults that are catastrophic to the system execution, however, we have already argued in the introduction, why such faults are less interesting for autonomous fault detection.

With this general approach we realized the exact scheduling of faults during each trial as follows (cf. Fig. 3): starting with the initial execution of the system state machine, the trial was separated into consecutive time slices of a fixed length, which were additionally separated by a fixed length pause. Within each of the slices, a single fault was scheduled for a fixed time interval. The fault instance was uniformly drawn from the available ones and the start time of fault within the slice was also determined using a uniform distribution[5]. This procedure was chosen to prevent accidental correlations between system states and the induced faults, e.g. through an operator manually triggering the faults. A uniform selection was chosen to provide the same statistical confidence for each fault type. To further reduce potential correlations the start of the first slice was

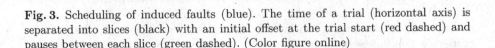

Fig. 3. Scheduling of induced faults (blue). The time of a trial (horizontal axis) is separated into slices (black) with an initial offset at the trial start (red dashed) and pauses between each slice (green dashed). (Color figure online)

---

[4] Up to the knowledge of an expert user.

[5] Limited so that the target fault time fits into the slice.

randomly offset after the state machine start using a uniform distribution up to 30 s. For the length of each fault 80 s was used. This time was determined by an expert developer of the system so that this person was able to detect each of the fault types from appropriate visualizations of the event communication and the performance counters[6], but without resulting in a catastrophic failure of the system due to the implemented resource leaks. The length of each slice was selected to be 160 s to provide sufficient variation for the fault occurrence and the pause time between slices was set to 20 s as the minimum acceptable distance between consecutive faults. This reflects the maximum time the recovery from any of the faults might take (after the signal to recover normal state, heuristically determined) so that instances are correctly separated in any case. The start of each slice is exposed via RSB to include scheduling information in the data set.

6   Corpus Content

With the explained method we have recorded a data set which consists of 10 executions of the system without induced faults as a baseline and 33 successful trials with induced faults. This number was chosen so that at least 10 complete instances of each induced fault were recorded during execution. The total time of recordings is 8:16 h, which results in an average of 11:33 min per trial. The recorded system, on average, exchanged middleware events with 113 Hz, including performance counters and fault scheduling information, which results in approx. 190 MB of raw event data per trial. For each RSB event, the middleware provides the following information, which is available in the data set:

Scope. Middleware channels the event has been sent to (string).
ID. A unique ID for the event including the sender ID.
Type. A string describing the data type of the contained data.
Data. The user-defined payload (binary, basic types or Google Protocol Buffers).
Method. An optional string describing a method call associated with the event.
Causes. A vector of other event IDs an event refers to.
Timestamps. Timestamps in µs precision describing event processing steps.
User-defined Infos. A user-defined set of string key-value pairs.

Performance counters have been recorded at 1 Hz, to align with the update rate for the counters from the Linux kernel and to prevent heavy load on the system. Only the network bandwidth information per process has been sampled at 0.5 Hz as it is more costly to generate this information. In detail, the following counters have been recorded:

CPU. Time spent in user and kernel mode.
Memory. Virtual and Resident Set Size.
Threads. Number of threads of the process.

---

[6] In appropriate situations. For instance, the grasping controller fault is only detectable in case the arm is used.

Decriptors. Open files and file descriptors.
I/O. Total reads and writes and disk-specific reads and writes.
Network. Receiving and sending bandwidth.

For each host the following information was recorded at 0.5 Hz:

CPU Total CPU usage per state for all virtual cores.
Load The 1 min, 5 min, and 15 min load average.
Memory Total and used system memory.

The trial recordings are available in the original `rsbag` format as well as in CSV files, where apart from the raw data the network bandwidth counters, the fault state of each component and relevant encoded events for each component have been conformed to the 1 Hz sampling of the other performance counters.

In addition to the trial data, machine-readable meta-data (CSV and JSON) describing the available trials (with contained faults, additional operator notes and results of the automatic validation), the structure of the system (component processes, relevant RSB scopes and distribution across the two laptops for each component, dependencies on other components), and the affected components for each fault is available.

Apart from the 33 trials of the core data set, 23 additional trials are available, which contain unexpected behaviors of the robot or system. These are separated from the core data set and can be used for explorative analyses based on manual annotations. Projects for the ELAN[7] annotation tool are included for this purpose, which include the system communication. The number of trials with undesired faults is that high because during the recording session an actual hardware issue appeared. A lose screw inside the gripper resulted in a situation where the sensors did not detect whether an object was inside the gripper or not. As a consequence, grasping frequently stopped at the point where the gripper was closed and did not recover. We have annotated these situations, which results in an additional fault type being included in the data set.

6.1 Data Set Examples

Figure 4 displays an exemplary trial of the data set for the central state machine component to visualize the kind of data available within the data set. In the upper part of the Fig. 3 of the available performance counters are shown. The lower part displays all events received and sent by the state machine. Each row of this plot relates to a single RSB scope and each vertical line in each row relates to a single event being communicated via this scope. The colored areas at the top of the first plot indicate the fault ground-truth information for this trial, where red areas indicate a fault that directly affects the state machine component whereas the orange area relates to an unrelated fault in other parts of the system. Finally, the dark shaded areas in all plots report the results of a fault detection approach which we have trained on the data set in order to verify the applicability for the intended purposes.

---

[7] https://tla.mpi.nl/tools/tla-tools/elan/.

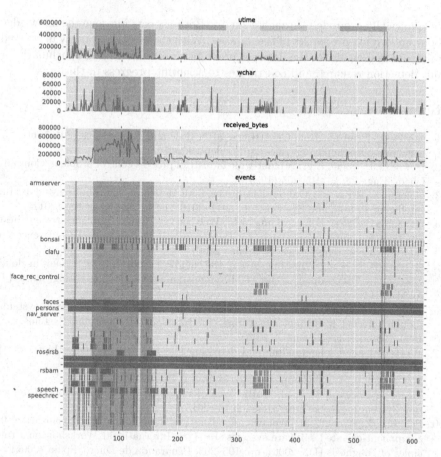

Fig. 4. Visualization of the recorded performance counters and events for the state machine component in one of the data set trials. The x-axis measures trial time in seconds. RSB scope names have been grouped by purpose.

7 Conclusion

With this contribution we have introduced a new publicly available data set for fault detection research on robotics systems. The data set specifically focuses on the overarching software system of event-based robots, an area currently lacking established benchmarks and publicly available data sets. It comprises a unique combination of performance information for system components and the complete system communication, which enables the application for a variety of purposes. We are actively using the data set for our own research, which validates its applicability. Our data set is generated in a challenging state of the art scenario and provides accurate ground-truth information and annotations. For this purpose, a novel mechanism of algorithmically inducing faults into a running system via the middleware has been proposed and the induced faults are empirically grounded through a survey. Both aspects improve the knowledge about underlying properties of the data set. With the available variety in covered faults and system states the data set presents a new challenging benchmark for fault detection research and contributes to scientific progress in this area.

References

1. van Beek, L., et al.: RoboCup@Home 2015: Rule and Regulations (2015). http://www.robocupathome.org/rules/2015_rulebook.pdf
2. Downs, J.J., Vogel, E.F.: A plant-wide industrial process control problem. Comput. Chem. Eng. **17**(3), 245–255 (1993)
3. Golombek, R., et al.: Online data-driven fault detection for robotic systems. In: Intelligent Robots and Systems, pp. 3011–3016. IEEE, San Francisco (2011)
4. Gonzalez-Bañales, D.L., Adam, M.R: Web survey design, implementation: best practices for empirical research. In: European and Mediterranean Conference on Information Systems, Valencia, Spain (2007)
5. Gunawi, H.S., et al.: What bugs live in the cloud?: a study of 3000+ issues in cloud systems. In: Proceedings of the ACM Symposium on Cloud Computing, pp. 1–14. ACM (2014)
6. Jiang, H., Elbaum, S., Detweiler, C.: Reducing failure rates of robotic systems though inferred invariants monitoring. In: Intelligent Robots and Systems, pp. 1899–1906. IEEE (November 2013)
7. Jin, G., et al.: Understanding and detecting real-world performance bugs. ACM SIGPLAN Not. **47**(6), 77–88 (2012)
8. McConnell, S.: Code Complete, 2nd. Microsoft Press (2004)
9. Meyer zu Borgsen, S., et al.: ToBI-Team of Bielefeld: The Human-Robot Interaction System for RoboCup@ Home 2015 (2015)
10. Peischl, B., Weber, J., Wotawa, F.: Runtime fault detection, localization in component-oriented software systems. In: 17th International Workshop on Principles of Diagnosis (DX 2006), pp. 195–203, Penaranda de Duero, Spain (2006)
11. Pettersson, O.: Execution monitoring in robotics: a survey. Robot. Auton. Syst. **53**(2), 73–88 (2005)
12. Siepmann, F., Wachsmuth, S.: A Modeling Framework for Reusable Social Behavior. In: Silva, R.D., Reidsma, D (ed.) Work-in-Progress Workshop Proceedings, pp. 93–96. Springer, Amsterdam (2011)
13. Steinbauer, G.: A survey about faults of robots used in robocup. In: Chen, X., Stone, P., Sucar, L.E., van der Zant, T. (eds.) RoboCup 2012. LNCS, vol. 7500, pp. 344–355. Springer, Heidelberg (2013)
14. Steinbauer, G., Wotawa, F.: Detecting and locating faults in the control software of autonomous mobile robots. In: Kaelbling, L.P. (ed.) International Joint Conference on AI, pp. 1742–1743 (2005)
15. Sydor, M.J.: APM Best Practices: Realizing Application Performance Management. Apress, New York (2010)
16. Wienke, J., Klotz, D., Wrede, S.: A framework for the acquisition of multimodal human-robot interaction data sets with a whole-system perspective. In: Multimodal Corpora: How Should Multimodal Corpora Deal with the Situation? Workshop Programme (2012)
17. Wienke, J., Wrede, S.: A middleware for collaborative research in experimental robotics. In: 2011 IEEE/SICE International Symposium on System Integration (SII), pp. 1183–1190. IEEE, Kyoto (2011)

euRathlon 2015: A Multi-domain Multi-robot Grand Challenge for Search and Rescue Robots

Alan F.T. Winfield[1]([✉]), Marta Palau Franco[1], Bernd Brueggemann[2],
Ayoze Castro[3], Miguel Cordero Limon[4], Gabriele Ferri[5], Fausto Ferreira[5],
Xingkun Liu[6], Yvan Petillot[6], Juha Roning[7], Frank Schneider[2], Erik Stengler[1],
Dario Sosa[8], and Antidio Viguria[4]

[1] Bristol Robotics Lab and Science Communication Unit, UWE Bristol, Bristol, UK
alan.winfield@uwe.ac.uk
[2] Fraunhofer FKIE, Bonn, Germany
[3] Oceanic Platform of the Canary Islands (PLOCAN), Canary Islands, Spain
[4] FADA Center for Advanced Aerospace Technologies, Seville, Spain
[5] NATO STO Centre for Maritime Research and Experimentation, La Spezia, Italy
[6] School of Engineering and Physical Sciences,
Herriot-Watt University, Edinburgh, UK
[7] Department of Electrical and Information Engineering,
University of Oulu, Oulu, Finland
[8] University of Las Palmas de Gran Canaria, Las Palmas de Gran Canaria, Spain

Abstract. Staged at Piombino, Italy in September 2015, euRathlon 2015 was the world's first multi-domain (air, land and sea) multi-robot search and rescue competition. In a mock-disaster scenario inspired by the 2011 Fukushima NPP accident, the euRathlon 2015 Grand Challenge required teams of robots to cooperate to map the area, find missing workers and stem a leak. In this paper we outline the euRathlon 2015 Grand Challenge and the approach used to benchmark and score teams. We conclude the paper with an evaluation of both the competition and the performance of the robot-robot teams in the Grand Challenge.

Keywords: Field robotics · Multi-robot systems · Land robots · Aerial robots · Marine robots · Benchmarking · Competitions

1 Introduction

A high-level aim of the three-year EU FP7 euRathlon project is to help speed-up progress towards practical, useable real-world intelligent autonomous robots through competitions; toward this aim euRathlon has created real-world robotics challenges for outdoor robots in demanding emergency response scenarios.

The euRathlon competitions aim to test the intelligence and autonomy of outdoor robots in demanding mock disaster-response scenarios inspired by the 2011 Fukushima accident. Focused on multi-domain cooperation, the 2015 euRathlon competition required flying, land and marine robots acting together to survey the disaster, collect environmental data, and identify critical hazards. The first

L. Alboul et al. (Eds.): TAROS 2016, LNAI 9716, pp. 351–363, 2016.
DOI: 10.1007/978-3-319-40379-3_36

(land) competition was held in 2013 in Berchtesgaden, Germany [1]. In September 2014, the second (sea) competition was held in La Spezia, Italy [2,3]. The final euRathlon Grand Challenge (air, land and sea) was held in Piombino, Italy, from 17th - 25th September 2015.

This paper proceeds as follows. First we outline the Grand Challenge concept then, in Sect. 3, we describe the location chosen for euRathlon 2015 and how the requirements of the Grand Challenge map to the physical environment. In Sect. 4 we outline the benchmarking/scoring schema developed for euRathlon 2015. The paper concludes in Sect. 5 by evaluating first the competition itself, including lessons learned, then the performance of the teams in rising to the Grand Challenge.

2   The Grand Challenge

Inspired by the 2011 Fukushima accident and the subsequent efforts to use robots to assess internal damage to the NPP buildings [4], we sought to develop a scenario which would - in some respects at least - provide teams with a comparable challenge. Clearly there were aspects that we could not replicate, in particular the radiological environment or chemical hazards – but we were able to offer significant challenges to radio communication. Other challenges included the weather, which reduced underwater visibility to less than 1m, the rough terrain for land robots, and obstructed access routes inside the building.

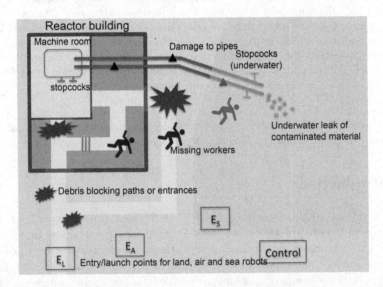

Fig. 1. Concept diagram for the euRathlon 2015 Grand Challenge scenario

Figure 1 shows the concept diagram for the Grand Challenge scenario. The key physical elements of the scenario are (1) a building on a shoreline which can act in the role of the 'reactor' building, with an internal mock 'machine room', (2) valves (stopcocks) in the machine room connected to pipes which lead out of the building and into the sea, with corresponding underwater valves, (3) damage or debris blocking paths or entrances outside or inside the building, (4) damage to the pipes and (5) missing workers. The Grand Challenge scenario comprised three mission objectives – outlined as follows.

- **Mission A**: Search for missing workers. Robots must search for two missing workers represented by mannequins dressed in orange suits, which could be inside the building, outside the building, floating on the sea surface near the coast, or trapped underwater. Teams received bonus points if a worker was found during the first 30 min of the Grand Challenge, because in a real scenario the probability of finding a missing person alive decreases rapidly with time.
- **Mission B**: Reconnaissance and environmental survey of a building. Robots must inspect a building to evaluate damage (represented by markers) and find a safe path to a machine room where valves were located. This required robots to survey the area, create a map of the building and the outdoor area surrounding it, and locate objects of potential interest (OPIs) in order to provide situational awareness to the team.
- **Mission C**: Pipe inspection and stemming a leak. Robots must localize four pipe sections on land, localize another four matching pipes underwater, look for damage to the land pipes and identify a contaminant leak (represented by a marker), reach the valves in the machine room and underwater, and close the two corresponding valves in synchrony.

In the published scenario description[1] we made it clear that the missions could be undertaken in any order, or in parallel. The Grand Challenge would be successfully met if all three mission objectives were met within 100 min, but importantly we did not specify how the challenge should be met, or with what robots (only limiting their number and kind).

3   Torre del Sale - the Competition Site

Securing a location for euRathlon 2015 was challenging given the requirements. We needed a suitable building on a shoreline and surrounding areas with safe access for land and flying robots, a safe shallow sea for marine robots and sufficient space for team preparation, organisers and spectators. Equally importantly we needed all of the necessary permissions to operate land, sea and air robots: for marine robots from the Port Authority and for flying robots from the Italian Civil Aviation Authority (ENAC).

The venue selected was an area in front of the ENEL (Italian National Company for Electricity) electrical power plant in Piombino, Italy. The location offered all the areas needed for the robots, space for hosting participants

---

[1] http://www.eurathlon.eu/index.php/compete2/eurathlon2015/scenarios2015/.

Fig. 2. The Torre del Sale, with the ENEL power plant in the background, and beach to the right

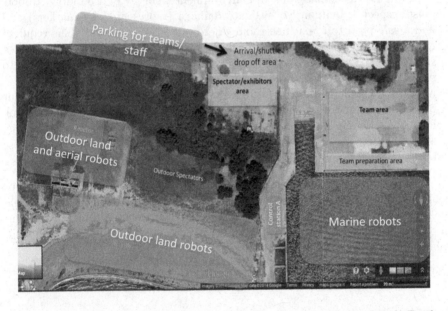

Fig. 3. Competition site, with the Torre del Sale at the left. Image: Google Earth

and public, and also offered a credible industrial context as a background for the competition. Permission was obtained from the State Property Authority to make use of a disused historical building on the sea shore, the Torre del Sale, as the mock reactor building with an internal room playing the part of the machine room. Figure 2 shows the Torre Del Sale building, and Fig. 3 shows a satellite image of the competition site, with the outdoor land, air and sea robot areas indicated.

4   Benchmarking and Scoring

Inspired by and adapted from the benchmarking approach of the RoCKIn Challenge [5] we developed a system-level benchmark (i.e. Task Benchmark) and module-level (i.e. Functionality Benchmark) for euRathlon 2015. The Task Benchmark evaluates the performance of the integrated robot systems while the Functionality Benchmark evaluates the performance of a specific module/functionality of the robot systems. Evaluating only the performance of integrated systems does not necessarily inform how the individual modules are contributing to global performance and which aspects of the module need to be improved. On the other hand, good performance at module level does not necessarily guarantee that systems integrating a set of well performing individual modules will perform well as an integrated system.

Focusing on module-level evaluations alone is also not sufficient to determine which robot system can achieve a specific task. Combining both system-level and module-level benchmarking enables us to perform a deeper analysis and gain useful insights about the performance, advantages and limitations of the whole robot system.

4.1   Matrix Approach to Task and Functionality Benchmarking

As discussed above, in order to perform a specific task which has a set of goals which must be reached a robot needs to execute a set of functionalities. The Functionality and Task Benchmarks can be represented in matrix form as in Fig. 4.

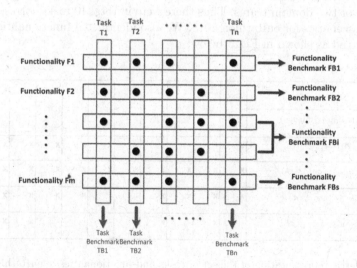

Fig. 4. Task (Vertical) and Functionality (Horizontal) Benchmarking illustration. Source: RoCKIn

Each task requires the effective implementation of several functionalities to be achieved successfully. Each functionality can be evaluated across different tasks or domains (e.g. Robot Navigation in both Land and Sea domains: indoor/outdoor/underwater navigation).

As illustrated in Fig. 4 suppose that for the competition we define N tasks $(T1, T2, ..., Tn)$ which correspond to the columns (vertical) and M functionalities $(F1, F2, ..., Fm)$ which correspond to the rows (horizontal), we will have N Task Benchmarks $(TB1, TB2, ..., TBn)$ and S Functionality Benchmarks (here $S \leq M$). Because we benchmark every task there will be the same number of benchmarks as the defined tasks. For some cases it is not quite necessary to evaluate each functionality in a task separately, for instance, a function of Obstacle Avoidance is an essential functionality of a robot but can be considered as part of the Navigation function, i.e., one Functionality Benchmark can evaluate more than one function at the same time. This is shown as Functionality Benchmark FBi above.

The Task benchmarks were used directly to score the competition results.

4.2   Functionality-Task Mapping for 2015 Scenarios

For the euRathlon 2015 competition, 10 scenarios across 3 domains (Land, Air and Sea) were defined. The 10 scenarios are categorised as *Trials* with 2 scenarios in each single domain (as shown in Fig. 5 below: L1, L2, S1, S2, A1 and A2), *Sub-Challenges* with 3 scenarios in combined two domains (L+A, S+A and L+S) and the *Grand Challenge* (GC) with 3 missions across all three domains. The purpose of the trials and sub-challenges was to, firstly, provide teams with practice in the competition environment and, secondly, provide judged events for single or two-domain teams. Thus there were in total 10 tasks corresponding to the 10 scenarios for euRathlon 2015. We also identified 4 functionalities to be benchmarked as shown in Fig. 5 below:

Tasks Functionalities (/Domain)	L1	L2	S1	S2	A1	A2	L+A	S+A	L+S	GC
2D Mapping (/L+A)	x	-	-	-	x	x	x	x	x	x
Object Recognition (/L+S+A)	x	x	x	x	x	x	x	x	x	x
Obstacle Avoidance (/L+S)	x	x	x	x	-	-	x	x	x	x
Object Manipulation (/L+S)	-	x	-	x	-	-	-	-	x	x

Fig. 5. Metric representation of the set of tasks and functionalities in euRathlon 2015. The /Domain indicates in which domains (Land, Air, Sea) the Functionalities are involved.

A set of ten detailed judging sheets (one per scenario) were devised for each single-domain trial, two-domain sub-challenge and the Grand Challenge, together with guidelines for judges. Data obtained directly by judges observing each event, when combined with analysis of data provided post-event by teams in standardised formats, provided the basis for both benchmarking and scoring.

The full benchmarking for tasks and functionalities are described in the document D3.2 "Benchmarks Evaluation (Part 2: Benchmarking and scoring for euRathlon 2015)[2].

5   Evaluation

5.1   The Competition

A total of 21 teams registered for euRathlon 2015 and, of these, 18 progressed successfully through the qualification process. Of those 18, two withdrew one week before the competition for different reasons; both teams did however attend euRathlon 2015 as visitors.

The 16 teams that participated in euRathlon 2015 are detailed in Table 1. They comprised a total of 134 team members from 10 countries with ~40 robots. A group photo is shown in Fig. 6. As shown in Table 1 there were 9 single domain teams, 2 two-domain teams and 3 three-domain teams. Through a team matching process we actively encouraged single- and two-domain teams to form combined air, land and sea teams. This process resulted in 3 new matched teams to complement the existing 3 multi-domain teams. Thus, of the 16 teams at euRathlon 2015, 10 were able to compete in the Grand Challenge scenario, as shown in Table 2.

Fig. 6. Group photo of euRathlon 2015 participants

[2] http://www.eurathlon.eu/index.php/benchmarking/.

Table 1. Teams with country of origin and domains of participation

Team name	Institution/company	Country	Land	Sea	Air
AUGA	ACSM	ES		X	
AUV Team TomKyle	University of Applied Sciences Kiel	DE		X	
AVORA	Universidad Las Palmas de Gran Canaria	ES		X	
bebot-team	Bern University of Applied Sciences	CH	X		X
B.R.A.I.N. Robots	B.R.A.I.N. Robots e. V	DE	X	X	
Cobham	Cobham Mission Systems	DE	X		
ENSTA Bretagne Team 1	ENSTA Bretagne (ex ENSIETA)	FR	X	X	X
ENSTA Bretagne Team 2	ENSTA Bretagne (ex ENSIETA)	FR	X	X	X
ISEP/INESC·TEC Aerial	ISEP & INESC TEC	PT			X
ICARUS	ICARUS FP7 Project	BE, DE, PL, PT, ES	X	X	X
Team Nessie	Ocean Systems Laboratory/Heriot Watt University	UK		X	
OUBOT	Obuda University	HU		X	
Robdos Team Underwater Robotics	Robdos SRL/ Universidad Politcnica de Madrid	ES		X	
SARRUS - Search And Rescue Robot of UPM & Sener	UPM SENER	ES	X		
UNIFI Team	University of Florence	IT		X	X
Universitat de Girona	Universitat de Girona	ES		X	

Table 2. Teams participating in the Grand Challenge, showing domains (L=Land, A=Air, S=Sea)

Grand Challenge Teams
AUV Team TomKyle (S) + bebot-team (L)(A)
B.R.A.I.N. Robots (L)(S) + UNIFI Team (S)(A)
Cobham (L) + Universitat de Girona (S) + ISEP/INESC TEC Aerial Team (A)
ENSTA Bretagne Team 1 (L)(S)(A)
ENSTA Bretagne Team 2 (L)(S)(A)
ICARUS (L)(S)(A)

The competition took place over 9 days. The first three days were for practice, then followed 2 days for single-domain trials, 2 days for two-domain sub-challenges, and the Grand Challenge in the final two-days. Including single-domain trials, sub-challenges and the Grand Challenge a total of 48 runs were judged. It should be noted that the position of missing workers, leaks, blocked routes and OPIs were randomised between GC runs, and at no time during the competition were teams allowed access into the Torre del Sale building or the machine room.

In parallel with the competition was a public programme, including evening lectures and public demonstrations in the Piombino city centre and at the competition site. Notably the programme included demonstrations from two finalists, including the overall winner, of the DARPA Robotics Challenge (DRC). A total of ~1200 visitors attended the competition and its public events, including several organised parties of school children, families and VIPs.

The logistics and local organisation work of euRathlon 2015 was considerable. The event was staffed by 78 people in total, including the organising staff, judging team, technical and safety team (including divers and safety pilots), media and film crew, stewards and volunteers; the judging team comprised 16 judges (12 from Europe and 4 from the USA). Despite the considerable challenges the event ran smoothly and – most importantly given the risks inherent in an outdoor robotics event – safely.

5.2   Grand Challenge Results

Using the methodology outlined in Sect. 4, the judges were able to assess the performance of the 6 Grand Challenge teams. As summarised in Fig. 7 scores were derived from 5 components: task achievements, optional task achievements, autonomy class, penalties and key penalties. A number of the task achievements were scored on the basis of judges witnessing an event, such as 'robot reaches the unobstructed entrance of the building' or 'robot enters the machine room'; others were scored following analysis of data supplied by teams after the run had been completed, such as map data or OPIs found. Optional achievements were bonus points awarded if, for instance, teams found both missing workers within 30 min, robots transmitted live video/image data during the run, or for direct robot-robot cooperation between domains. The autonomy class was judged on the basis of observing teams, with 1 point awarded for full autonomy, 0.5 for semi-autonomous operation and 0 for tele-operation. Penalties were marked for

Summary (Total Scores/Numbers)						
Teams	COBHAM + ISEP/INESC + UDG	ENSTA 2	ENSTA 1	B.R.A.I.N + UNIFI	BEBOT + TOMKYLE	ICARUS
Achievement(A)	40.5	3.3	21	7	31.5	26
Optional Achievement (OA)	8	4	3	3	9.5	9
Autonomy Class (AC)	9	3	2	1.2	5.5	15.565
Key Penalty (KP)	0	16	3	5	2	2
Penalties (P)	3	0	0	0	0	0
T (Time)	100	100	100	100	76	100
Overall Result						
*S=A+OA+0.5*AC-KP	53	0	22	5.6	41.75	40.7825
Rank	1st	--- -	3rd	4th	2nd=	2nd=

Fig. 7. Grand Challenge scores and ranking

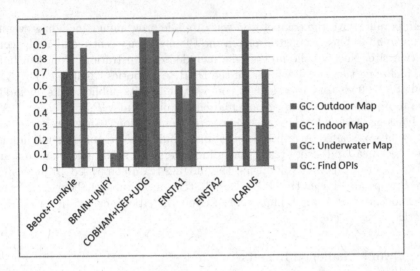

Fig. 8. Functionality benchmarks for the Grand Challenge (Color figure online)

each manual intervention per achievement, or key penalties for mission critical errors such as closing the wrong valve.

Within two hours of completion of the Grand Challenge teams were required to provide vehicle navigation data, mission status data, map information and object recognition information, all using the Keyhole Markup Language (KML) format. This allowed judges to load KML files into Google Earth for evaluation[3].

Figure 8 shows the functionality benchmarks for the Grand Challenge. Of the functionality benchmarks proposed in Fig. 5 we were unable to evaluate obstacle avoidance and object manipulation because of insufficient data. However, we had good data to compare mapping in all three domains, and object recognition (finding OPIs). In Fig. 8, 1.0 is a perfect score, and it is notable that overall winners Cobham+ISEP+UDG achieved 1.0 for finding OPIs, and 0.95 for indoor and underwater maps, however a weakness was outdoor mapping at 0.56. Team ICARUS however achieved a perfect score for the outdoor map, but failed to produce an indoor map. Team Bebot+TomKyle on the other hand produced a perfect indoor map, and was very successful in finding OPIs (0.87) but did not produce an underwater map. As an example Fig. 9 shows the outdoor map generated by fusing the data from air and land robots by team ICARUS.

In euRathlon, because of the unstructured nature of the environment and changes in conditions between events the benchmarks are relatively coarse. However, our Benchmarking and Scoring methodology proved to be very successful in allowing a thorough and transparent evaluation of the performance of teams during the euRathlon 2015 competition. Perhaps the best indicator of the success of the approach was the fact that teams were clearly differentiated in both task and functionality benchmarks; notably no scores were appealed. The detailed scores

---

[3] See http://www.eurathlon.eu/index.php/compete2/eurathlon2015/results2015/.

Fig. 9. The fused map obtained by the ground and aerial vehicles of the ICARUS team during the Grand Challenge. Credits: team ICARUS.

exposed strengths and weaknesses, both between teams and of the state of the art as represented by competing teams and their robots. The overall winners of the Grand Challenge, scoring 53 out of a maximum achievable of 75 points, were team ISEP/INESC TEC (Air), Team Cobham (Land) and Universitat de Girona (Sea), shown with their robots in Fig. 10. This was a particularly impressive outcome given that these three teams had not worked together until arriving at euRathlon 2015. However, of the teams entering the Grand Challenge five

Fig. 10. Overall winners of the euRathlon 2015 Grand Challenge: ISEP/INESC TEC (Air), Team Cobham (Land) and Universitat de Girona (Sea)

achieved creditable performance in mapping, finding missing workers and closing valves in a complex search and rescue scenario that placed great demands on both the robots and the teamwork needed to coordinate those robots.

5.3   Lessons Learned

By all measures euRathlon 2015 was a very successful event. We attracted a larger number of teams than originally planned, and the team matching process proved to be very successful. Indeed perhaps the most significant outcome of not just euRathlon 2015 but the whole project was in bringing together air, land and sea robotics domains to create a new community. We estimate that we have, through workshops and competitions trained ~200 roboticists in outdoor multi-domain robotics.

From a technical point of view we were impressed by the performance of teams in the Grand Challenge noting however that there were a number of common difficulties that all teams experienced. The first was radio communication. Most teams expected to use WiFi networks to maintain communication with land robots, and despite some innovative approaches to overcoming range limitations, such as dropping repeaters or using several land robots as a multi-hop network, all teams experienced challenges. The second was human-robot interfaces – many teams had poorly designed interfaces with their robots which severely tested those operating or supervising robots from inside hot control tents. The third limitation was human-human interaction. We did not specify how the teams communicated between land, sea and air control stations, but it was clear that the most successful multi-domain teams were those who established and rehearsed clear channels and protocols for human-human coordination between the domains. The real challenges are often not technical but human.

Acknowledgements. The euRathlon project was funded within the EU FP7 programme, grant agreement number 601205.

References

1. Winfield, A., Palau Franco, M., Brueggemann, B., Castro, A., Djapic, V., Ferri, G., Petillot, Y., Roning, J., Schneider, F., Sosa, D., Viguria, A.: euRathlon outdoor robotics challenge: year 1 report Advances in Autonomous Robotics Systems: 15th Annual Conference, TAROS 2014. Springer, Birmingham(2014)
2. Ferri, G., Ferreira, F., Sosa, D., Petillot, Y., Djapic, V., Franco, M.P., Winfield, A., Viguria, A., Castro, A., Schneider, F., Roning, J.: euRathon 2014 marine robotics competition analysis. Eurocast 2015 Workshop on Marine Sensors and Manipulators, Las Palmas de Gran Canaria (2015)
3. Petillot, Y., Ferreira, F., Ferri, G.: Performance measures to improve evaluation of teams in the euRathlon 2014 sea robotics competition. IFAC-PapersOnLine **48**(2), 224–230 (2015)

4. Nagatani, K., Kiribayashi, S., Okada, Y., Otake, K., Yoshida, K., Tadokoro, S., Nishimura, T., Yoshida, T., Koyanagi, E., Fukushima, M., Kawatsuma, S.: Emergency response to the nuclear accident at the Fukushima Daiichi nuclear power plants using mobile rescue robots. J. Field Robot. **30**(1), 44–63 (2013)
5. Amigoni, F., Bastianelli, E., Berghofer, J., Bonarini, A., Fontana, G., Hochgeschwender, N., Locchi, L., Kraetzschmar, G., Lima, P., Matteucci, M., Miraldo, P., Nardi, D., Chiaffonati, V.: Competitions for benchmarking: task and functionality scoring complete performance assessment. IEEE Robot. Autom. Mag. **22**(3), 53–61 (2015)

An Hybrid Online Training Face Recognition System Using Pervasive Intelligence Assisted Semantic Information

Dongfei Xue[✉], Yongqiang Cheng, Ping Jiang, and Martin Walker

Department of Computer Science, University of Hull, Hull, England
D.Xue@2014.hull.ac.uk,
{Y.Cheng,P.Jiang,Martin.Walker}@hull.ac.uk

Abstract. In face recognition, the large sizes of training databases can place a heavy burden on computing resources and may produce unsatisfactory results due to significant amount of irrelevant features for target screening. We adopt the technology of wireless sensor networks by storing semantic information in wireless tags to assist grouping of candidates. The semantic information of nearby people such as gender and race is provided to the robot and help it narrows its search to a smaller subset of the database. Hence the face recognizer can be simplified by training the selected subset samples that makes online training possible. Furthermore, the feature space can be constantly adjusted benefiting from online training to distinguish faces with higher accuracy and the resolution of training samples can also be adjusted based on the camera and target distance. In order to further improve the correct rate, permutation post processing has been employed. The proposed hybrid approach has been validated in experiments with a promising low error rate. Compared to other face recognition systems, our system is better suited to work on a human-machine interactive robot which needs to detect targets under different illumination conditions and from different distances.

Keywords: Face recognition · Semantic information · Pervasive intelligence

1 Introduction

The size of a human's social network is estimated to be 150 [1]. It is suggested that the number of neocortical neurons in our brain limits the organism's information-processing capacity, and this then limits the number of relationships that an individual can monitor simultaneously. So, even if a person claims he or she has thousands of friends on Facebook, only about 150 of those relationships are meaningful in their head [2].

Though sometimes we are bothered by bad memory, few of us like to record a person's information in a contact book in detail. People are used to describing each other with more abstract words, such as gender, race, age, color of hair, etc. This semantic information is helpful and can easily guide us in picking a person out of a crowd.

Just as people use a contact book, a robot has all information stored in its database. A face recognition system is a computer application that automatically verifies an

L. Alboul et al. (Eds.): TAROS 2016, LNAI 9716, pp. 364–370, 2016.
DOI: 10.1007/978-3-319-40379-3_37

identity from a digital image or a video frame. The system recognizes faces by comparing the selected facial features from a facial database [3]. Though large memory is not a big problem for a robot today, it still cannot recognize faces as well as we do. That is because the world model installed in the robot cannot be completely consistent with its observed real world. The results of face recognition are easily affected by the noises of illumination, face pose and expressions [4].

In recent years, many efforts have been made to bridge the gap between models and the real world, but most of them place heavier burdens on the available computing resources. For example, images of multiple different poses per person can be pre-installed in the robot database to match the target in different poses, but that requires a larger database size of database and means more images to compare [5]. Through deep-learning methods, faces can be represented with more expressive high-level features, but it needs 3 days to finish training [6].

We think that the face recognition task can be simplified in a pervasive intelligence environment. Just like humans focus their interest on a limited number of acquaintances, we can have a robot focus its attention only on the nearby people.

Compared with other face recognition systems, our system has the following features:

1. Small database size: we use only one picture per person as a training sample.
2. Low cost computation: the algorithms are simple so calculations cost less.
3. Online training: the classifier adjusts itself depending on who is nearby, and can get higher accuracy.

In our system, it becomes possible to train a face recognizer online. The errors caused by the resolution difference between models and observed faces are restricted.

We will introduce the system structure in Sect. 2, then discuss about the algorithm in Sects. 3 and 4, and finally will compare the results on different resolution images in Sect. 5.

2   Overview of the Method

As discussed above, searching and training the whole database places a heavy burden on computing resources for the robot, and the result often cannot reach a satisfactory level in real practice [7]. To solve this issue, our proposal is to separate the whole database into pieces, train the classifier in real time, and enable only a part of the database each time.

On the other hand, there always exist some inconsistences between database models and robot observation which are hard to estimate. Simple features such as HOG [8] may not be enough for a robot to distinguish a large group of people (say 10000), but they still can perform well if limit the size of candidates to a small number (under 10).

The main problem is how to focus robot's attention on only the most probable candidates and ignore the others. In other words, we want to supply some prior knowledge about who is nearby, then the robots can correctly match the target from a subset of the whole database. The technology we use here is a Wireless Sensors Network (WSN).

2.1 Wireless Sensors Network

WSN comprise two parts: the dispersed sensors in the environment and the integrated robots working within it [9].

In our experiment, all people working in the office will wear a Bluetooth Communication Module (Tag). The Tag stores the identity of each person and accepts a query from robot. And we have a robot to work in the office which is installed with a camera and is able to communicate with the Tags in its vicinity. When it approaches a group of people, the Tags on people will inform the camera about their identities. The camera adjusts its algorithm depending on how many people and who is in the vicinity of it.

2.2 System Structure

With the query to the nearby tags, the robot can limit the candidates to those nearby people. And these people will be further separated into smaller groups according to the semantic information such as their gender, race, and age. Finally, the robot will match the target within one of the small groups.

The workflow of our system can be found in Fig. 1.

1. By constantly querying nearby tags, the whole face database does not need to be taken into calculation. Only the nearby face samples are chosen from database according to Tag information and used to compare with the image from the camera. The 2D face images are encoded using HOG features.
2. In order to get higher accuracy, the candidates are separated into smaller groups. Several binary classifiers are trained according to the candidates' semantic information such as gender and race. The training method used here is the Fisher-LDA algorithm, which will be introduce in the next section.
3. In preprocessing, all 2D faces are encoded and installed in the robot's database in the form of feature vectors. The face database used in our experiments is Extended Yale B [10]. Each original face image size is 192×168 pixels and will be projected into a 37 bit vector before further computing.
4. After getting the scope of identities who are nearby and correspond to the semantic information, these candidates' face images will be collected from the database and be used to train the classifier.
5. Detect the face image from the camera and project it into the feature space.
6. The result is determined by calculating Euclidean distance between the observed face from the camera and each face vector in the subset of the database.

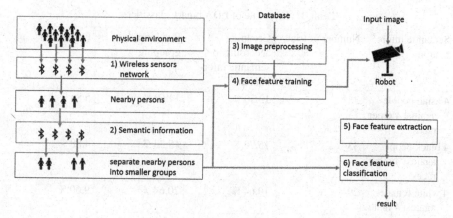

Fig. 1. Workflow of the system

3   Preprocessing

In the experiments, we are using "The extended Yale Face Database B" [10]. We collect only one face image of each identity (suffix: P00A + 000E + 00) for training purposes. All of the 2432 images are used to check the accuracy rate of the classifiers.

In preprocessing, each face image is resized to 64×56 pixels, then encoded into a 1512 dimensional vector of HOG features [8] (8×8 cell size, 2×2 block size, 0 to 180 degrees with 9 histogram channels), finally it will be compressed to n-1 dimensional vector through the PCA method (n is the number of classes) [11].

The results of our experiments shows the error rate of the HOG features stays below 15 % if the size of database is under 12, and it reaches 20 % when the database contains 20 identities. To reach a higher accuracy (e.g. above 90 %), then only 4 people can be distinguished using HOG features.

4   Candidate Screening

From the above experiment, we know that it can only distinguish 4 people with 90 % accuracy. If there are more than 4 people, we need to separate them into small groups of less than five. We find it shows better geometric distribution if we separate the candidates according to their semantic information (genders, races), and the classifier can then get higher accuracy. The results can be found in Table 1.

The result can be furtherly improved by means of permutation and combination. As described in Sect. 2, the number of candidates is controlled to a small size by Tag information, because only the nearby people are calculated (see algorithm 1).

D. Xue et al.

Table 1. Error rate of LDA binary classifiers

Semantic information	Number of trials	Grouping according to semantic information	Randomly grouping	Permutation
4 asian people against 4 other races	32	13.50 %	14.69 %	5.88 %
4 black people against 4 other races	33	15.79 %	18.33 %	4.88 %
4 white female against 4 male	29	19.64 %	20.64 %	9.60 %
4 white male against 4 other races female	29	12.31 %	16.92 %	3.74 %

Algorithm 1: permutation

1) Assume there are $I_1 + I_2$ person nearby, I_1 of them belongs to class 1, and the others I_2 belongs to class 2.
2) Calculate the Euclidian distance d_{ij} between each observed faces i and each means of classes j. We have the whole set of results as below.

$$D = \{d_{ij} | i = [1 \cdots I_1 + I_2], j = [1,2]\}$$

3) Choose $d_{ij} = \min (D)$, if j = 1, then we say $x_i \in class$ 1; if else we say $d_{ij} \in class$ 2
4) Update $D = D - \{d_{i1}, d_{i2}\}$
5) Repeat step 3) and 4) until $D \in \emptyset$

5   Candidate Matching

After candidate screening, the number of candidates is reduced to less than 5. Finally, the target can be located by comparing Euclidian distance, and the result can also be improved by permutation algorithm as described in Sect. 4. The experiment results are listed below Table 2.

Table 2. Error rate of candidate matching

Semantic information	Number of trials	Error rate
4 asian people against 4 other races	32	8.68 %
4 black people against 4 other races	33	7.68 %
4 white female against 4 male	29	12.90 %
4 white male against 4 other races female	29	6.36 %

In the above experiment, only one sample image per person is used in training classifiers. For a group of 8 people, we separate them into two smaller groups first, then match it in one of the small groups. In total, 123 groups of identities are tested and the average error rate is 8.85 %.

It also shows that the error rate rises when the resolution of the model is different to that of the observed face image (see Table 3). For a pre-trained face recognizer, the observed face image should be resized to the same size of model in database, but image resizing and cropping brings noises and uncertainties.

Table 3. Error rate on different resolution

Samples image resolution		192×168	96×84	64×56
HOG cell size		24×24	12×12	8×8
Feature number		1512	1512	1512
Observed image resolution	192×168	13.65 %		
	96×84	11.94 %	8.56 %	
	64×56	13.61 %	9.38 %	7.24 %

An online trained face recognizer can be trained after it detects faces from the camera. It can adjust the resolution of the model to match the observed face image.

6 Conclusion

We have a robot to work in a pervasive intelligence environment, where everyone wears a Bluetooth Tag which keeps communicating with nearby robots. With the help of this WSN, a robot can estimate who is in its vicinity and focus its attention on a subset of its face image database. The semantic information stored in each Tag helps the robot to separate the nearby people into smaller groups. Therefore the calculation is simplified and makes it possible to train the face recognizer in real time.

Unlike a pre-trained face recognizer, our system constantly adjusts its feature space to suit a subset of the database and consequently has higher accuracy. The gap between the pre-installed model and the observed world is narrowed, because more prior knowledge is provided to the robot in advance to narrow the search space. The resolution of models can be adjusted in training according to the distance to the target.

References

1. Hill, R.A., Dunbar, R.I.: Social network size in humans. Hum. Nat. **14**(1), 53–72 (2003)
2. NPR: Don't Believe Facebook; You Only Have 150 Friends (2011)
3. Sonka, M., Hlavac, V., Boyle, R.: Image Processing, Analysis, and Machine Vision. Cengage Learning, Boston (2014)
4. Wagner, A., et al.: Toward a practical face recognition system: robust alignment and illumination by sparse representation. IEEE Trans. Pattern Anal. Mach. Intell. **34**(2), 372–386 (2012)
5. Beymer, D.J.: Face recognition under varying pose. In: Proceedings of the 1994 IEEE Computer Society Conference on Computer Vision and Pattern Recognition, 1994, CVPR 1994. IEEE (1994)

6. Taigman, Y., et al.: Deepface: closing the gap to human-level performance in face verification. In: 2014 IEEE Conference on 2014 Computer Vision and Pattern Recognition (CVPR). IEEE (2014)

7. Delac, K., Grgic, M., Grgic, S.: Independent comparative study of PCA, ICA, and LDA on the FERET data set. Int. J. Imaging Syst. Technol. **15**(5), 252 (2005)

8. Dalal, N., Triggs, B.: Histograms of oriented gradients for human detection. In: IEEE Computer Society Conference on Computer Vision and Pattern Recognition, 2005, CVPR 2005. IEEE (2005)

9. Li, X., et al.: Servicing wireless sensor networks by mobile robots. IEEE Commun. Mag. **50**(7), 147–154 (2012)

10. Lee, K.-C., Ho, J., Kriegman, D.J.: Acquiring linear subspaces for face recognition under variable lighting. IEEE Trans. Pattern Anal. Mach. Intell. **27**(5), 684–698 (2005)

11. Jolliffe, I.: Principal Component Analysis. Wiley Online Library, Hoboken (2002)

Performance Analysis of Small Size and Power Efficient UWB Communication Nodes for Indoor Localization

Reza Zandian$^{(\boxtimes)}$ and Ulf Witkowski

Electronics and Circuit Technology Department,
South Westphalia University of Applied Science, Soest, Germany
{zandian.reza,witkowski.ulf}@fh-swf.de

Abstract. In this paper, the localisation capabilities of ultra-wideband (UWB) communication devices are evaluated. A test platform is designed to perform experiments in indoor environments, to record the data for different node set-ups and to evaluate the results in practice. The platform development comprises hardware design of the anchor and tag node, development of the PC software for communication to the nodes, collecting the measured distance data and performing localisation algorithms. At the end, some experiments are performed in both line of sight (LOS) and non-line of sight (NLOS) cases with blocking and non-blocking barriers. The experiment results confirm a distance measurement accuracy of 10 cm in LOS conditions for 85 % of the measured distances in the range of 0.5 m to 30 m. In NLOS cases an additional offset can be observed in the measurement results causing a higher relative error for short distances.

Keywords: UWB · ToA · TDoA · Indoor localisation · Ranging · LOS · NLOS

1 Introduction

Advancements in the localisation technology that resulted from developing cheaper radio modules and more accurate localisation devices, shifted the applications of localisation techniques from military to civil purposes. Among these applications, tracking, navigation, health care, search and rescue, entertainment, etc. can be mentioned which promise a large market volume and ongoing research activities to improve quality and applicability. Several different technologies have been used so far to localise an object or a person. The GPS system and other similar technologies have shown great potential for localisation when the target is outdoor. However, the efficiencies of these devices drop drastically when the target is located in an indoor area. Problems are blocking of GPS signals in indoor applications and low accuracy of measurements which might be in the range of several meters. Among the applied approaches, radio based solutions proved to be more successful in indoor areas as they can overcome problems such as lack of light, no line of sight to target, barriers or walls around the transmitter, sound interferences, and others. Unlike the development of outdoor localisation solutions based on GPS, GLONAS and other methods, the pace of advancement in indoor localisation techniques is gentle. Several radio based localisation techniques and standards have been proposed and tested by scientists. These include RFID, Wi-Fi, Bluetooth,

© Springer International Publishing Switzerland 2016
L. Alboul et al. (Eds.): TAROS 2016, LNAI 9716, pp. 371–382, 2016.
DOI: 10.1007/978-3-319-40379-3_38

ZigBee, RF radios and also Ultra-Wideband (UWB) as measurement approach. These technologies have been evaluated and compared in [1–3]. In this paper we concentrate on UWB technology to realize an indoor localization system, because it can provide high accuracy in indoor distance measurements as basis for an indoor localization system.

This paper is organized as follow: After introducing the UWB technology in Sect. 1 the hardware implementation of UWB nodes is discussed in Sect. 2. Based on the role of a node, this can be tag or anchor, different software versions are required. Details are given in Sect. 3. In order to characterize UWB based distance measurements several experiments have been performed. Results are discussed in Sect. 4 and finally Sect. 5 concludes the paper.

1.1 State of the Art in Localisation Using UWB

Among the aforementioned techniques, UWB has gained more attention due to its wide range of frequency band which reduces the multipath effect and has resulted excellent performances when time based algorithms are applied for distance measurement. The wide frequency band results in narrow pulses in the time domain allowing accurate time of flight measurements of emitted pulses. Considering the development of UWB devices, the first aim was to transfer large amount of data in short time, benefiting from the large bandwidth of the system. However, difficulties such as antenna design, needs of fast processing devices and complexity of signal processing leaded to loss of interest by manufacturers. Up to recently, it was not cost effective to use UWB technique for localisation and measurement purposes. Also the performance was not good. But advances in designing high performance processors in small packages as well as improvements in chip level design techniques paved the way to get the UWB technology available with modest budget requirements. In context of localisation techniques and related algorithms, ToA (Time of Arrival), TDoA (Time Difference of Arrival), RSSI (Received Signal Strength Indication), AoA (Angle of Arrival), etc. can be named which have been evaluated in many scientific papers [4–7, 15]. Some scientists have performed experiments to improve the accuracies, signal to noise ratio and reduce error probabilities by applying filtering techniques [8], combining different localisation techniques or radios [9, 16] and mathematical models [10]. Common in all reports is that they have achieved successful results up to some extents, however, challenges such as multipath effects, multi transmitter signals, degradation of signal strengths by barriers, noise effects and many other factors usually limit the rate of location sampling and its accuracy up to few decimetres. With latest implementation of UWB signal generation and processing in silicon, higher accuracy can be achieved at moderate power consumptions and reasonable costs.

1.2 Recent Advances in the UWB Hardware

DecaWave company is currently leading the market of UWB devices producing the cheapest UWB modules with good accuracy (~10 cm) [11]. According to specifications, the modules are suitable for ToA or TDoA topologies depending on the users defined internal data processing program. The DecaWave radio chip is able to localize an object with the data update frequency of 1–10 Hz (Fig. 1a). Some manufacturers developed a module based on the DecaWave hardware which is combined with many other motion detection sensors (gyro, accelerometer, magnetometer, altimeter, temperature and humidity sensor, etc.) that result in increase of the accuracy up to 10 cm (Fig. 1b) and may benefit from data fusion approaches in terms of accuracy or availability.

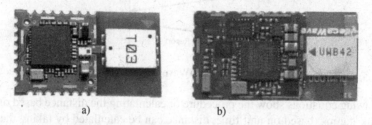

a)   b)

Fig. 1. (a) DecaWave's DWM1000 module [14], (b) RTLS1000MCU includes ARM-M4, gyro, accelerometer, compass, pressure, temperature and humidity sensors [12]

1.3 Possible Localisation Techniques

Due to the high degree of integration and configuration options, the DecaWave module has been used for hardware design and localisation tests. Several different connection topologies are applicable on DecaWave modules. A few of them are shown in the following figures.

In point-to-point structure, one node acts as anchor which is connected to a host system (Fig. 2a). The other node will be the tag element that communicates with the anchor. In this type of connection only the distance of the two nodes can be estimated. The advantage is that the nodes do not need to be synchronised. In order to estimate the distance, the clock of both devices will be synchronised by a set of transmit and receive procedures as depicted in Fig. 3.

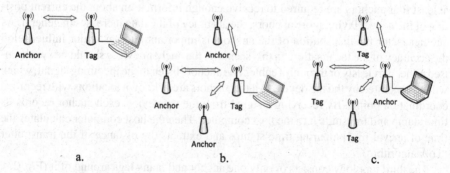

a.   b.   c.

Fig. 2. Supported connection topologies for distance measurement and object localization based on the DecaWave module [12–14]

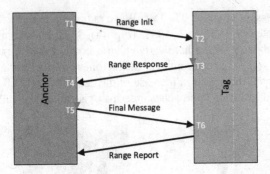

Fig. 3. Synchronization procedure in DecaWave modules based on ToA technique [12]

Following equations show the procedure of calculating the distance based on measured time stamps (based on unit time, distance can be calculated by taking the propagation velocity of the radio signals into account):

$$T_{FG} = \left[(T_4 - T_1) - (T_3 - T_2) \right] + \left[(T_6 - T_3) - (T_5 - T_4) \right] \qquad (1)$$

$$T_{FE} = \frac{T_{FG}}{4} \qquad (2)$$

In these equations T_{FG} is the overall time required to send a signal between two modules for the whole process and T_{FE} is the time that the signal requires to travel between the modules. At first, the anchor device sends its time stamp to the tag device. After a certain amount of time, the message will be returned back to the anchor. The anchor measures this time and sends the new time stamp to the tag device again. Having the total time that a signal needed to travel between two modules, the distance value can be calculated. This procedure is graphically demonstrated in the Fig. 3.

In the second connection topology (Fig. 2b), several anchor devices are used to estimate the location of the tag device. For 2D localisation at least three anchors and for 3D at least four anchors are required to receive enough information about the current position of the node. Having more anchors, the accuracy of the data increases in general. As another factor the distribution of the anchors is important and has major influence on the accuracy of the localisation. In this scenario, the anchor devices should be synchronised either wirelessly or through a cable. The anchors listen to the incoming signal, which will be generated by the tag device. All the anchors are in known locations with recorded coordination points. By receiving a signal from the tag device, each anchor records its time stamp and transmits it to the host computer. Then the host computer calculates the time of arrival by comparing time stamps and extracts the distance of the transmitter (ToA algorithm).

The third topology consists of only one anchor and many tags around of it (Fig. 2c). The application of this topology is mostly detection of the tags in a certain perimeter. Therefore the exact location of the tag cannot be estimated rather only the number of the tags and their distances to the anchor. Real applications of this topology are searching for items carrying tags, supervising children, health care services, navigation in supermarket, primitive security of machines, etc.

2 Hardware Implementation of UWB Nodes

Due to the features of the decaWave UWB module, such as cost, accuracy, simplicity and minor required implementation area, we decided to setup our own UWB localisation system based on this module. The first prototype of the localisation module is already provided which contains the decaWave UWB module with all required antenna tuning and matching circuit, voltage regulators as well as a low energy consumption family of ARM microcontroller from ST Microelectronics. The block diagram of the UWB node is shown in the Fig. 4.

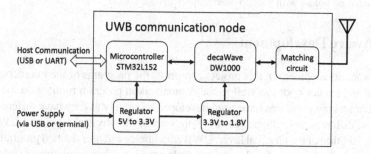

Fig. 4. Block diagram of the implemented elements in the tag and anchor nodes

The microcontroller on board enables the users to implement a certain program to decide on the role of a node (anchor or tag), to choose connection topology and to perform calculations of the measurements. The achieved results can be transmitted to the host PC for further analysis or storage purposes over UART or USB connections. The antenna circuit can be mounted externally based on the user requirements. In this project, an antenna circuit suggested from decaWave is used. The anchor node is equipped with a backup power supply and a holder which makes it easy to transport and install the anchors at different positions.

The first prototypes of the tag and anchor nodes are shown in Fig. 5. In the next version of the prototype, it is planned to implement IMU (Inertial Motion Unit) sensors (gyroscope, accelerometer and magnetometer) to acquire the motion data and merge it with the available data from UWB systems using data fusion techniques. The final solution is called a hybrid node, which is able to deliver more precise location data. Colombo et al. [18] have implemented this technique in a wearable platform combining localization node with IMU sensors with good results. Using this technique, in addition of distance information, the location of the tag node in cardinal directions can be estimated which will be useful for navigating robots to a certain direction. The performance of the system will be evaluated in a test platform described in [17].

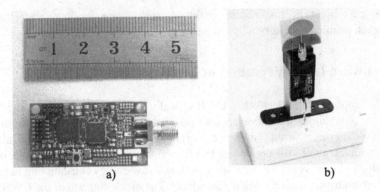

Fig. 5. (a) Developed hardware platform used for both anchor and tag modules, (b) The anchor node mounted on holder with backup powersupply in base socket

3   Software Development

The software development in this project comprises the programs of the microcontrollers in the tag and anchor nodes as well as the demonstration program running on the server. The anchor and tag programs have been developed separately as they have different tasks and follow different procedures. However, they share the physical layer and UWB chip library. In addition of the physical layer, UWB chip library and user defined program which are common in both types of nodes, the tag node contains a data communication layer for transferring the data and commands between the tag node and the server PC. These layers are demonstrated graphically for tag and anchor nodes in the Fig. 6.

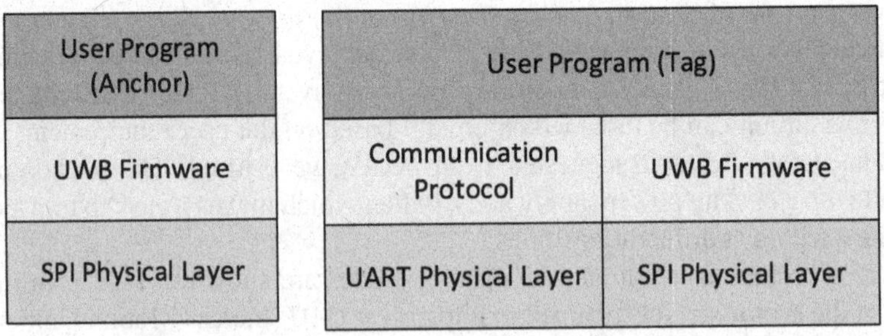

User Program (Anchor)	User Program (Tag)	
UWB Firmware	Communication Protocol	UWB Firmware
SPI Physical Layer	UART Physical Layer	SPI Physical Layer

a.   b.

Fig. 6. Software layer structure of the (a) anchor node, (b) tag node

The tag program starts its routine by selecting an anchor number. The next step is to check the pairing status of the nodes. If the nodes are not paired, the pairing procedure begins. In case of successful pairing, the measurement procedure begins which is identical to procedure introduced in Sect. 1.3 and Fig. 3. In case the pairing procedure fails or the measurement procedure has been performed, another anchor will be selected and the whole routine will be repeated until all anchors are polled once, see Fig. 7.

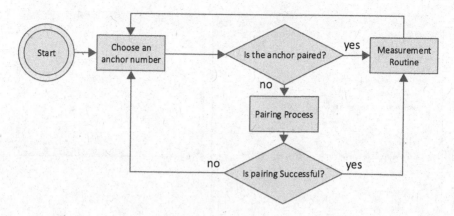

Fig. 7. The tag node measurement algorithm with anchor polling mechanism

For demonstration purposes, an application for a PC has been developed which simplifies monitoring and configuration of the distance measurement module as well as implementation of localisation algorithms. With the help of this software, users are able to setup the configuration parameters of the module, address of the tag and anchor modules in the tag device and finally read and demonstrate the measurement data and its related parameters. The location data of the tag and anchors as well as pairing status of the nodes can be displayed graphically. The screen shots of the configuration tab (Fig. 8) as well as graphical visualization of the node locations (Fig. 9) are depicted below.

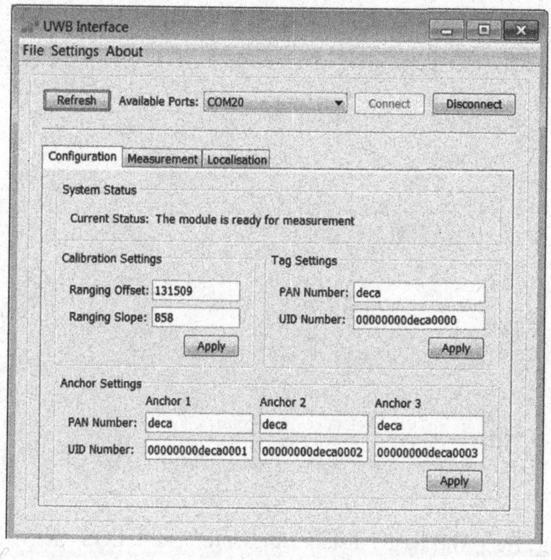

Fig. 8. Screen shot of the configuration tab in the server side program

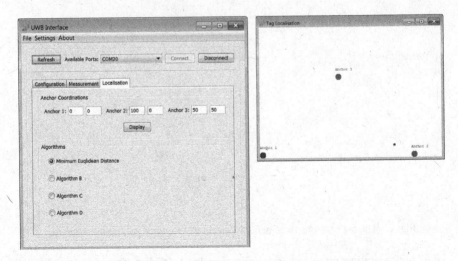

Fig. 9. Screen shot of the localization tab and the graphical display of the node locations

4 Results and Discussion

In order to evaluate the performances of the designed nodes, several experiments have been carried out and evaluated. The first experiment is performed in LOS conditions inside a long corridor with no barriers between two nodes. Aim was to measure the distance between the nodes for distances in the range of 0.5 m to 30 m. The results of this experiment is depicted in Fig. 10.

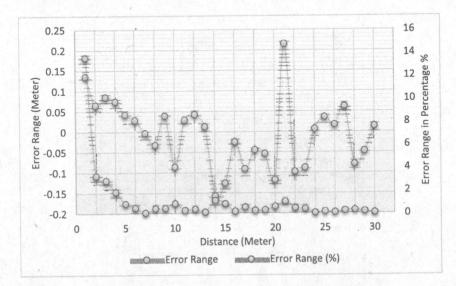

Fig. 10. Error range for UWB based distance measurement in LOS case (Color figure online)

The blue dots in this figure show the average error of the measurements performed at different distances. For each distance, 50 samples are taken which takes about 5 s. The range of the absolute error never exceeded 25 cm and more than 85 % of all measurements were in the tolerance range of maximum 10 cm. The red dots show the same results in percentage. Considering the 10 cm tolerance of the measurement, a larger percentage error can be expected for shorter distances, however for large distances of more than 5 meters the accuracy of the distance measurement is better than 1 percent error.

The second experiment was to evaluate the performance of the module in NLOS conditions. In order to perform this, several measurements are taken in the laboratory with barriers in between the nodes. The barriers in this case were non-blocking items, so the radio signal may also reach the receiver going through a reflective path. This cause some interference in the final result and the accuracy rate decrease in comparison to LOS measurements. Figure 11 shows the corresponding results. The measurements are performed in four different distances with and without barriers. The anchor is located at coordination of zero and the tag module has been moved to spotted distances. The diversities of the measured values are also depicted which shows the accuracy range of the measured values. The experiment at the distance 6.5 m is performed with two barriers in between of the nodes each with the thickness of 70 centimetres made of wood and metal pieces. The maximum observed error for all measurements is +50 cm caused from the two barriers. The effect of a barrier in the signal path is constant offset to final result of the distance measurement.

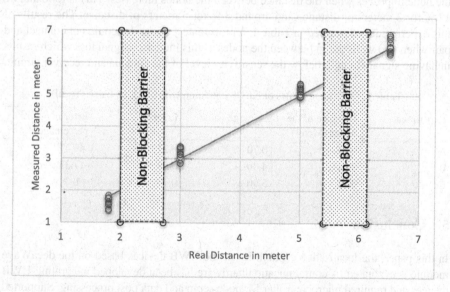

Fig. 11. NLOS measurement with non-blocking barriers in between of the nodes

The last experiment is performed again in NLOS condition but with a wall made of stones between the nodes (blocking barrier). Three different distances have been tested. One node is placed in front of the wall that has a thickness of 41 cm. The other node has been placed directly after the wall (see Fig. 12a), in 3 meters of distance behind the wall (part b), and at 5 meters of the distance (part c).

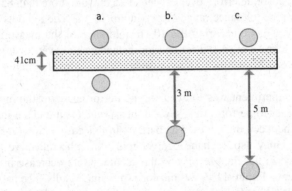

Fig. 12. NLOS condition with wall between tag and anchor UWB nodes

The presence of the wall between tag and anchor node attenuates the signal and causes an additional transit time through the wall. Therefore the wall produces an offset of the measured distance. When the measurement is carried out at a short distance right after the barrier, the accuracy of the nodes decreases drastically (50 %) as the radio signal needs a longer time to travel through the wall however the (relative) performances of the node improves when the distance between the nodes increases. This is for example 17 % for 3 meters of distance and around 8 % for 5 meters of distance. The results of this experiment is described in Table 1. Unfortunately the nodes could not connect and pair when two walls are in between the nodes as this increases signal loss which results in having signal strength below the sensitivity level of the antenna and receiver circuit.

Table 1. Measured values of NLOS condition with blocking barrier (wall)

Experiment	Distance of the nodes (Meter)	Average value	Error	Error in percentage %
A	0.47	0.70	0.232	49.47
B	3.41	4.00	0.59	17.33
C	5.41	5.90	0.49	8.3

5   Conclusion

In this paper, the localisation capabilities of the UWB devices based on the decaWave module is evaluated. A communication hardware has been developed integrating UWB devices and required microcontroller for node setup and data post processing. Supported interfaces are UART and USB to be able to easily integrate the communication devices into mobile vehicles and stationary infrastructure nodes for indoor applications. The developed software for node's microcontroller, supports anchor and tag role for the hardware. A test setup is used to perform the experiments and record the results in practice.

For this, a PC software for communication to the nodes, collecting the distance data and performing localisation algorithms has been developed. At the end several experiments are carried out in both LOS and NLOS conditions with blocking and non-blocking barriers. The experiment results prove the accuracy of about 10 cm in LOS conditions for 85 % of the measured distances in the range between 0.5 m and 30 m. The measured error never exceeded 25 cm. In NLOS cases the accuracy may be deviate in case the distance of the barriers to the nodes are too short however this effect reduce when the overall distance of the nodes increases. It has been observed that barriers appear as constant offset in the measurement results. A blocking barrier also significantly reduces maximum communication range.

In the next step of this work, the number of anchors should be extended to more than five and different localisation algorithm should be applied for 2D and 3D localisation. The results of this work can be used in localisation scenarios to implement projects in real life such as search and rescue scenarios, elderly assisting applications, healthcare, object monitoring and security primitive of devices, navigation of robots and drones in indoor areas and other applications.

References

1. Farid, Z., Nordin, R., Ismail, M.: Recent advances in wireless indoor localization techniques and system. J. Comput. Netw. Commun. **2013** (2013). Article ID 185138, doi:10.1155/2013/185138
2. Stojanovic, D., Stojanovic, N.: Indoor localization and tracking methods, technologies and research challenges. Facta Univ. Ser. Autom. Control Robot. **13**, 57–72 (2014). ISSN 1820-6425
3. Jinhong, X., Zhi, L., Yang, Y., Dan, L., Xu, H.: Comparison and analysis of indoor wireless positioning techniques. In: 2011 International Conference on Computer Science and Service System (CSSS), pp. 293–296, 27–29 June 2011
4. Basem, A., Aboelmagd, N.: A survey of recent indoor positioning systems using wireless networks. Int. J. Electron. Commun. Comput. Eng. **5**(5), 1197–1204 (2014). ISSN (Online): 2249–071X, ISSN (Print): 2278–4209
5. Singh, S., Shakya, R., Singh, Y.: Localization techniques in wireless sensor networks. (IJCSIT) Int. J. Comput. Sci. Inf. Technol. **6**(1), 844–850 (2015)
6. Sven, A.: Evaluation of Different Radio-Based Indoor Positioning Methods, Master thesis, Linköping, 9 April 2014
7. Mišić, J., Milovanović, B., Vasić, N., Milovanović, I.: An overview of wireless indoor positioning systems. INFOTEH-JAHORINA, vol. 14, March 2015
8. Chung, H.Y., Hou, C.C., Chen, Y.S.: Indoor intelligent mobile robot localization using fuzzy compensation and kalman filter to fuse the data of gyroscope and magnetometer. IEEE Trans. Industr. Electron. **62**(10), 6436–6447 (2015)
9. Xiong, Z., Song, Z., Scalera, A., Ferrera, E., Sottile, F., Brizzi, P., Tomasi, R., Spirito, M.A.: Hybrid WSN and RFID indoor positioning and tracking system. EURASIP J. Embed. Syst. **2013**(1), 1 (2013)
10. Qamar, A., Faruq, U.: Modelling and simulation of UWB radar system for through the wall imaging and doppler detection. Int. J. Eng. Trends Technol. (IJETT) **17**(7), 325–330 (2014)
11. DecaWave Presentation Slides. DecaWave's ScenSor DW1000: The World's Most Precise Indoor Location and Communication CMOS Chip IdTechEx USA November 2013

12. Sewio Company. Photo available online at: http://www.sewio.net/uwb-sniffer-2/analyzing-decawave-two-way-ranging-twr/. Accessed 25 Jan 2016
13. Prophet, G.: Featured. Articles in EDA Europe. Two-way ranging, real-time location evaluation kit, 11 March 2015. http://www.edn-europe.com/en/two-way-ranging-real-time-location-evaluation-kit.html?cmp_id=7&news_id=10005966#VqYDsfkrLmE
14. DecaWave Company. http://www.decawave.com/products. Accessed 25 Jan 2016
15. Laaraiedh, M., Yu, L., Avrillon, S., Uguen, B.: Comparison of hybrid localization schemes using RSSI, TOA, and TDOA. In: Wireless Conference 2011 - Sustainable Wireless Technologies (European Wireless), 11th European, pp. 1–5, 27–29 April 2011
16. Kabir, M.H., Ryuji, K.: A hybrid TOA-fingerprinting based localization of mobile nodes using UWB signaling for non line-of-sight conditions. Sensors **12**(8), 11187–11204 (2012). PMC. Web. 25 Jan. 2016
17. Kemper, P., Tetzlaff, T., Witkowski, U., Zandian, R., Mamrot, M., Marchlewitz, S., Nicklas, JP., Winzer, P.: Small size robot platform as test and validation tool for the development of mechatronic systems. In: Fira Conference, South Korea (2015)
18. Colombo, A., Fontanelli, D., Macii, D., Palopoli, L.: Flexible indoor localization and tracking based on a wearable platform and sensor data fusion. IEEE Trans. Instrum. Meas. **63**(4), 864–876 (2014)

Author Index

Printed in the United States
By Bookmasters

Printed in the United States
By Bookmasters